Environmentally Conscious Materials and Chemicals Processing

Environmentally Conscious Materials and Chemicals Processing

Edited by
Myer Kutz

John Wiley & Sons, Inc.

Library of Congress Cataloging-in-Publication Data:

Environmentally conscious material and chemical processing / edited by
Myer Kutz.
        p. cm.
    Includes bibliographical references and index.
    ISBN 978-0-471-73904-3 (cloth)
        I. Kutz, Myer.

Printed in the United States of America

10 9 8 7 6 5 4 3 2 1

To W. Bradford Wiley, Mike Harris, and Bob Polhemus, in fond memory.

Contents

Contributors   ix

Preface   xi

1   Engineering of Mineral Extraction   1
Michael G. Nelson

2   Waste Reduction in Metals Manufacturing   33
Dr. Carl C. Nesbitt

3   Life-cycle Evaluation of Chemical Processing Plants   59
David Brennan

4   Waste Reduction for Chemical Plant Operations   89
Ali Elkamel, Cheng Seong Khor and Chandra Mouli R. Madhuranthakam

5   Industrial Waste Auditing   125
C. Visvanathan

6   Reverse Production Systems: Production from Waste Materials   155
I-Hsuan Hong, Jane C. Ammons and Matthew J. Realff

7   Environmental Life-cycle Analysis of Alternative Building Materials   179
Bruce Lippke and Jim L. Bowyer

8   Wastewater Engineering   207
Say Kee Ong, Ph.D

9   The Environmental Impacts of Packaging   237
Eva Pongrácz

10   Aqueous Processing for Environmental Protection   279
Fiona M. Doyle, Gretchen Lapidus-Lavine

11   Solid Waste Disposal and Recycling   307
Georgios N. Anastassakis

12   Processing Postconsumer Recycled Plastics   357
Gajanan Bhat

13   Supercritical Water Oxidation   385
Philip A. Marrone and Glenn T. Hong

Index   455

Contributors

Jane C. Ammons
Georgia Institute of Technology
Atlanta, Georgia

Georgios N. Anastassakis
National Technical University
    of Athens (N.T.U.A.)
Athens, Greece

Gajanan Bhat
The University of Tennessee
Knoxville, Tennessee

James L. Bowyer
University of Minnesota
St. Paul, Minnesota

David Brennan
Monash University
Victoria, Australia

Fiona M. Doyle
University of California at Berkeley
Berkeley, California

Ali Elkamel
University of Waterloo
Waterloo, Ontario, Canada

Glenn T. Hong
General Atomics
San Diego, California

I-Hsuan Hong
National Chiao Tung University
Hsinchu, Taiwan

Cheng Seong Khor
University of Waterloo
Waterloo, Ontario, Canada

Gretchen Lapidus-Lavine
Universidad Autónoma
    Metropolitana—Iztapalapa
México

Bruce Lippke
University of Washington
Seattle, Washington

Chandra Mouli
University of Waterloo
Waterloo, Ontario, Canada

R. Madhuranthakam
University of Waterloo
Waterloo, Ontario, Canada

Philip A. Marrone
Science Applications International
    Corporation
Newton, Massachusetts

Michael G. Nelson
University of Utah
Salt Lake City, Utah

Carl C. Nesbitt
Michigan Technological University
Houghton, Michigan

Say Kee Ong
Iowa State University
Ames, Iowa

Eva Pongrácz
University of Oulu
Oulun Yliopisto
Finland

eva.pongracz@oulu.fi
08-553-2345

Matthew J. Realff
Georgia Institute of Technology
Atlanta, Georgia

Chettiyappan Visvanathan
Asian Institute of Technology
Klongluang Pathumthani,
Thailand

Preface

Many readers will approach this series of books in environmentally conscious engineering with some degree of familiarity with, knowledge about, or even expertise in one or more of a range of environmental issues, such as climate change, pollution, and waste. Such capabilities may be useful for readers of this series, but they aren't strictly necessary. The purpose of this series is not to help engineering practitioners and managers deal with the *effects* of man-induced environmental change. Nor is it to argue about whether such effects degrade the environment only marginally or to such an extent that civilization as we know it is in peril, or that any effects are nothing more than a scientific-establishment-and-media-driven hoax and can be safely ignored. (Other authors, fiction and nonfiction, have already weighed in on these matters.) By contrast, this series of engineering books takes as a given that the overwhelming majority in the scientific community is correct, and that the future of civilization depends on minimizing environmental damage from industrial, as well as personal, activities. However, the series goes beyond advocating solutions that emphasize only curtailing or cutting back on these activities. Instead, its purpose is to exhort and enable engineering practitioners and managers to reduce environmental impacts—to engage, in other words, in *environmentally conscious engineering*, a catalog of practical technologies and techniques that can improve or modify just about anything engineers do, whether they are involved in designing something, making something, obtaining or manufacturing materials and chemicals with which to make something, generating power, or transporting people and freight.

Increasingly, engineering practitioners and managers need to know how to respond to challenges of integrating environmentally conscious technologies, techniques, strategies, and objectives into their daily work, and, thereby, find opportunities to lower costs and increase profits while managing to limit environmental impacts. Engineering practitioners and managers also increasingly face challenges in complying with changing environmental laws. So companies seeking a competitive advantage and better bottom lines are employing environmentally responsible design and production methods to meet the demands of their stakeholders, who now include not only owners and stockholders, but also customers, regulators, employees, and the larger, even worldwide community.

Engineering professionals need references that go far beyond traditional primers that cover only regulatory compliance. They need integrated approaches centered on innovative methods and trends in design and manufacturing that help them focus on using environmentally friendly processes and creating green products. They need resources that help them participate in strategies for designing environmentally responsible products and methods, resources that provide a foundation for understanding and implementing principles of environmentally conscious engineering.

To help engineering practitioners and managers meet these needs, I envisioned a flexibly connected series of edited handbooks, each devoted to a broad topic under the umbrella of environmentally conscious engineering, starting with three volumes that are closely linked—environmentally conscious mechanical design, environmentally conscious manufacturing, and environmentally conscious materials and chemicals processing.

The intended audience for the series is practicing engineers and upper-level students in a number of areas-mechanical, chemical, industrial, manufacturing, plant, and environmental-as well as engineering managers. This audience is broad and multidisciplinary. Some of the practitioners who make up this audience are concerned with design, some with manufacturing, and others with materials and chemicals processing, and these practitioners work in a wide variety of organizations, including institutions of higher learning, design, manufacturing, and consulting firms, as well as federal, state, and local government agencies. So what made sense in my mind was a series of relatively short handbooks, rather than a single, enormous handbook, even though the topics in each of the smaller volumes have linkages and some of the topics (*design for environment, DfE,* comes to mind) might be suitably contained in more than one freestanding volume. In this way, each volume is targeted at a particular segment of the broader audience. At the same time, a linked series is appropriate because every practitioner, researcher, and bureaucrat can't be an expert on every topic, especially in so broad and multidisciplinary a field, and may need to read an authoritative summary on a professional level of a subject that he or she is not intimately familiar with but may need to know about for a number of different reasons.

The Environmentally Conscious Engineering series is composed of practical references for engineers who are seeking to answer a question, solve a problem, reduce a cost, or improve a system or facility. These handbooks are not research monographs. The purpose is to show readers what options are available in a particular situation and which option they might choose to solve problems at hand. I want these handbooks to serve as a source of practical advice to readers. I would like them to be the first information resource a practicing engineer reaches for when faced with a new problem or opportunity—a place to turn to even before turning to other print sources, even any officially sanctioned ones, or to sites on the Internet. So the handbooks have to be more than references or collections of background readings. In each chapter, readers should feel that they are in the hands of an experienced consultant who is providing sensible advice that can lead to beneficial action and results.

The third title in the Environmentally Conscious Engineering Series, the *Handbook of Materials and Chemicals Processing*, ranges widely for a volume with only thirteen chapters. This handbook covers the making and recovery-from-wastes of a variety of industrial materials, including metals, chemicals, plastics, building materials, and packaging. Included are discussions not only of the extraction, production, and recycling of metals, but also methods for reducing waste and preventing pollution in chemical plants. The handbook covers methods for auditing industrial wastes, as well as techniques for disposing of, recovering, and recycling materials from industrial wastes, end-of-life products, and packaging. Two chapters in the handbook show how one of the key tools of environmentally conscious engineering—life-cycle evaluation or life-cycle analysis—can be used to minimize waste and pollution in chemical plant operations and to improve the performance of building materials. In addition, water and wastewater quality are treated in several chapters.

I asked the lead contributors, who are from four continents (eight are from the United States, plus one each from Canada, Finland, Greece, Thailand, and Australia), to provide short statements about the contents of their chapters and why the chapters are important. Here are their responses:

Michael G. Nelson (University of Utah in Salt Lake City, Utah), who contributed the chapter on **Engineering of Mineral Extraction**, writes, "Mineral extraction provides most of the materials and fuels needed for the comfortable and productive lives desired by all the world's people. As developing countries raise living standard for their inhabitants, it is particularly important that mineral extraction take place with full consideration the protection of the environment. This chapter describes the methods used for the extraction of solid minerals from the earth, and details the procedures to be followed to ensure that the environment is protected."

Carl C. Nesbitt (Michigan Technological University in Houghton), who contributed the chapter on **Waste Reduction in Metals Manufacturing**, writes, "The metals industry is a mainstay of the economy of the world; a primary indicator of growth and strength of any nation's economy can be tied to the consumption or production of metals. However, waste production is unavoidable within this large industry. This chapter summarizes the processes and strategies implemented by the industry to lessen the impact of wastes from metal production and recycle, and illustrates the challenges of reducing wastes at all of the production stages."

David Brennan (Monash University in Victoria, Australia), who contributed the chapter on **Life-cycle Evaluation of Chemical Processing Plants**, writes, "Maximizing resource efficiency and minimizing waste over the life cycles of products and their processing plants are globally accepted as urgent priorities. Achieving these objectives requires an ability to understand and define complex systems, to evaluate system performance using both economic and environmental criteria, and to project system performance over extended time horizons. This chapter addresses some key concepts, strategies, and methodologies required to develop the necessary skills and understanding for tackling these tasks."

Ali Elkamel (University of Waterloo in Ontario, Canada), who along with Cheng Seong Khor, Chandra Mouli, and R. Madhuranthakam, contributed the chapter on **Waste Reduction for Chemical Plant Operations**, writes, "The paradigm of green engineering that advocates environmentally conscious practices and operations has emerged as the perennial driving force of the industrial sector of the global economy. This chapter aims at providing an integrated treatment of objectives and methods for identifying and implementing waste reduction and pollution prevention (P2) initiatives in the chemical industry. We consolidate a set of tools, techniques, and best practices that would ensure the successful development of an economically sustainable pollution prevention program with a multiscale survey and categorization of these available approaches at the levels of macro-, meso-, and microscale."

C. Visvanathan (Asian Institute of Technology in Klongluang Pathumthani, Thailand), who contributed the chapter on **Industrial Waste Auditing**, writes, "As new opportunities to reuse and recycle industrial waste are explored, complete profiling of waste is essential. The first step to exploring such an opportunity is to undertake comprehensive waste auditing. This chapter discusses industrial waste audit techniques in an easy, simple, step-by-step approach that can even be understood by those who are new to the subject."

Jane C. Ammons (Georgia Institute of Technology in Atlanta, Georgia), who along with Hsuan Hong and Matthew J. Realff contributed the chapter on **Reverse Production Systems: Materials Production from Wastes**, writes, "Reverse production systems to handle return flows of supply chain and production wastes, packaging, and end-of-life products are evolving rapidly due to increasing pressures from economic opportunities, legislation requirements, and global environmental concerns. Key to the profitability, prevalence, and long-term growth and stability of these recycling/reuse systems are decisions associated with network design, growth, and market adaptation. This chapter compares the performance of centralized versus decentralized approaches for strategically planning reverse production systems so that economic viability can be maximized and material flow secured."

Bruce Lippke (University of Washington in Seattle), who along with James L. Bowyer contributed the chapter on **Environmental Life Cycle Analysis of Alternative Building Materials**, writes, "Environmental improvement is a universal goal. Companies want to improve their environmental performance. Builders and architects want to produce environmentally responsible buildings. Governments want to minimize the burden of their purchases. Environmental life-cycle assessment, or LCA, has become the tool of choice for identifying environmental improvement. This chapter describes the tools, and provides examples of how life-cycle assessment can be used to improve environmental performance in building construction through design, material, and process selection."

Say Kee Ong (Iowa State University in Ames), who contributed the chapter on **Wastewater Engineering**, writes, "Regardless of the waste reduction/pollution prevention efforts employed in many industrial processing facilities, there will be a certain quantity of industrial wastewater generated that needs to be treated before it can be discharged safely into the environment. This chapter outlines some of the legislative constraints on wastewater discharge and current technology and practices for the treatment of industrial wastewater from industrial facilities."

Eva Pongrácz (University of Oulu in Oulun Yliopisto, Finland), who contributed the chapter on **Environmental Impacts of Packaging**, writes, "Packaging is a part of modern life, offers convenience, ensures safe delivery of the product to the consumer while preserving the designed and processed values of the product. The concern about the effect of packaging on the environment derives from the relatively high percentage of packaging waste in household waste (about one third). Recovery of packaging wastes through recycling is thus strongly promoted by legislative bodies. However, the amount packaging waste is also a signal of the level of consumerism, and its recovery requires the involvement of the public in both collective and individual levels."

Fiona M. Doyle (University of California at Berkeley), who along with Gretchen Lapidus Lavine contributed the chapter on **Aqueous Processing for Environmental Protection**, writes, "Aqueous processing is already well established for the primary processing of many minerals, materials, and chemicals. However, it has enormous potential that is still relatively untapped for treating waste materials, and replacing existing processing technologies whose environmental impact is becoming recognized as unacceptable. This chapter summarizes many of the unit operations used in aqueous processing, with the goal of providing the information needed to modify these operations for different situations, while achieving appropriate water quality objectives."

Georgios N. Anastassakis (National Technical University of Athens in Greece), who contributed the chapter on **Solid Waste Disposal and Recycling**, writes: "This chapter deals with the unit operations that can be used to separate the streams of solid waste into recyclable single constituents and combustible products minimizing the quantity of material to be landfilled. This chapter is a valuable tool to waste processing engineers, as it describes in details the working principles of the various separation methods as well as the auxiliary processes that secure a smooth operation of the units (feeding, handling) and marketable products (agglomeration, baling, storage). In each case, the available equipment and the prerequisites for their application are presented."

Gajanan Bhat (The University of Tennessee in Knoxville) who contributed the chapter on **Processing of Post-Consumer Recycled Plastics**, writes, "Plastics continue to replace traditional materials in many applications and are being used in several new products. Many advantages of plastics in their performance as well as in the ease of their fabrication have contributed to their growth. Since large shares of the products are used as one-time or short-time usage products, they end up in the waste stream. The continuing growth of plastics consumption has created problems in municipal solid waste (MSW) disposal. Thus, it is important to find a suitable way of dealing with plastic waste. The problems, issues and opportunities in the post-consumer recycling of plastics are discussed in this chapter. Successful recycling of post-consumer plastics involves collection, cleaning, separation and then processing to produce value added products from them. A brief overview of plastics market, waste generation from plastics, current level of recycling, and products produced from recycled plastics is also provided."

Philip A. Marrone (SAIC/FOCIS Division in Newton, Massachusetts), who along with Glenn Hong and Kevin Downey contributed the chapter on **Supercritical Water Oxidation (SCWO)**, writes, "Supercritical water oxidation is an organic waste destruction technology that provides extremely clean and environmentally benign byproducts. This chapter describes the science and technology behind this elevated pressure and temperature process, with a focus on practical aspects for various applications." That ends the contributors' comments. I would like to express my heartfelt thanks to all of them for having taken the opportunity to work on this book. Their lives are terribly busy, and it is wonderful that they found the time to write thoughtful and complex chapters. I developed the handbook because I believed it could have a meaningful impact on the way many engineers approach their daily work, and I am gratified that the contributors thought enough of the idea that they were willing to participate in the project. Thanks also to my editor, Bob Argentieri, for his faith in the project from the outset. And a special note of thanks to my wife Arlene, whose constant support keeps me going.

Myer Kutz
Delmar, New York

CHAPTER 1

ENGINEERING OF MINERAL EXTRACTION

Michael G. Nelson
University of Utah Salt Lake City, Utah

1	INTRODUCTION	1
2	THE RANGE OF MINING ACTIVITIES	1
3	TRADITIONAL MINING PRACTICES	2
4	ENVIRONMENTALLY CONSCIOUS PRACTICES	3

4.1	Exploration	4
4.2	Development	8
4.3	Extraction	12
4.4	Reclamation	20
4.5	Closure	26
5	CASE STUDY	29

1 INTRODUCTION

Mining may be defined as the removal from the earth of solid material, to be sold at a profit. Note that liquid and gaseous products, such as petroleum and natural gas, are excluded from this definition. Table 1 shows one way of classifying materials that are recovered by mining, including examples for each category. Mining of stone and sand is sometimes referred to as quarrying.

Activities associated with mining have direct and lasting effects on both the physical and the human environments. This chapter considers only the former. The social, economic, and political effects in the latter category are addressed as *sustainability* issues, and are extensively discussed elsewhere,[1] as are issues of worker health and safety.[2]

2 THE RANGE OF MINING ACTIVITIES

Mining includes the following activities:[2]

1. *Exploration:* Economic deposits are identified and their characteristics are determined to allow recovery.
2. *Development:* Preparations are made for mining.
3. *Extraction:* Valuable material is removed for sale or processing.
4. *Reclamation:* Disturbances caused by any of the preceding activities are corrected or ameliorated.

Table 1   Classifications of Mined Resources

Metals	Industrial Minerals	Stone and Sand	Fuels
Copper	Potash	Dimension stone	Coal and peat
Gold	Phosphate	Sculptural stone	Oil shale
Iron	Salt	Natural gravel	Tar sand
Lead	Trona	Crushed rock	Oil sand
Silver	Kaolin	Paving stone	
Platinum	Bentonite	Building stone	
Tin	Limestone		
Zinc			

5. *Closure:* Activity ceases and the area is abandoned or returned to another use.

Mining occurred in prehistoric times with the extraction of flint and other rocks for use in tools, and of red ochre, a hydrated iron oxide, for use in coloring the bodies of the deceased, possibly to give an appearance of continued vitality.[3] Mining activity has occurred on every continent except Antarctica, where it is prohibited by international treaty.

Mining is a basic activity, underlying all industrial and commercial activity. In 2003, mineral production in the United States was valued at over $57 billion, approximately one-half of one percent of the gross domestic product. Mining is an important industry in 51 developing countries—accounting for 15 to 50 percent of exports in 30 countries and 5 to 15 percent of exports in another 18 countries. It is important domestically in three other countries.[1]

3   TRADITIONAL MINING PRACTICES

In the recent past, mining was done with little concern for its effects on the environment. The consequences of this approach often resulted in significant damage to the natural environment, including (but not limited to) the following:[4]

- Unreclaimed mine pits, shafts, tunnels, and refuse piles may result in landslides and large amounts of blowing dust.
- Surface and groundwater may be contaminated by solid particulates and chemical contaminants released by active and abandoned workings.
- There are potential hazards to humans, livestock, and wildlife from falling into abandoned pits, shafts, and tunnels.
- Erosion, with its consequent loss of soil and vegetation in and around unreclaimed workings can be a significant problem.
- Abandoned, underground workings may cause with surface subsidence.

Examples of these kinds of mining-induced damages are readily found in any historic mining district. In fact, water contamination and abandoned waste piles left by ancient Roman mining activity in Spain resulted in rediscovery of the orebody on which Rio Tinto, one of the largest mining companies in the world, was founded.[3]

In most cases, the environmental damage caused by mining was accepted by society because of the economic benefits that derived from mineral extraction. There were some notable exceptions, such as the lawsuit by downstream farmers against gold miners in California's Mother Lode district. Those miners used a method called *hydraulicking*, in which entire hillsides were washed away with a powerful stream of water, so the gold-bearing gravels could be processed. This resulting debris clogged streams and rivers and flooded meadows and fields, causing serious damage to agriculture.[5]

The public's expectations of the mining industry began to change in the 1950s, and by the end of the 1970s, governments in developed countries had enacted broad environmental laws that had direct bearing on all industrial activities, including mining.[6] Although environmental standards still vary among countries, almost all major mining companies now state as policy that they will operate all their mines, regardless of location, to *first-world standards* of environmental protection and worker health and safety.

4   ENVIRONMENTALLY CONSCIOUS PRACTICES

This section describes the processes in each of the mining-associated activities already listed, then describes the procedures and precautions required to ensure that each process is completed in an environmentally conscious manner.

Before discussing the individual activities involved in mining, it is important to point out that, to be truly environmentally conscious, all activities must be conducted with the future in mind. This will not only minimize the environmental effects of each activity, but will also result in significant cost savings. Three examples will illustrate this principle:

1. Design and location of roads developed for exploration should consider future needs for mining and processing. This will minimize unnecessary road building and decrease effects on the local ecosystem.
2. When containment ponds are required for drilling fluids or cuttings, consideration should be given to the future use of those ponds for tailings impoundment, stormwater catchment, or settling ponds.
3. The overall mine plan should include consideration of the requirements of reclamation and restoration. Placement of waste piles, tailings ponds, and similar materials should be carefully planned to minimize rehandling of material.

The mining operation should be thoroughly planned, in consultation with government agencies and people living in the region. Historic, cultural, and biological resources should be identified, and plans made for their protection. This is especially important in areas where indigenous people have little exposure to the technologies used in mining. Mining companies should take the appropriate steps to ensure that the indigenous people understand the mining plans. The indigenous peoples' values and land use practices must be understood, and steps must be taken to protect them. In some locations, this will necessitate the involvement of anthropologists, sociologists, and other experts. It may also require a planning and approval process that differs from those to which the company is accustomed. For example, some indigenous peoples make decisions by group consensus, necessitating large community meetings that may last for several days.

4.1   Exploration

Exploration includes three activities: reconnaissance, sampling, and geophysical surveys.

Reconnaissance

Reconnaissance often begins with aerial surveying and photogrammetry. This is followed by work on the ground, in which topography, surface geology, and geologic structures are mapped in detail. Obviously, the ground work requires that geologists, surveyors, and other personnel have access to the prospective mining property, with the ability to conveniently move about.

The primary environmental concern during reconnaissance is the effect of personnel movement and habitation. The following precautions should be observed:

- When personnel are transported to exploration sites by helicopter or airplane, landing strips and pads should be carefully selected to avoid damage to sensitive habitat and endangered or threatened species.
- When personnel travel around the exploration site in motorized vehicles, care should again be taken to ensure that sensitive habitat and endangered or threatened species are not damaged by tire tracks, or disturbed by engine exhaust and noise.
- Campsite locations should be carefully chosen to minimize effects on habitat, flora, and fauna.
- Low-impact camp practices should be observed. All trash should be collected and stored for removal when the camp is vacated. In especially sensitive areas, such as permafrost tundra and designated wetlands, solid human waste should also be stored and removed.
- Metal and plastic survey markers should be used carefully, and should be removed if exploration ceases and no further development occurs.

Sampling

Sampling includes geochemical sampling and material sampling by drilling and excavation. *Geochemical samples* are relatively small samples of surface water and soil, which are analyzed for a suite of trace elements whose presence in certain concentrations can indicate the presence of a mineral deposit. Geochemical sampling often occurs concurrently with reconnaissance, and the environmental concerns and precautions are identical to those noted for reconnaissance.

In *drilling and excavation*, larger amounts of material are removed for analysis and process testing. Thus, the environmental concerns are greater. Power equipment, often large, must be delivered to the site and moved to each sampling location. Some small drills can be delivered by helicopter, but most drills are either self-powered or moved by truck or tractor. Drills must be set up and operated on a level, stable *drill pad*, which is usually prepared with powered excavating equipment. The process requires water as a drilling fluid, and may use additives in the fluid. Sample excavations may be small test pits sunk with a backhoe, trenches cut with a bulldozer, or large, bulk sample pits developed by drilling and blasting.

The environmental consequences of sampling vary considerably among the various activities of sampling. All sampling activities require that personnel and equipment be moved to and around the exploration site; thus, all of the precautions listed in the previous section apply to this activity as well. In addition, the following steps should be considered:

1. When any material is removed for samples, the disturbed surface should be restored as nearly as possible to its original condition. For small samples, taken for geochemical analysis, this will likely require nothing more than smoothing the surface with a shovel and carefully replacing any small plants removed in sampling. For larger samples, backhoe pits, dozer trenches, and large test pits should be filled, recontoured, and revegetated with an appropriate mix of seeds and planting, as described in more detail in Section 4.4 on Reclamation. The length of time for which large sample sites may be left open is usually stated in the exploration permit, based on the applicable government regulations.

2. Roads and drill pads should also be reclaimed as just described. In some cases, the jurisdictional government may wish exploration roads to be left in place.

3. Drilling fluids and drill cuttings not removed as samples should be disposed of properly. These materials should not be placed into surface waters. They should be placed in stable configurations, with embankments and diversions provided to prevent contamination of runoff water where necessary.

4. In the development of pits for bulk samples, more extensive design for environmental compliance is required:

    a. Sample pits will likely require reclamation similar to that required for full-scale mining pits.

    b. It will likely be necessary to strip and stockpile any topsoil, for use in reclamation. It may also be necessary to divert surface water flow, both in streams and runoff, so that water will not flow through the pit and be contaminated.

    c. If the pit is deep enough to encounter groundwater, the groundwater monitoring procedures described in Section 4.3 on Extraction may be required.

    d. If bulk sample pits are developed by drilling and blasting, correct procedures for the handling and use of explosives must be observed.

Geophysical Surveys

Geophysical surveys involve the measurement of certain of the earth's physical properties, anomalous values of which may indicate the presence of valuable minerals. These properties include electromagnetic and gravitational field strength, electrical conductivity or resistivity, seismic reflectivity, and natural or induced radiometric emission. Preliminary electromagnetic and gravitational geophysical surveys are usually conducted using airborne instruments, and thus have very few environmental consequences.

Electromagnetic and gravitational surveys on the ground require that the appropriate sensor or sensors be moved over the site following a surveyed grid, so that the sensor readings are correlated with location. Some sensors are small enough to be carried in a backpack, while others are carried on a motorized vehicle.

Electrical and seismic surveys require the delivery to site and emplacement of a number of sensing instruments, which are usually connected by cables to a survey vehicle. The survey vehicle generates required output signals, electrical or seismic, then receive and log the signals returned by the sensors. Seismic surveys require the inducement of high-amplitude, seismic waves into the ground. This is done using explosives, projectiles, or large, hydraulically actuated *pads* that directly contact the surface of the ground.

Radiometric surveys measure naturally occurring or artificially induced ionizing radiation. Measurement of naturally occurring radiation is made for two purposes. The first is to locate ores of uranium or other radioactive elements. Initial surveys of this type are often airborne, and are followed up by traversing the surface and taking readings with a Geiger counter or similar device, as is the practice with electromagnetic or gravitational surveys. The second is to locate strata or other geologic members that are known to emit ionizing radiation. For example, the marine shales that often overlie coal seams emit gamma radiation, because small amounts of radon daughters are concentrated there by the depositional process. This type of survey is almost always conducted by measuring the radiation in a drill hole, as a function of depth. This may be done by using an instrumented drill string, or by lowering a radiation sensor into a previously drilled hole.

Surveys of artificially induced ionizing radiation are based on the fact that bombardment with activated neutrons causes many elements to emit gamma rays. The various elements present in the activated material are identified by the characteristic wavelengths of the emitted gamma rays. These surveys are almost always conducted by measuring the induced radiation in a drill hole, as a function of depth. This may be done by using an instrumented drill string that includes a neutron source and gamma detector, or by lowering such an assembly into a previously drilled hole.

The environmental consequences of geophysical surveys vary considerably among the various activities just described. When geophysical surveys are conducted on the ground, seven precautions should be observed:

1. When personnel are transported to survey sites by helicopter or airplane, landing strips and pads should be carefully selected to avoid damage to sensitive habitat and endangered or threatened species.
2. When personnel travel around the exploration site by motorized vehicles, care should again be taken to ensure that sensitive habitat and endangered or threatened species are not damaged by tire tracks, or disturbed by engine exhaust and noise.
3. Campsite locations should be carefully chosen to minimize effects on habitat, flora, and fauna.
4. Low-impact camp practices should be observed. All trash should be collected and stored for removal when the camp is vacated. In especially sensitive areas, such as permafrost tundra and designated wetlands, solid human waste should also be stored and removed.
5. Metal and plastic survey markers should be used carefully, and should be removed if exploration ceases and no further development occurs.
6. All sensors, wires, and other electronic equipment should be removed from the survey site. This may not be an issue for electromagnetic and gravimetric surveys.
7. Roads and vehicle parking pads should also be reclaimed. For electrical, seismic, or radiometric surveys, it will likely be necessary to strip and stockpile any topsoil, for use in reclamation. It may also be necessary to divert surface water flow, both in streams and runoff, so that water will not by flowing across roads and pads. In some cases, the jurisdictional government may wish exploration roads to be left in place.

There are a few other precautions specific to the type of survey used. The effects of the induced seismic waves used in seismic surveys should be carefully analyzed, for potentially harmful effects on wildlife and nearby human structures. Also, when conducting radiometric surveys, drilling fluids and drill cuttings not removed as samples should be disposed of properly. These materials should not be placed into surface waters. They should be placed in stable configurations, with embankments and diversions provided to prevent contamination of runoff water where necessary.

4.2 Development

Development is the preparation the facilities, equipment, and infrastructure required for extraction of the valuable mineral material. It includes land acquisition, equipment selection and specification, infrastructure and surface facilities design and construction, environmental planning and permitting, and initial mine planning.

Land Acquisition

Land acquisition is the activity that results in attaining the required level of control over those properties that are necessary for the timely development of a mining project. It may include outright purchase, rental, or leasing of the pertinent land parcels. In some cases, a mining company may purchase or lease the rights to extract a particular mineral commodity, while another party may have rights to extract another commodity, or to use the land for some other purpose. For example, one company may have the right to mine coal under a tract of land, while another may have the right to extract oil and gas. Similarly, the respective rights to extract minerals and to harvest timber may be separately owned or leased, or one entity may have mining rights while another has rights to graze livestock on the surface.

In land acquisition, the following environmental considerations apply:

1. *The land acquired should include all areas that will be used in extraction of the mineral deposit.* This must include area for tailings ponds, waste piles, topsoil stockpiles, access roads, surface facilities, and so on. Care must be exercised to ensure that an inadequate land position does not force installation of surface facilities in unsuitable locations—a waste dump located too close to a river, for example.

2. *Acquisition of land parcels with preexisting environmental contamination may result in the obligation of the acquiring company to remove or ameliorate the contamination.* For example, a company may acquire an old mine site, with the intention of exploiting previously uneconomic material, and find that existing buildings on the site contain asbestos insulation. The new owner may be required to remove that asbestos, an expensive process.

3. *In areas where endangered species are known, those species must be protected.* It may be necessary to acquire additional land to provide replacement habitat for animals or plants disturbed by mining activity, or to allow design of mining operations that will disturb existing habitat.

Equipment Selection and Specification

All equipment to be used in the mine is selected and specified in the development stage. This includes production and support equipment. Two environmental considerations apply:

1. *The source of power for mobile equipment should be carefully analyzed.* This choice affects the gaseous emissions from engines in the equipment, and also the overall energy consumption balance of the mine. If electricity is readily available, electric or electric-assisted haulage equipment may be preferred. At high altitudes, internal combustion engines require modifications to ensure efficient and clean operation.

2. *Requirements for air quality control should also be analyzed.* These may influence the selection of power sources for mobile equipment, and will also determine the number and type of water trucks required for dust control on the surface.

Infrastructure and Surface Facilities Design and Construction

Infrastructure and surface facilities may include all the following:

- Roads and railroads
- Electric power lines and substations
- Fuel supply lines and tank farms
- Water supply lines, water tanks, and water treatment plants
- Sewage lines and sewage treatment plants
- Maintenance shops
- Storage sheds and warehouses
- Office buildings, parking lots, shower facilities, and changehouses
- Worker accommodations (housing, cafeteria, infirmary, and recreation facilities)
- Houses for pumps, fans, and hoists
- Waste piles and impoundments
- Ponds for catchment of surface and ground water, and mine drainage

Six general environmental considerations apply to the design and construction of all surface facilities and infrastructure:

1. *Select locations of surface structures and infrastructure to minimize the effects of their construction and utilization on surface water, groundwater, plant and animal ecosystems, and nearby human habitation.* Surface

structures and infrastructure include roads, railroads, and power lines, for example.

2. *Remove and store topsoil for future use from areas that will be later reclaimed.* This applies only to facilities and infrastructure that will be razed at the end of the mine's life.

3. *Control runoff so that all water exiting the boundaries of the permitted mine site is captured and treated as required to meet applicable discharge standards.* Many jurisdictions permit mines as *zero-discharge* facilities, meaning that all water or solid waste is contained within the permitted mine area.

4. *Protect surface water and wetlands, using buffer zones or stream relocation designed to protect streams and wetlands.* Because these features may exist directly over mineral reserves, their relocation may be necessary.

5. *Revegetate disturbed areas.* Recontour the land to a defined standard and revegetate with an approved mix of seeds and plantings.

6. *Control emissions of dust and noise to meet the requirements of local regulations and the reasonable expectations of persons living nearby.* In particular, if explosives are used, control air blast and ground vibration as required.

Specific considerations for certain types of surface facilities or infrastructure include the following:

- Maintenance shops should be designed to avoid contamination of soil and water by spilled fuel and lubricants.
- Surface facilities should be designed to minimize energy consumption, utilizing solar water heating and alternative electric power generation wherever possible.
- Surface facilities should also be designed architecturally to harmonize with the natural surroundings in the locale.

Environmental Planning and Permitting

The forward-thinking approach mentioned earlier is especially important in environmental planning and permitting. In the *permitting process*, careful planning is imperative, so that information is gathered efficiently, the applications submitted to regulators meet all requirements, including timely submission, and mine planning costs are minimized by avoiding redesign required after agency review. As pointed out by Hunt, "One of the most serious deficiencies encountered by many operators in the permitting process is to fail to adequately plan the time required to collect baseline data and to provide for public and agency review of applications."[7] Hunt goes on to list the major tasks and activities required during early phases of environmental planning:

- Identification of the regulatory agencies that will be responsible for permitting the proposed mining operation
- Acquisition of all necessary permitting forms and copies of applicable regulations
- Identification of possible legal or technical restrictions that may make approval difficult or require special attention in the permitting process
- Identification of environmental resources information that will be required and the development of a plan and schedule for collection of the required data[7]

After preliminary planning, permit applications must be completed. Many jurisdictions require a detailed environmental audit, which is then described in a report, often called an *environmental impact statement*, or EIS. Depending on local regulations, the EIS may be issued for public review and comment. Topics covered in an EIS usually include the following fifteen areas:

1. Permits required and permitting status
2. Ownership of surface and mineral rights
3. Ownership of the company or companies that will be operating the mine, including contractors
4. Previous history of the entities described in Item 3
5. Information on archaeological, historical, and cultural resources within the planned mine site and adjacent areas
6. Hydrology of the mine site and adjacent areas, including the presence of potential pollutants and plans for dealing with them
7. Potential for dust and air pollution, and plans for mitigation
8. Plans for use and control of explosives
9. Plans for disposal of mine waste, including methods for controlling dust generation, slope stability, and seepage of contaminated water
10. Plans for impoundments, and methods for preventing failure or overtopping
11. Plans for construction of backfilled areas during mining, and for backfilling of mine excavations when mining is finished, including methods for preventing slides or other instability
12. Potential for surface subsidence, both immediate and long-term, and proposed methods of preventing subsidence or compensating for its negative effects
13. Potential for fires in mine workings, at outcrops, and in spoil piles, and methods for prevention and control of such fires
14. Planned use of injection wells or other underground pumping of fluids

15. Information on previous mine workings, plans for reclaiming previous workings, and description of how the planned workings will interact with previous workings

Mine Planning

It may be said that all of the development activities described to this point constitute *mine planning*. However, the term as used here refers specifically to the planning of the mine workings, including gaining access to the mineral deposit, removing the valuable mineral, and handling any waste material produced. The activity as here comprehended also includes tasks described as *mine design*. Although the details of these processes are described in the next section, planning for them is begun in the development stage, and is thus discussed at this point.

In mine planning, the specific, day-to-day operations of the mine are set out, analyzed, and documented. Mine planning continues throughout the life of every mine, to account for changing geologic and economic conditions. The objective of mine planning at each point in time is to determine the manner in which the mine should be operated to optimize the goal(s) of the operating entity, usually the return on investment.

Although mine planning varies with site-specific conditions, the following points must be determined in the initial plan, and continually reevaluated throughout the life of the mine:

- The extent of the mineral deposit
- The market for anticipated products of the mine
- The manner in which the deposit will be accessed for extraction
- The equipment and personnel that will be used for extraction
- The operating sequence for extraction
- The disposition of extracted material, both waste and valuable mineral
- The interactions of mine workings with adjacent or overlying areas and properties

Initial mine planning will obviously take place concurrently with the environmental planning and permitting, and the consideration of environmental consequences of the initial mine plan will be included in this process.

4.3   Extraction

Mining methods may be classified according to a variety of schemes. Here it is convenient to distinguish *surface mining*, *underground mining*, *aquatic* or *marine mining*, and *solution mining*.

Solution mining is the removal of valuable minerals from in-place deposits, by dissolving the mineral in a suitable liquid that is then removed for recovery of the desired constituent. It may also be called *in-situ mining* or *in-situ leaching*. Solution mining uses techniques similar to those used in the extraction of petroleum and natural gas, and is thus not discussed further in this chapter. The operational details and environmental considerations of surface mining, underground mining, and aquatic or marine mining are discussed separately.

Surface Mining

Surface mining is the removal of material from the earth in excavations that are open to the surface. In some cases, material is removed directly from the earth's surface; for example, sand and gravel may be removed directly from deposits of those materials put in place by ancient lakes or rivers. In other cases, material that has no economic value (*overburden*) covers the valuable material, and must be removed.

A typical operational cycle for surface mining, described in detail by Saperstein, is as follows:[8]

1. Install erosion and sedimentation controls.
2. Remove topsoil from areas to be mined.
3. Prepare the first drill bench by leveling the bench with a bulldozer, inspecting and scaling the highwall as required, and laying out the blast holes.
4. Drill the blast holes.
5. Blast the rock.
6. Load the fragmented material.
7. Haul the fragmented material, waste to the waste dumps and product to loadout for sale or to subsequent processing.
8. Manage the waste dumps as required by contouring waste piles to stable configuration or returning waste to mine workings for use in reclamation.
9. Prepare mine workings for reclamation by
   a. Recontouring to the original contour, or to another, approved and stable configuration.
   b. Returning the stored topsoil to the mine site, and spreading it uniformly on the recontoured surfaces.
10. Reclaim the prepared surfaces by
    a. Revegetating with an approved mixture of seeds and plantings.
    b. Irrigate and maintain revegetated areas as required until a stable condition is reached.
11. Remove temporary drainage controls and stream diversions.

The environmental considerations that must be addressed during surface mining include the following:

1. Control drainage from mine workings to avoid contamination of surface water, groundwater, and the existing ecosystem. Treatment may be required to remove particulates or chemical contamination.

2. Control inflow or surface water and storm runoff so these waters are not contaminated by passing through the mine workings.

3. Carefully analyze the groundwater regime in and around the mine workings, and provide monitoring wells as required. In some cases, it may be advisable to minimize the interaction of the mine workings with groundwater by pumping out the aquifers in and around the mine, and discharging the pumped water to constructed wetlands on the surface.

4. Design and construct waste dumps to minimize erosion and protect surface and groundwater.

5. Control emissions of dust and noise to meet the requirements of local regulations and the reasonable expectations of persons living nearby. In particular, if explosives are used, control air blast and ground vibration as required.

6. When mining is done at night, consider the effects of artificial light on wildlife and humans living near the mine.

7. Control slope stability and landslides. Of course, this is also necessary for successful mining operation, but inadequate control of slope stability may also lead to environmental damage. Slope failure may block or contaminate streams, damage wildlife habitat, and cause flooding that endangers wildlife and humans living near the mine.

8. Consider the effects of increased traffic to and from the mine site, especially when product haulage will result in a marked increase in heavy truck traffic.

Reclamation and closure of surface mines are discussed in Sections 4.4 and 4.5.

Underground Mining

Underground mining is the removal of material from the earth in excavations below the earth's surface. Access to such underground workings may be gained through a *drift* or *adit*, a *shaft*, or a *slope*. Drifts and adits are horizontal tunnels, usually in a hillside, that connect to the mineral deposit. In the case of minerals like coal that occur in *seams* (near horizontal deposits of fairly uniform thickness and considerable areal extent), the drift may be developed in the mineral itself. Travel in drifts is by rail or rubber-tired vehicles. Shafts are a vertical tunnels developed from the surface to access mineral bodies below. Travel in shafts is by cages or cars, which are lowered and raised by a mechanism on the surface, similar to elevators in tall buildings. Slopes are tunnels neither vertical nor horizontal, and may be lineal or spiral. Travel in lineal slopes may be by rubber-tired vehicles or hoists, but rarely by rail. Travel in spiral slopes (also called *ramps*) is by rubber-tired vehicles.

Again, Saperstein has described in detail a typical operational cycle for underground mining in which material is fragmented by drilling and blasting, as summarized:[8]

1. Enter the workplace after the previous blasting round is detonated.
2. Ensure that the workplace is in good condition and safe for continued work by checking that ventilation is adequate and that blasting fumes have been removed; providing for dust suppression; checking for the presence or hazardous gases; and inspecting for and removing loose material.
3. Load the fragmented material.
4. Haul the fragmented material to the appropriate location.
5. Install ground support as required.
6. Extend utilities as required: ventilation, power (electricity or compressed air), and transportation.
7. Survey and drill blast holes for the next round.
8. Load the explosives and connect the detonation system.
9. Leave the workplace and detonate the round.

This cycle is typical for the mining of narrow, steeply dipping veins (typical of metal ores), and for massive deposits of limestone, salt, and dimension stone, where the vertical extent of the valuable mineral is 6 meters or higher.

The operational cycle for methods in which material is mechanically fragmented and removed in a continuous process may be similarly summarized:

1. Enter the workplace after the required ground support has been installed.
2. Ensure that the workplace is in good condition and safe for continued work by checking that ventilation is adequate and that blasting fumes have been removed; providing for dust suppression; checking for the presence or hazardous gases; and inspecting for and removing loose material.
3. Cut and load the fragmented material until the limit for advance is reached.
4. Concurrently, haul the fragmented material to the appropriate location.
5. Remove fragmentation, loading, and hauling equipment from the workplace.
6. Install ground support as required.
7. Extend utilities as required: ventilation, power (electricity or compressed air), and transportation.
8. Survey as required to ensure that the mined opening is maintaining the required dimensions and directional orientation.

This cycle is typical for mining in near-level seams of relatively soft material that have considerable areal extent and are thinner than 6 meters, such as coal, trona, phosphate, and potash.

Seven environmental considerations must be addressed during underground mining:

1. Locate ventilation fans and hoist houses to minimize the effects of noise on wildlife and humans living nearby.
2. Whenever possible, dispose of mine waste in underground mine workings.
3. When waste dumps are constructed on the surface, design them to minimize erosion and protect surface and groundwater.
4. Control drainage from mine workings to avoid contamination of surface water, groundwater, and the existing ecosystem. Treatment may be required to remove particulates or chemical contamination. Control drainage from mine workings to avoid contamination of surface water, groundwater, and the existing ecosystem. Treatment may be required to remove particulates or chemical contamination.
5. Control inflow or surface water and storm runoff so these waters are not contaminated by passing through the mine workings.
6. Carefully analyze the groundwater regime in and around the mine workings, and provide monitoring wells as required. In some cases, it may be advisable to minimize the interaction of the mine workings with groundwater by pumping out the aquifers in and around the mine, and discharging the pumped water to constructed wetlands on the surface.
7. Analyze and predict the subsidence likely to result from mine workings. Design workings to minimize the effects of subsidence on surface structure, utilities, and important natural features such as lakes, streams and rivers, and wildlife habitat.

Reclamation and closure of underground mines are discussed in Sections 4.4 and 4.5.

Aquatic or Marine Mining

Aquatic or marine mining is the removal of unconsolidated minerals that are near or under water, with processes in which the extracted mineral is moved by or processed in the associated water. This type of mining may also be referred to as *alluvial mining* or *placer mining*. The two major types of aquatic or marine mining are *dredging* and *hydraulicking*.

Materials typically recovered by aquatic or marine mining have usually been deposited in fluvial, aeolian, or glacial environments, and are thus unconsolidated. They include aggregate (sand and gravel), and materials deposited because of their relatively high specific gravities. The latter deposits are called *placers*, and include native (naturally occurring) precious metals, tin (as the oxide cassiterite), heavy mineral sands (oxides of zirconium, hafnium, titanium, and others), and precious stones. In some cases, placer deposits are mined by methods described in the section on surface mining; in such cases, the environmental precautions given for those methods apply.

Dredging is the use of a powered mechanism to remove unconsolidated material from a body of water. The mechanism is almost always a type of bucket or shovel. In the simplest case, it may be a metal bucket moved by chains or steel cables that are attached to a pole. The bucket is dropped through the water and into the solid material on the bottom. As the bucket is retracted, its weight and trajectory force its leading edge into the solid material, and the bucket fills. When the bucket comes to the surface, it is emptied. This type of dredge is usually installed on the shore, near a body of water that covers a valuable mineral deposit.

More complex dredge mechanisms attach several buckets to a wheel or a *ladder*. A ladder is structure designed to support a series of buckets attached to a chain, which moves continuously in a loop. Both mechanisms are usually installed in a floating vessel, the *dredge*. While moving, the bucket wheel or the dredge ladder is lowered into the unconsolidated material below the surface, picking up that material and returning it to the dredge, where it is either further processed or transferred to the shore for further usage. The dredge vessel may operate in a natural body of water or in an artificial body called a *dredge pond*. A typical operational cycle for dredging is shown below:

1. If dredging in a dredge pond, complete the steps a–g; if not, proceed to Step 3.
   a. Locate a source of water for filling the pond.
   b. Install erosion and sedimentation controls, and divert surface water as required.
   c. Prepare dikes or dams required for the dredge pond. The perimeter of the dredge pond must extend beyond the extent of the mineral to be recovered, to allow for deposition of overburden or tailings removed in the first pass of the dredge.
   d. Fill the dredge pond.
   e. Remove vegetation.
   f. Remove and stockpile topsoil.
   g. If necessary, remove overburden with excavating equipment, and place overburden in stable piles.
   h. Proceed to Step 3.
2. If dredging from the shore, prepare the site for installation of the dredge mechanism.
3. Install the dredge.
4. Remove by dredging any overburden that was not removed in Step 1 g.

    a. When dredging in a natural body of water, the overburden will be deposited under the water, in an area from which the valuable mineral has already been removed, or in an area under which there is no valuable mineral.

    b. When dredging in a dredge pond or from the shore, the overburden will be deposited beside or behind the dredge, in an area from which the valuable mineral has already been removed, or in an area under which there is no valuable mineral. When mining in a dredge pond, the overburden will for a time be at least partially under water; when mining from the shore, it will not. However, in both cases, the overburden will over time drain and be left exposed. It will thus require reclamation as described below.

5. Remove the valuable material by dredging.

6. Process the valuable mineral as required, to concentrate the valuable constituent(s). This processing is almost always done on the dredge, and is integrated with the dredging of the material.

    a. When dredging in a natural body of water, the tailings will be deposited under the water, in an area from which the valuable mineral has already been removed, or in an area under which there is no valuable mineral.

    b. When dredging in a dredge pond or from the shore, the tailings will be deposited beside or behind the dredge, in an area from which the valuable mineral has already been removed, or in an area under which there is no valuable mineral. When mining in a pond, the tailings will for a time be at least partially under water; when mining from the shore, it will not. However, in both cases, the tailings will over time drain and be left exposed. They will thus require reclamation, as described below.

7. Remove the valuable constituent(s) for sale or further processing off site.

8. Deposit the tailings under the water, in an area from which the valuable mineral has already been removed, or in an area under which there is no valuable mineral.

9. Continue dredging until reaching the limits of the dredge's cables and other connections, the limits of the pond, or the boundary of the deposit.

10. If dredging in an artificial pond, prepare and fill the next pond as in Step 1, and transfer the dredge. If dredging in a natural body of water, move the dredge to the next location.

11. Begin dredge operation in the new location by returning to Step 3. Simultaneously, and in accordance with the approved reclamation plan, reclaim overburden and tailings piles from previous work, according to Step 12 and 13.

12. Prepare mine workings for reclamation:
    a. Recontour to a stable the original contour, or to another, approved and stable configuration.
    b. Return the stored topsoil to the mine site, and spread it uniformly on the recontoured surfaces.
13. Reclaim the prepared surfaces:
    a. Revegetate with an approved mixture of seeds and plantings.
    b. Irrigate and maintain revegetated areas as required until a stable condition is reached.
14. Remove temporary drainage controls and stream diversions.

There may be considerable variation in the cycle, depending on how the dredge is transported between sites.

Hydraulicking is a method of mining placer deposits that was used extensively in the past, but has fallen into disfavor because of the potential for serious effects on the stability of the remaining surface and on nearby surface waters.

Hydraulicking required a large deposit of auriferous alluvium and a source of water that would provide sufficient volume and pressure head. Overburden and pay gravel were both removed by high-pressure water that flowed through a *monitor* — a large nozzle that could be rotated horizontally and vertically. Material removed by the flow from the monitor moved to the bottom of the valley, where the flow was controlled so that overburden passed directly into the stream channel and pay gravel flowed through a sluice or similar recovery device for recovery of the gold. Tailings from the sluice also flowed into the stream channel.

As already mentioned, hydraulicking of the placer gold deposits in the famous California Gold Rush of 1849 led to severe contamination of the rivers that drained the area of the gold deposit. The solids from the hydraulicking operations filled rivers with solids, which were eventually deposited downstream, with serious consequences for agriculture in the downstream areas.

Hydraulicking is still a very inexpensive method for moving large volumes of unconsolidated material, and is still used on a limited basis, with five precautions:

1. Material discharge from a hydraulicking operation must be captured and treated to remove solids, including, in particular any fine, suspended solids. This will require a settling pond, and chemical flocculants may also be necessary. Even water that appears to be clear must be tested by appropriate methods, and treated on the basis of those tests.
2. Tailings, overburden, and deactivated settling ponds must be reclaimed appropriately, usually be recontouring and revegetation. It may be necessary to remove and stockpile top soil for use in reclamation. Coarse gravel, which may accumulate in separate piles, should receive special attention to ensure that it is reclaimed properly.

3. The area of the mining operation should be isolated from the flow of surface streams and runoff, to prevent contamination of those waters.

4. Discharge of water from the hydraulicking operation should be managed to prevent interference with the existing flow regimes in the drainage. The quantities and velocities of discharge should not modify the existing stream flow in a manner that will cause erosion, undercutting of banks, or flooding.

5. Highwalls and embankments produced by hydraulicking should be reclaimed, again by recontouring and revegetation. Even when top soil is removed and stockpiled, there may not be enough topsoil for the reclamation required. In such cases hydroseeding will likely be required.

4.4   Reclamation

In the most general sense, reclamation is the gaining or recovery of land for a purpose that is perceived as higher or more beneficial, as in the recovery of low-lying coastal land by the building of dikes, or the recovery of desert lands by irrigation. However, as applied to mining activities, "... *reclamation* is a response to any disturbances to the earth and its environment caused by mining activity."[9]

The terms *restoration* and *rehabilitation* are often used synonymously with *reclamation*. However, *restoration* implies that a given site is returned to the exact conditions that existed before the disturbance, and *rehabilitation* indicates that the disturbed land is returned to conditions that conform to a prior use plan (often agricultural). In fact, it is often neither possible nor desirable to return disturbed land to its previous condition or use.

Reclamation is often considered only in association with surface mining, but, as reflected in the previous discussion, the activities of underground mining may also necessitate reclamation. However, the reclamation activities required in association with underground mining are almost always less numerous and less extensive.

Reclamation practice varies widely, depending on the type of mine, its location, and the applicable legal requirements. In all cases, it is important from the beginning to include reclamation, and the required planning for reclamation, in the mining plan. As summarized by Ramani et al. and Riddle and Saperstein, there are four required planning steps:[10,11]

1. Make an inventory of the premining conditions.
2. Evaluate and decide on the postmining requirements of the region, consistent with legal requirements and the needs and desires of the affected groups and individuals.
3. Analyze alternative mining and reclamation schemes to determine how the postmining requirements can be most easily met.

4. Develop an acceptable mining, reclamation, and land-use scheme that is optimal for the technical, social, economic, and political or legal conditions.

Detailed descriptions of the reclamation procedures and requirements for all situations are beyond the scope of this chapter. Instead, a detailed description will be given for surface coal mining, with additional comments for other cases that are considered noteworthy.

Premine Planning and Permitting

The following description of reclamation requirements for surface coal mining is based on requirements in the United States, as given in the Surface Mining Control and Reclamation Act (SMCRA) of 1977.[12]

Under SMCRA, specific requirements are given for obtaining a mining permit. A formal Reclamation and Operations Plan is required to document the information and conclusions of the planning process. The reclamation plan should include descriptions of the following:

- The uses of the land at the time of the permit application, and if the land has a history of mining activities, the uses that predate any mining activity
- The capability of the land, prior to any mining, to support a variety of uses, based on soil and foundation characteristics, topography, and vegetation
- The proposed use for the land following reclamation, including a discussion of the utility and capacity of the reclaimed land to support a variety of alternative uses
- The manner in which the proposed postmining land use is to be achieved, and the activities required to achieve this usage
- The consideration given to making mining and reclamation operations compatible with existing surface owners' plans and applicable, governmental land-use plans
- The consideration given to developing the reclamation plan in a manner consistent with local physical, environmental, and climatological conditions

The Reclamation and Operations Plan must specifically address the following:[13]

- Air pollution control
- Protection of fish and wildlife
- Protection of the hydrological balance
- Postmining land uses
- Ponds, impoundments, banks, dams, and embankments
- Surface water diversions
- Protection of public parks and historic places
- Relocation or use of public roads

- Disposal of excess spoil and mine waste
- Transportation facilities
- Development of new workings near existing underground workings
- Subsidence control
- Maps, plans, and cross-sections

A large amount of information is required for preparation of the Reclamation and Operations Plan. Table 2 summarize those requirements.[9]

Land-Use Analysis

The Reclamation and Operations Plan also requires an analysis of suitable uses for the land after reclamation. It is not always necessary to restore land to its premining condition and utilization. Often, a utilization of higher value is acceptable. For example, a previously forested hillside adjoining pastureland might be acceptably restored as pasture. In areas with high relief, mountaintop removal and valley-fill methods may be used to create new flat areas, which have positive use for agriculture and residential development.

SMCRA identifies 10 categories of postmining land use: crop land; pasture or land occasionally cut for hay; grazing land; forest land; residential development; industrial or commercial development; recreation; fish and wildlife habitat; developed water resources; and undeveloped land. The most suitable postmining land use will be determined in cooperation with local residents and interest groups, and government agencies.

Reclamation Operations

Reclamation operations are complex and extensive. The time sequence for reclamation activities in surface coal mining is summarized as follows:[10]

1. During site preparation:
    a. Install control measures (diversion, sediment traps and basins, etc.).
    b. Clear and grub; market lumber if possible; stockpile brush for use in filters or chip brush for use a mulch.
    c. Stabilize areas around temporary facilities such as maintenance yards, power station, and supply areas.
2. During overburden removal:
    a. Divert water away from and around active mining areas.
    b. Remove topsoil and store as required.
    c. Selectively mine and place overburden strata as required.
3. During coal removal:
    a. Remove all coal insofar as possible.

Table 2   Information Needed to Prepare Reclamation and Operations Plan

Natural Factors

1. Topography
   a. Relief
   b. Slope
2. Climate
   a. Precipitation
   b. Wind patterns and intensity
   c. Humidity
   d. Temperature
   e. Climate type
   f. Growing season
   g. Microclimatic characteristics
3. Altitude
4. Exposure (aspect)
5. Hydrology
   a. Surface hydrology
      • Watershed considerations
      • Flood plain delineations
      • Surface drainage patterns
      • Runoff amounts and qualities
   b. Groundwater hydrology
      • Groundwater table
      • Aquifers
      • Groundwater flow amounts and qualities
      • Recharge potential
6. Geology
   a. Stratigraphy
   b. Structure
   c. Geomorphology
   d. Chemical nature of overburden
   e. Coal characterization

7. Soils
   a. Agricultural characteristics
      • Texture
      • Structure
      • Organic matter content
      • Moisture content
      • Permeability
      • pH
      • Depth to bedrock
      • Color
   b. Engineering characteristics
      • Shrink-swell potential
      • Wetness
      • Depth to bedrock
      • Erodibility
      • Slope
      • Bearing capacity
      • Organic layers
8. Terrestrial ecology
   a. Natural vegetation
      • Characterization
      • Uses and survival needs
   b. Crops
   c. Game animals
   d. Resident and migratory birds
   e. Rare and endangered species
9. Aquatic ecology
   a. Aquatic animals
   b. Aquatic plants
   c. Aquatic life systems
      • Characterization
      • Uses and survival needs

Cultural Factors

1. Location
2. Accessibility
   a. Travel distances
   b. Travel times
   c. Transportation networks
3. Site size and shape

4. Surrounding land use
   a. Current
   b. Historical
   c. Land-use plans
   d. Zoning ordinances

(*continued overleaf*)

Table 2   (*continued*)

5. Land ownership	**h.** Transportation/utilities
a. Public	**i.** Water
b. Commercial or industrial	**7.** Population characteristics
c. Private or residential	**a.** Population
6. Type, intensity, and value of use	**b.** Population shifts
a. Agricultural	**c.** Population density
b. Forestry	**d.** Age distribution
c. Recreational	**e.** Number of households
d. Residential	**f.** Household size
e. Commercial	**g.** Average income
f. Industrial	**h.** Employment
g. Institutional	**i.** Educational levels

    b. To control postmining groundwater flows, manage the condition of the strata immediately below the coal seam. In some conditions, it may be desirable to break these strata, in others to preserve them intact.

4. Immediately after coal removal:

    a. Seal the highwall if necessary.

    b. Seal the lowwall if necessary.

    c. Backfill or bury toxic materials and boulders.

    d. Dispose of waste.

    e. Ensure compaction.

5. Shortly after coal removal:

    a. Rough grade and contour, taking into account the following:
- Time required for grading, as related to advance of mining and seasonal constraints
- Steepness of slopes
- Length of uninterrupted slope
- Compaction required
- Reconstruction of underground and surface drainage patterns

    b. If necessary, amend mine spoil to condition the root zone for revegetation, taking into account the following:
- Type of amendment required (fertilizer, limestone, flyash, sewage sludge, etc.)
- Depth of application
- Top layer considerations, including temperature and color, water retention (size, consist, and organics), mulching, and tacking

6. Immediately prior to the first planting season:
   a. Fine-grade and spread topsoil, accounting for seasonal fluctuations
   b. Mechanically manipulate soil as required—rip, furrow, deep-chisel or harrow, or construct dozer basins
   c. Mulch and tack
7. During the first planting season, seed and revegetate, considering time and method of seeding and choice of species.
8. At regular, frequent intervals, monitor and control slope stability, water quality (pH, chemical content, and particulate content), and vegetation growth.

Backfilling and Grading. In most cases, backfilling and grading must return the land to the contour that existed before mining, eliminating highwalls, spoil piles, and depressions. The resulting slopes must be stable, each with the appropriate angle of repose. Another important objective is the elimination to minimize erosion and water pollution. Mining often leaves exposed coal seams, acid-forming or other toxic materials, and combustible materials, all of which must be buried to eliminate interaction with surface waters.

Erosion is controlled by grading to create terraces and diversions that decrease slope lengths and direct runoff water to safe outlets. During backfilling, the permeability of the fill can be controlled to achieve the desired degree of water percolation. The requirements for backfill and grading in reclamation operations should be considered in the selection of mining method.

Soil Reconstruction. All topsoil must be removed as a separate layer and segregated. When the topsoil is less than 6 inches (15 centimeters) thick, the unconsolidated layer beneath the topsoil must be removed and treated as topsoil. When the topsoil cannot be redistributed immediately, it must be stored according to approved methods. When topsoil is redistributed, it must have a uniform thickness, not be excessively compacted, and be protected from erosion by wind and water.

The postmining land-use objective must be selected with the characteristics of the topsoil in mind. Chemical soil properties are more easily amended than physical properties such as texture. Thus, when adequate topsoil is not available, the substitute material must be selected with special attention to texture, coarse-fragment content, and mineral content. Bulk density and soil strength must be considered during soil handling. Methods and equipment used for redistribution of topsoil must be selected to minimize compaction, which can inhibit root penetration and movement of air and water.

Revegetation. The vegetative cover resulting from reclamation must be diverse, effective, and permanent. The extent of the new vegetation must be at least as great as that of the natural, premining vegetation. The soil must be stabilized from erosion. Normally, planting must be done during the first normal season suitable for planting after redistribution of topsoil. Specific standards defining successful revegetation depend on the approved postmining land use; the general requirement is the achievement of 90 percent of the established standard. This cannot be achieved in one successful growing season; thus, the responsibility period is 5 years in areas that receive more than 26 inches (66 millimeters) of average annual precipitation, and 10 years in areas that receive 26 inches (66 millimeters) or less.

Postmining land use must also be considered when planning revegetation, taking into account species selection, soil amendments, and methods of planting. For example, when reforestation is planned, a major concern is competition from herbaceous plants, which must be controlled by scalping or proper use of herbicides. Similarly, when reclaiming land for pasture or hayland, important variables are water control, species selection, seeding time, fertilizer selection, seedbed preparation, seeding method, and use of mulches.

Reclamation of Mine Waste Disposal Sites.   Mine waste piles will include material resulting from mining, such as overburden, waste rock from mining, and tailings, slurries, and slimes from processing. Also included may be waste from treatment of air and water, ordinary garbage, debris from construction, and discarded equipment. However, many jurisdictions require that garbage, construction debris, and discarded equipment be disposed of in a separately sited and approved sanitary landfill. Although this may initially be more expensive, it will greatly simplify the reclamation of true *mine waste piles*, and is thus recommended even where not required.

Generally, about twice as much waste results from mining as from processing, with most of the waste being overburden. The stability and proper handling of mine waste are influenced by its chemical and physical properties. Chemical properties are primarily responsible for potential hazards to public health and effects on the physical and biological environment. Chemical properties also determine the suitability of the waste for further processing. Physical properties determine the appearance of the waste during and after reclamation, and may strongly influence the ultimate form of the postmining land surface.

4.5   Closure

All mines eventually close, though there are different degrees of closure. Mines that are not open for production but could be reopened may be referred to as *temporarily closed* or *semi-permanently closed*; in this chapter, only *permanent closure* is considered.

Permanent closure occurs when a mine ceases operation because of economic conditions, depletion of reserves, political conditions, or any other reason. Closure generally includes the following three steps:

1. Sealing of underground mine openings
2. Removal of surface facilities
3. Reclamation of surface mines and surface areas of underground mines[14]

By definition, *sealing* is the securing of mine entries, drifts, stopes, adits, shafts, and boreholes to protect against ingress and gas and water emissions for the safety of the public. Similarly, *abandonment* is the voluntary act of abandoning and relinquishing a mining claim or the intention to mine. This differs from *forfeiture*, which is involuntary surrender of mining claims or the right to mine by neglect.[15]

Mine Facility Removal

All surface facilities that do not serve a purpose in the postmining use plan must be removed. These include buildings, conveyors, rail lines, transfer stations and loading facilities, electrical lines and substations, pipelines, roadways, drainage ponds, and drainage channels. Any hazardous material must be removed from the site, and disposed of in an approved facility. Sanitary landfills and other waste disposal areas must be reclaimed.

Removal of buildings and structures is often done by a company specializing in such activities. Demolition should be accomplished using standard construction equipment wherever possible, exercising due care not to endanger life or property. After demolition, all debris must be removed from the area, including concrete, concrete block, timber, metal scrap, and so on. All foundations must be removed to a depth below the soil horizon as specified in the reclamation plan, usually 2 to 3 feet (0.6 to 1.0 meters). After removal of structures, the surface should be graded and backfilled as specified in the reclamation plan.

As soon as a decision for closure is reached, a complete inventory of all equipment, parts, and supplies should be made. Using this inventory, plans can be made for appropriate disposition of each item. Many items can be resold or transferred to another mine owned by the company. In the case of large underground mining equipment, such as coal longwall units, it may be less expensive to abandon the equipment underground than to recover it to the surface.

Sealing of Underground Mines

Seals for underground mines must be designed in accordance with good engineering practice, and in conformance with applicable regulations. A complete hydrological analysis is required to determine if there will be a hydraulic head on the mine seal. The shutdown of the mine's ventilation system must be planned to prevent accumulation of hazardous gasses, and to ensure safe working conditions throughout closure and sealing. During sealing, areas near the seal point must be continuously monitored for accumulation of explosive or flammable gasses, which might be ignited by construction activities. Temporary measured may be necessary, such as the erection of temporary stoppings and direct supply of fresh air to the area where the seal is being installed.

Methods of sealing are described in detail by Gray and Gray.[14] *Dry seals*, used where development of a hydraulic head is not expected, are constructed by placing cement blocks or another suitable material in the mine opening to prevent the entrance or air, water, and persons. *Wet seals* are constructed with water traps so that water can flow from the mine but mine gasses are retained. *Hydraulic seals* are designed to stop the discharge of water from the mine, and must be constructed to support the developed hydraulic head. A hydraulic seal will cause the mine to flood, excluding air and thus retarding the oxidation of sulfide materials that results in acid mine drainage.

Boreholes must also be sealed. In one typical method, the surface casing and protective cap are removed to a few meters below the proposed final surface elevation. A plug is then installed in a competent stratum that is as close as possible to the intersection of the borehole with the mine roof. Finally, the borehole is filled with a nonshrinking cement grout, usually to within 2 feet (0.6 meters) of the surface, and the remaining length is filled with dirt.

Sealing of shafts requires extensive planning. In almost all cases, the shaft is completely filled with inert material. In coal mines, or other mines where flammable or explosive gasses may be present, approximately 50 feet (15 meters) of the shaft at the bottom must be filled with noncombustible material. After filling, shafts are capped with at least 6 inches (15 centimeters) of concrete. Normally, the shaft cap is fitted with a 2-inch (50-millimeter) vent pipe that discharges 15 feet (4.5 meters) above the surface. Hydraulic shaft seals can also be installed, with no provision for ventilation or discharge. These are designed with using the same methods used in designing a surface dam. After sealing, the shaft should be fenced to prevent access by unauthorized individuals.

Maintenance

After closure, almost all mine sites will require several years of care and maintenance. Periodic inspections should be conducted to verify that all seals are intact and functioning correctly, and to ensure that water discharges are within the limits specified by the permits. It may also be necessary to clean out water diversion and containment structures, and to support revegetation by regrading, reseeding, applying fertilizer, and irrigating.

Postmining Liability

Even after reclamation and closure are completed, the mining company should take ongoing responsibility for water treatment and subsidence mitigation. The need for water treatment may continue indefinitely, and mine operations should be planned to support these activities. Similarly, subsidence may not occur until many years after mining has ceased. If necessary, the company should post a bond or provide some other means of assurance to regulatory agencies that water treatment will continue as long as necessary, and that subsidence damages will be properly dealt with.

5 CASE STUDY

The Island Copper Mine in British Columbia, Canada, operated from 1970 through 1995. The reclamation and closure of the mine represent an excellent example of environmentally conscious mining practice. Welchman and Aspinall provided a case history of the mine, which is summarized here.[16]

Island Copper was operated by BHP Minerals Canada Ltd., a subsidiary of BHP International of Australia. When the mine ceased operation, BHP carried out a comprehensive plan to achieve mine closure, while also making a transition to commercial and industrial use of the site.

At the peak of its operation in 1980, the mine had 900 employees. The orebody was mined by conventional open-pit, truck-and-shovel methods. A total of 400 million dry short tons (364 million metric tons) was removed, and processed for recovery of copper and molybdenum. An additional 600 million short tons (545 million metric tons) of overburden were excavated, resulting in an oval pit that was 7,900 feet (2,408 meters) long, 3,500 feet (1,067 meters) wide, with a bottom elevation 1,322 feet (403 meters) below sea level. The operation had two unique features: Mill tailings were deposited below the ocean surface in Prince Rupert Inlet, and a plastic concrete seepage barrier, 4,000 feet (1,219 meters) long, 108 feet (33 meters) deep, but only 2.8 feet (0.86 meters) wide, was installed along the original shoreline of the inlet.

Island Copper instituted a comprehensive water management program to control runoff from waste rock dumps, maintain pit dewatering, and recycle all site drainage through the concentrator. Reclamation of disturbed land and waste rock deposits began early in the mine's life—in the early 1970s—and continued through mine closure.

At startup, the mine purchased 100 acres (40 hectares) of municipal land for development of residential subdivisions for mine employees. The mining company paid all the costs for installation of roads, water, sewage, and other utilities to the new housing sites. The company also paid for an extensive upgrade to the municipal water system, including an upgrade of the existing dam, and a new trunk line and storage tanks. Increased tax assessments and an issue of municipal debentures purchased by the company permitted construction of a new sewage treatment plant. The mine also assisted in construction of a new hospital, ice arena, swimming pool, theater, and several parks.

Planning for closure began when the site was being cleared, in 1969. Five million short tons (4.5 million metric tons) of overburden and glacial till were stockpiled for use in recontouring and revegetation of the waste rock deposits that would be produced over the next 25 years.

The mine closure plan included four main areas: the open pit, the waste rock deposits, the marine environment, and the physical plant. The initial plan was submitted for discussion to the provincial authorities in 1990, and a revised plan was submitted in 1994. Copies of the plan were also distributed to government buildings and public libraries throughout the region, and several open houses and bus tours were held for the public.

In 1996, the open pit was decommissioned by flooding it with seawater from the adjacent Inlet. This created a 528-acre (214-hectare) lake, which stabilized the pit walls and provided an effective receptacle for the moderate amount of acid drainage from the waste rock. The acid drainage is diluted by the large volume of the lake, and heavy metals are precipitated to the bottom by the action of sulfate-reducing bacteria. Precipitation and surface drainage have formed a cap of fresh to brackish water on the surface of the lake.

Land reclamation included recontouring and planting of four waste rock deposits covering 500 acres (202 hectares), and a 650-acre (263-hectare) landfill created by deposition of waste rock in Rupert Inlet. Revegetation had been ongoing throughout operations, with the planting of 600,000 alder and lodgepole pine seedlings. This will eventually produce a cedar-hemlock forest of a type typical to the region. The reclaimed area will be monitored until it is clearly seen that the new forest is sustainable, after which the property will be returned to the provincial government.

The waste rock landfill in Rupert Inlet was graded to the low-tide mark. To provide a varied habitat for marine organisms, six bays were sculpted into the shoreline. This area is visually indistinguishable from a reference station selected earlier. Benthic recolonization and populations of key marine species have returned to premining levels. Physical, chemical, and biological monitoring of the marine environment continued through December 1998.

Finally, the mine's physical assets were removed and disposed of. All fuel, chemicals, and designated special wastes were removed and disposed of according to applicable regulations. Contaminated areas were rehabilitated. Buildings, machinery, and equipment were sold through an international firm specializing in disposal of such items. Structures not purchased for ongoing use on site were dismantled and their sites were reclaimed as wildlife habitat.

Welchman and Aspinall noted eight lessons to be learned in regards to reclamation and closure:[16]

1. *Shutdown, closure, and postclosure costs should be a capital expenditure item beginning with the prospect-approval feasibility study.* While in the present value sense it may be inconsequential, nevertheless the importance of early environmental planning for reclamation and closure must be spotlighted.

2. *The direct closure impact in terms of tailings handling, disposal of chemicals and hydrocarbons, and pit shutdown should be an integral part of*

the design criteria employed during the detailed engineering phase of the project. Specifically, locations of dumps should not only be dictated by transportation costs but also by reclamation and final closure suitability/plans. Ongoing reclamation and the last stage of production should be integral parts of the final closure plan.

3. *The selection of key reclamation and closure alternatives should be backed up with a full suite of technical information generated by in-house experts and competent credible consultants.* Supporting information should be open for inspection and both company and consulting experts should be available to provide corroborating testimony.

4. *At times, a Catch-22 condition may arise in that regulators can require a finalized closure plan to approve a mine project's startup.* However, company management tends to be reluctant to authorize detailed engineering of any facet of a project without prior regulatory approval.

5. *Closure is a distinct phase of the mine life.* During construction and operations it is taken for granted that plans must be modified to allow for changing conditions. Similarly, the closure plan, no matter how detailed, should be sufficiently flexible to allow minor adjustments and even major changes as the unfolding circumstances warrant.

6. *Constant communications with regulators are essential, and all efforts should be made to keep them fully informed of decisions and problem.* They must also be protected from either surprise and/or embarrassment. Whenever possible, it pays to take the initiative before being asked to do so by the regulators. Island Copper's frequent response was: "Been there, done that, here's the report." This attitude cannot be bettered.

7. *Before the fact, constructive criticism should be actively sought from environmentalists.* There are several reasons for that:
   - They may notice anomalies while the project is still in the planning stage, therefore saving on costly retrofitting.
   - It is better to have people working with you than against you.
   - It provides background credence to the effort.
   - It supplies a good public relations image.

8. *A sufficient closure fund/bond should be established when either construction starts or production begins.* This should be done even if it is not mandated.

REFERENCES

1. *Breaking New Ground—The Report of the Mining, Minerals, and Sustainable Development Project*, Earthscan Publications Ltd., London, 2002, p. 410.
2. H. L. Hartman (ed.), *S.M.E. Mining Engineering Handbook*, 2nd ed., Society for Mining, Metallurgy, and Exploration, Littleton, Colorado, 1992, vol. 1 and vol. 2.

3. R. Raymond, *Out of the Fiery Furnace*, The Pennsylvania State University Press, University Park, Pennsylvania, 1984, pp. 90–91.

4. J. R. Craig, D. J. Vaughan, and B. J. Skinner, *Resources of the Earth: Origin, Use, and Environmental Impact*, Prentice Hall, Upper Saddle River, NJ, 2002, p. 520.

5. P. Shabecoff, *A Fierce Green Fire*, Hill and Wang, New York, 1993, pp. 35–36.

6. L. M. Kaas and C. J. Parr, "Environmental Consequences," in *S.M.E. Mining Engineering Handbook*, 2nd ed., H. L. Hartman, (ed.), Society for Mining, Metallurgy, and Exploration, Littleton, Colorado, 1992, pp. 174–201.

7. D. K. Hunt, "Environmental Protection and Permitting," in *S.M.E. Mining Engineering Handbook*, 2nd ed., H. L. Hartman (ed.), Society for Mining, Metallurgy, and Exploration, Littleton, Colorado, 1992, pp. 502–519.

8. L. W. Saperstein, "Basic Tasks in the Production Cycle," in *S.M.E. Mining Engineering Handbook*, 2nd ed., H. L. Hartman, (ed.), Society for Mining, Metallurgy, and Exploration, Littleton, Colorado, 1992, pp. 174–201.

9. R. J. Sweigard, "Reclamation," in *S.M.E. Mining Engineering Handbook*, 2nd ed., H. L. Hartman, (ed.), Society for Mining, Metallurgy, and Exploration, Littleton, Colorado, 1992, pp. 1181–1197.

10. R. V. Ramani et al., "Pre-mining Planning for Environmental Control in Surface Coal Mines," Preprint No. 77-F-387, SME, New York, 1977, p. 25.

11. J. M. Riddle and L. W. Saperstein, "Premining Planning to Maximize Effective Land Use and Reclamation," Chapter 13, *Reclamation of Drastically Disturbed Lands*, American Society of Agronomy, Madison, Wisconsin, 1978, pp. 223–240.

12. *Surface Mining Control and Reclamation Act*, 30 U.S.C. 1231, *et seq.*, http://www.osmre.gov/smcraindex.htm, 1977.

13. M. L. Clar and M. J. Arnold, "Exploration and Permit Application Review/Approval Procedures for Coal Mining Operations Under a Federal Program," Final Report, Hittman Associates, U.S. Office of Surface Mining Reclamation and Enforcement, Washington, DC, 1981.

14. T. A. Gray and R. E. Gray, "Mine Closure, Sealing, and Abandonment," in *S.M.E. Mining Engineering Handbook*, 2nd ed., H. L. Hartman, (ed.), Society for Mining, Metallurgy, and Exploration, Littleton, Colorado, 1992 pp. 659–674.

15. P. W. Thrush (ed.), *A Dictionary of Mining, Mineral, and Related Terms*, U.S. Bureau of Mines, Superintendent of Documents, Washington, D.C., 1968, p. 1269.

16. B. Welchman and C. Aspinall, "Island Copper Mine: A Case History," in *Mine Closure and Sustainable Development*, T. Khanna (ed.), Results of the workshop organized by the World Bank Group Mining Department and the Metal Mining Agency of Japan, Mining Journal Books, Ltd., London, 2000, pp. 659–674.

CHAPTER 2

WASTE REDUCTION IN METALS MANUFACTURING

Dr. Carl C. Nesbitt
Department of Chemical Engineering, Michigan Technological University,
Houghton, Michigan

1 **WASTES AT THE MINE SITES**   35
  1.1 Tailings Impoundments— Planning and Operating   36
  1.2 Hydrometallurgical Wastes   39
  1.3 Acid-Rock Drainage (ARD)   41

2 **CHEMICAL METALLURGY WASTES**   43
  2.1 Pyrometallurgical Processing Wastes   44
  2.2 Hydrometallurgical Processing Wastes   50

3 **CONCLUSIONS**   57

The metals industry is one of the most important sectors of the world economy. In the United States alone, the consumption of steel products approaches 120 million tonnes each year, while the consumption of aluminum averages 6.5 million tonnes per year and the copper consumption averages over 4 million tonnes each year. The production of metals comes from primarily two routes—production of metal from mineral resources or recycle of spent metal products. For example, on average approximately 67 million tonnes of steel are produced from recycled scrap (~55%), 2.9 million tonnes of aluminum were made from old and new scrap (~52%), and 1.3 million tonnes of copper were produced from scrap (~30%) in the United States each year. Almost 75% of the lead used in the United States is from secondary sources and recycling. Recycling metals directly saves our society by reducing the amount of waste generated, and energy consumed, to produce new metal products. Improving economies of individual nations typically are preceded by an increased demand for metals and metal products. An increased consumption of construction products (especially steel and copper) is one of the leading indicators that an individual country's economy is growing—as has occurred in the China and Asia markets.

The mining of ores is still the primary means of producing new metals for consumption. Every state of the United States has mining of mineral deposits (sand, gravel, salt, clays, etc.), but the majority of metal mines are in the western states.

In the world, the biggest metal mining countries are in the Pacific Rim (including most South American countries, Australia, Indonesia, America, Canada, Mexico, and China), Africa (principally in South Africa, Ghana, Zimbabwe, Tanzania, Zambia, and the Democratic Republic of the Congo), and several former Soviet Republics (Kazakhstan, Uzbekistan, Tajikistan, and Turkmenistan) ores are mined, comminuted, and beneficiated to produce concentrates made of just the minerals with metal value. These concentrates are then processed by extractive processes (such as smelting or leaching) to produce a low-grade metal that can be refined or alloyed.

Recycled metals are typically separated by the consumers or the recyclers. More complex materials, as from spent automobiles and appliances, require complete shredding and an initial separation of the materials into general types (magnetic and nonmagnetic). Then the individual metal types can be melted and refined to produce metal for market. These processes, for the most part, have not changed in over 100 years. However, without exception, the amount of metal recovered from scrap has increased across the board—that is, nearly every metal consumed contains at least 10 percent and as much as 70 percent of its mass from recycled metals. That trend has risen over the years, and should continue to increase as metal demand increases in Asia and North America, as energy conservation is sought, and as environmental concerns rise. Take, for instance, the recycling of automobiles. European automobile manufacturers are now required to take ownership of old or damaged automobiles for recycle into new models. The U.S. automobile industry may face a similar fate in the next few years. Currently, there are more than 300 automobile shredding plants throughout the United States, each of which is capable of shredding up to 1,000 cars per day.

The discovery and development of new reserves, the recycling of metal scrap, and the production of consumer metals is not without waste. Generally speaking, all waste can be classified by its form—solid, liquid, or vapor—and the processes (or locations) that generates it—at the mine, mill, smelter, or refinery. By far, the largest mass and volume of waste produced in metal production is in the form of solid wastes. Mining produces waste in the form of overburden that has to be removed to expose the mineral value beneath, and waste material from the separation and recovery of mineral value in concentrates has to be collected and disposed. Slags are also a necessary byproduct of making pure metal ingots and sheets at smelters and refineries. Some processes, such as hydrometallurgical processing, will produce solutions rife with dissolved metals that must be treated before impounding or release. Vapors, including dusts and combustion products, are also formed in the metal production processes. While the amounts of metal wastes vary with the metal of interest, this chapter will discuss the types of wastes produced, steps to mitigate the release of wastes, and new developments into reuse and recycle of the wastes as valued products.

1 WASTES AT THE MINE SITES

The U.S. Environmental Protection Agency classifies mining and the minerals industry as the largest sector of waste production in the country. Just in the past few generations, mining techniques and processing techniques have improved to a state at which lower grade ores can now be more economically treated. No industry demonstrates this better than the gold industry. As long ago as the California Gold Rush of the 1850s, gold ores with 10 to 20 troy ounces per ton (310–620 g/t) were economically processed from placer deposits. However, since the adoption of cyanidation to recover gold in the early 1900s and the advent of heap leaching in the late 1960s, gold ores with as low as 0.025 troy ounce per ton (0.8 g/t) can now be treated at a near break-even point.

This trend in treating lower-grade ores is generally seen in all metal mines. Long gone are the 1 to 5 percent copper ores, or the 60 percent iron ores. Metal concentration steps must now be incorporated to liberate and recover the valuable metal. This is not without cost. Look at the gold mining example. At 0.03 troy ounce per ton, all of the rock will be turned into waste as the miniscule quantity of gold is extracted. As the processes to selectively recover the metal value improved, so too the techniques of waste handling had to be improved.

Environmental stewardship has been relegated to mining companies, especially in recent years. In some countries and states (most notably Wisconsin and Michigan), securities in the forms of bonds or cash payments must be made before licenses or permits will be approved for new operations. These securities are based on the expected impact of the waste piles (i.e., the probability of acid drainage generation, toxicity of contents, and total mass of earth moved). The bond is kept by the local governments and could be used to help mitigate any catastrophic failure of impoundments, or any long-term environmental problems that could arise from the mining of the ores, such as acid drainage or wind-blown dust.

When mining companies consider developing a mineral ore deposit, as much as 10 years of planning and permitting is conducted prior to loading the first shovel of dirt. Local and federal regulatory agencies require completed environmental impact assessments, risk analyses, waste storage plans, well permits, water permits, and discharge permits for the life of the mine. As part of that planning the mine operators must have a firm plan in place for storing the mine overburden and *poor rock*. The mining technique (open pit or underground) will also be decided based on the size and depth of the ore body, the mechanics of the surrounding rock formations, and so on. Underground mines are by definition a more selective means of extracting the ore and rarely require poor rock or waste rock piles. The mine plan is established to specifically extract ore, and little poor rock is excavated. In open pits, the poor rock piles are created by using dump trucks and bulldozers. The trucks back to the edge and end dump.

The dozers push the dumped rock over the side, and *berm* the edge to give the truck drivers a sense of where the edge is. This action causes the level piles (similar to large peninsula mesas) to grow progressively outward from the mine in a planned strategic method. By the end of the mine life these piles may be shaped to more closely resemble natural mountains, and are reseeded with native flora. Although this practice is fairly modern (began in earnest in the 1980s), the impact of long-term mining projects has been lessened by the practice.

The regulatory agencies (MSHA, local and federal EPA, etc.) and regulatory laws (such as RCRA, "zero discharge" permits, etc.) have been empowered to ensure that no form of waste can be released from an individual mine site in any form (solid, liquid, or vapor) by any means (land, air, or water). During mine operations, airborne losses are mitigated through dust-control measures, which are of constant concern to the mining companies that have to transport ores and waste by haul trucks over dirt roads. As ore bodies are mined, roads are considered temporary. A road used to haul overburden or ore may be removed in the future, so paving the haul roads would not be a logical option. Today's haul roads are often tomorrow's waste rock or ore. Water trucks are a part of the operating fleet in mines, especially during the hot, dry months of the summers. Often, additives such as tree saps or polymers are added to the water to cover haulage roads to give longer-term dust suppression without harming the value or waste-storage ability of the rock underneath the roads. Solid wastes from the mine (i.e., waste rock or overburden) are either stacked next to the mine or sometimes used in construction of dams and dikes for tailings impoundments.

Mine properties have to be much larger than the area of the deposit, by necessity. Open-pit mines are built to include the natural angle of repose of the material to prevent slides into the mine. This practice requires that the actual mine size be a large area beyond the outlying portions of the ore deposit. Also, overburden disposal areas, space for processing plants, ore storage piles, warehousing, the maintenance garage for the large equipment, and office space must be built on the property. But by far the largest footprint belongs to the massive tailings impoundments that must sequester the mineral wastes into perpetuity. These specially designed (and permitted) areas act as the mining company's personal landfills of rock and liquid wastes.

1.1 Tailings Impoundments—Planning and Operating

All of the waste of a milling operation is sent to the tailings impoundment for sequestration. Mineral beneficiation requires that the gangue minerals be separated from the valued minerals. In most operations, the mass flow rate of solids in the tailings stream is not much different than the feed rate to the plant. For instance, in a gold milling operation, for every ton of ore fed to the plant, 0.1 troy ounce of gold is extracted—that's a net reduction in the feed mass of 0.0000003 tons per feed ton of ore for such an operation. The waste material is then sent to the tailings impoundment for permanent storage.

As mineral processing techniques have improved, the head grade of treatable ores has diminished. In essence, the mining industry is generating more tons of tailings per ton of product, but more efficiently and more economically. This means that for the same metal production, the tailings impoundment must be larger for today's operations than similar operations from several years ago.

The technology for handling the tailings has dramatically improved as well. The construction of tailings dikes has evolved to be an on-going task of the engineers and operators of the mine site. Compaction testing of dikes and dams, density measurements, ground permeabilities, and downstream well monitoring are all part of the added duties of today's process engineers and operators. In the past, these tasks were given to consulting companies who engineered the starter dikes, and built the extensions. Now, in-house engineers are trained and certified to perform these duties as part of the permitting plan of the operation.

The term *tailings* is generally defined as the solid-bearing waste product of a mineral concentrating process. More specifically, the tailings of an operation are a mixture of solid particles ranging from coarse granular (25 mm) to fine particles (300–50 μm), or even colloidal particles (<30 μm). However, it should be understood that processing ores requires very large volumes of water. Water is essential in the success of mineral separation processes, it is the best medium to transport large quantities of solids, and it eliminates dust. While tailings thickeners are used to recover a portion of the water from the waste material, the slurries that report to the tailings impoundments are typically 55 to 60 percent solids. Tailings ponds are therefore required to house the solids, as well as a large amount of process water. This conglomeration of particles and water make tailings a complex material to handle. Add to the physical complication that the waste rock of a mine can consist of literally any mineral from quartz to clays, and it is easy to see that this material is inconsistent from mine to mine. All tailings impoundment areas must be designed specifically for the operation—that is, one size does not fit all.

In some mining processes, the tailings water may contain reagents that have to be treated before the slurries can be pounded. For instance, free cyanide or dissociable cyanide species must be treated to immobilize them in the tailings impoundment. The addition of ferrous sulfate solutions to cyanide-laden waste streams will sequester free cyanide in a stable precipitate known as *Prussian blue*, $Fe_4[Fe(CN)_6]_3$. More stringent treatment methods, such as Caro's acid (concentrated sulfuric acid and hydrogen peroxide), destroy most cyanide species as ammonia or nitrogen and CO_2.

Waste impoundment design begins well before the operation begins construction of the plant. The size, height, and type of impoundment are predetermined by a number of factors. Designers of the tailings impoundment must keep in mind the following aspects:

- Characteristics of the tailings and liquid wastes of the processing facilities

- Proximity to the ore body itself
- Permeability of the soil in the area
- Soil stability to support a large dike
- Topography of the area
- Volume of the holding area for the life of the mine
- Excess volume or flood diversion trenching to handle a 100-year flood scenario
- Applicability of installing a water retrieval system

Properly designed tailings impoundments will not only sequester the solids, but will also hold the water. As a precaution, all tailings ponds require monitor wells downstream of and along side the dikes. These wells constantly monitor the condition of the aquifer using conductivity probes and thermistors to measure any changes in salinity or temperature that could indicate seepage or leaks from the impoundment. Any perturbation to these measurements will send a signal to the operators that the water behind the tailings dike is leaking through the soil. Initial design of the tailings impoundment area requires testing of the permeability of the soil and ground to determine if the chosen site is proper for impoundment. In some circumstances, the ground may be made impermeable using geomembranes, clays, or other materials prior to start-up of the milling operation. However, the wells are safeguards mandated to watch for problems. Should a leak be detected, loads of clays, such as bentonite, can be dumped on dikes to stop seepage. These clays swell as they absorb water and are excellent additives to stop the seepage through a dam. Similar methods are used in ground seepage, as well.

At start-up, a *starter dike* is in place to begin collecting the tailings and wastewater at the startup of the operation. Typically, starter dikes are undersized to allow for optimal construction of the lowest portion of the diking system. Also, the total amount of tailings that will need to be impounded was predicted based on the reports of the exploration geologists who have estimated the size and grade of the ore in the deposit. Occasionally, this estimate is changed as the mine is developed, leading to an increased life beyond the life of the impoundment area. For these reasons, expansions to the tailings diking system are necessary, and are most often planned in advance.

It is understood that the initial dike is not at the final height, but will be the base for future expansions, or lifts, to be constructed as the operation proceeds. Often, the subsequent dikes are constructed by mixing of native soils with the coarse fraction of the tailings (separated by hydrocyclones on the dikes themselves). The new dikes are laid directly above the starter dike, allowing the level of tailings to grow subsequently. Figure 1 shows one operating strategy of this type.

The initial design of the dike must take into account this eventuality. The ground that will be used to support the dikes must be stable enough to not slide or degrade under the weight of the tailings and the dike. Typically, the ground

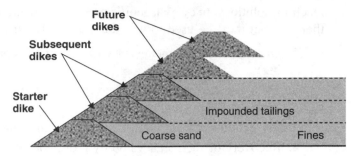

Figure 1 Upstream method for tailings dike construction. Tailings is naturally fractionated on the face of the dikes. Once the volume is full, another lift is placed above the starter dike.

is prepared by removing all plants and roots, and loose material. Then layers of gravel, sand, and clays are compacted into the area to make the area more stable to support the initial and subsequent dikes to ensure tailings impoundment stability.

All lifts of the dikes must meet specifications of earthen dams. The amount of mechanical compaction, the permeability, and the size fractions of the components are carefully monitored by the engineers in charge of the lift construction.

1.2 Hydrometallurgical Wastes

Hydrometallurgy has grown in importance since 1940. Before that gold was principally the only metal recovered using aqueous chemical processing. The adoption of cyanidation in the 1900s caused the dramatic rise in production from hundreds of ounces per year in the late 1800s to the 2005 estimated production of 2,500 metric tons per year. Likewise, copper production by leaching of ores has risen from a few thousand tons in the early 1950s ($\sim$1% of the world production) to the current estimate of 3.9 million metric tons per year (over 24% of the world copper production).

Hydrometallurgical processing has been used to treat lower-grade ores more economically than the traditional concentration/smelting operations. The processes have improved from vat leaching of copper ores to autoclaving finely ground ores to recover minute quantities of gold. The advent of heap and dump leaching in the late 1960s has dramatically lowered the grade of ores that can be processed economically.

The processing strategy used is directly dependent on the grade of ore and the ease of dissolving the metal of interest. Lixiviants range from basic cyanide solutions to strong sulfuric acid solutions. Bioleaching of ores, or ferric oxidation from bacteria, can also be used to treat ores that would have been relegated to waste heaps in the past. The lixiviant used is directly dependent on the metal to be leached. Although these reagents are typically not selective, subsequent processes have been implemented to not only increase the concentration of the metal in aqueous solutions, but also eliminate the unwanted metals present in the leaching solutions. In every leaching process, there are five steps (or combinations thereof) that must be used to win the metal value from ore:

1. Dissolution of the metal value to the aqueous phase (also known as leaching)
2. Solid-liquid separation to recover the aqueous phase apart from the spent ore
3. Purification/concentration of the metal of value sought
4. Recovery of the metal value by reduction or precipitation
5. Solid-liquid separation to recover the metal apart from the solution

Although the overall processes have evolved for specific metals, and leachants, there are principally only two leaching techniques that provide the *pregnant leach solution* for metal recovery: percolation leaching and agitated tank leaching.

Percolation leaching techniques generally include both the dissolution step and the separation of the liquid from the ore in one unit process. Percolation leaching includes in situ leaching, dump, and heap leaching. In situ leaching, most commonly used in the uranium industry, involves the direct dissolution of ore where it resides in the ore body. The lixiviant solutions are pumped into the ground to directly leach the metal-bearing minerals then recovered by pumping the solution out of the ground. Figure 2 shows a typical five-well pattern used to ensure that all leach solution (injected in the center ejection well) is recovered through four adjacent withdrawal wells. Persistent rinsing of the area after the metal is recovered ensures that these wells will not remain contaminated. This process is considered only for ore bodies that meet special guidelines, such as no proximity to aquifers, readily soluble metals with weak lixiviant solutions, and impermeable host rock. Typically, there are no solid wastes from the processes. No actual solids are mined, so there are no materials to remove. Treatment methods may be required to clean the solutions for reuse or disposal.

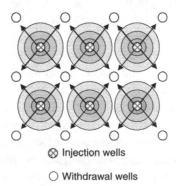

⊗ Injection wells

○ Withdrawal wells

Figure 2 Typical arrangement of a five-hole pattern for in situ leaching of ores. The central hole is used to inject the leachant, which flows radially to the outer holes from which the metal-laden solutions are withdrawn.

In dump leaching, the ore is mined and stacked on an impermeable pad. It is processed by the direct application of leach solutions onto the top of the piles. Heap leaching is similar to dump leaching with the exception that the ore is mined and then crushed to produce a more uniform-sized material for the leach pad. In dump and heap leaching, the spent piles must be rinsed thoroughly to remove any toxic materials and unreacted leachants (especially cyanide in precious metal operations), and to neutralize the pH of the pile. The rinse solution is collected and treated to recover valuable metals that may remain in the piles, and to remove any other metal ions species. The solution is then neutralized for reuse in rinsing the heaps, or prior to discharging the water. Once completely rinsed, the piles are typically bulldozed into a more natural shape, covered with topsoil and seeded with natural flora of the region. Often, these leached piles appear to be small hills on the landscape.

In *agitated tank leaching* systems, finely ground ore must be mixed with the lixiviant until the reaction is complete. Liquid/solid separation processes are used to recover the solution and prepare the waste for disposal in a tailing impoundment. Unlike percolation leaching operations, these operations require a large impoundment area to not only handle the solids, but also to neutralize the chemistry and pH of the solutions used to extract the metal value.

Economically speaking, the inexpensive percolation leaching processes are reserved for lower graded ores (that is, ores with lower net metal value.) The major costs of these processes are mining, stacking, and crushing of the ore. In agitated tank leaching operations, the ore must be mined, crushed, and finely ground. This added expense will give a better recovery of metal, but will also cost considerably more than the simple percolation processes. So these processes are reserved for the higher-grade ores, which will return more revenue per ton. Tailings impoundments must handle the solids, plus treatment methods must be included to neutralize the chemistry of the solutions prior to sequestering the tailings in the ponds. Also, every mine operator is responsible for preventing animals and birds from being harmed by these large manmade lakes. Great care is given to maintain fences and other methods to prevent birds from landing in the open water.

1.3 Acid-Rock Drainage (ARD)

Sulfide mining has been under attack for the perceived assumption that all sulfide mines (including base metal mines and high-sulfur coal mines) lead to the automatic generation of acidic mine drainage from the tailings and/or poor rock piles. In fact, acid-rock drainage and acid-mine drainage are the result of microorganisms that convert the soluble forms of iron and sulfur into high oxidizing solutions at low pH. *Acidithiobacillus ferrooxidans* are the primary culprit in acid-rock generation (ARD), but other microorganisms have been isolated and contribute to the problem.

The mechanism by which these bacteria produce ARD is still debated, but the fact that *Acidithiobacillus ferrooxidans* is a chemolithoautotropic microorganism that relies on oxidizing ferrous ions (Fe^{+2}) to ferric ions (Fe^{+3}) is well documented. The bacteria are aerobic, and they catalyze the oxidation by the following reaction:

$$O_2(aq) + 4Fe^{+2} + 4H^+ \rightleftharpoons 2H_2O + 4Fe^{+3}$$

It should be noted that the bacterial presence is absolutely necessary for this reaction to proceed at low pH. As written, the reaction is thermodynamically spontaneous to the right and rapid at pH above 3.5; that is, ferric will be formed by simply aerating a ferrous solution at pH >3.5. However, at a pH above 2.4, the solubility of the ferric ion is reduced by the rapid and complete formation of ferric hydroxide ($Fe(OH)_3$) precipitate. The bacteria are required to catalyze the formation of ferric at a pH below 2 so that soluble ferric ions are produced. This is the primary reason why the presence of bacteria is required for acid rock formation.

Understand that the bacteria will not survive, or will have very slow growth rates, in rock that does not contain the proper elements—namely, iron and sulfuric acid. The ferric that is generated by the bacteria will oxidize most sulfide minerals to solubilize the metal ions and will also oxidize the sulfide components into sulfates (hence, acid is generated from the sulfide minerals). But the bacteria first require a source of iron for their growth. That is why it is most common to find acid-rock drainage generated in sulfide mineral outcrops that contain any of the common iron sulfide minerals, such as pyrite, marcasite, pyrrhotite, and chalcopyrite. The natural weathering of these minerals (albeit quite slow kinetically) will produce a small concentration of soluble ferrous ions that will allow for the natural bacteria to begin growth. As the bacteria cultures mature and expand, the net result is full-fledged production of a water that is high in oxidation potential (from the ferric ions) at a low pH (caused by the direct and indirect oxidation of the sulfides to sulfuric acid). This product water will then proceed through the rock mass and liberate all other metals and sulfides to produce the byproduct of ARD, acid mine drainage (AMD), and acid drainage.

For many ores, pyrite is considered a waste mineral; that is, its value is too low to be economically recovered. The presence of pyrite in copper ores, gold ores, and coal, among others, can greatly reduce the value of the ores. For instance, in copper sulfide ores, a small amount of iron is required for optimal smelting. But too much (as would be provided by a small amount of pyrite in the concentrate) would mean a higher energy cost to smelt the concentrates. Pyrite in gold ores directly interferes with the cyanidation process. Often, the gold will actually be found in the pyrite minerals of the ore. The pyrite is readily solubilized by cyanide, so it consumes much of the cyanide, leaving none for precious metal extraction. In coals, pyrite and other sulfide minerals are the primary source of the sulfur in the coal. Most eastern coals in the United States contain high sulfur content, almost 60 to 70 percent of which comes from the mineral pyrite. Desulfurization techniques that selectively remove inorganic sulfide minerals are the most successful at sulfur reduction of coals.

ARD is common wherever the presence of pyrite is greatest. Poor rock piles of coal mines and metal mines are notorious sites for acid generation. The Bunker Hill mine of northern Idaho, the Berkeley Pit of Montana, and the Leviathan Mine of eastern California, are all notorious ARD sites that have pyrite in the deposits that have become exposed to the atmosphere in rock piles or the mines workings itself.

The mining companies that are harvesting the value from these mines are quite cognizant of the long-term costs that ARD poses to their business. International Network for Acid Prevention (INAP) is an organization of international mining companies dedicated to the reduction of the generation of acid from waste rock. This group is instrumental in sharing technologies for preventing and treating acid-rock runoff as it happens. Most of the successes in prevention involve preventing oxygen from contacting the waste. Storing pyrite-laden solid waste in impoundments that are covered with water, soils, plants, or impermeable geomembranes has been successfully used by some of the mining companies to keep acid generation to a minimum. Deep-water storage of wastes reduces the optimal conditions of bacterial activity. Dissolved oxygen is depleted by the aerobes themselves. As long as the flow of water is diverted so that the oxygen cannot be replenished, the area becomes anaerobic naturally. Anaerobic bacteria are known for their sulfide reduction chemistry which produces H_2S from $SO_4^=$. These bacteria directly reduce the acid content and precipitate any soluble metal species as stable metal sulfides.

Several processes have been used to combat acid-rock drainage that has already been formed. Metal precipitation and acid neutralization with lime or limestone is the most common technique used to limit the environmental impact of an acidic, metal-bearing solution from waste material. Metal hydroxide formation is quite simply achieved with moderate changes in the pH of the solution. Metal sulfide formation is typically preferred because the sulfide precipitates are more stable, more quickly formed, and much easier to remove by filtration or sedimentation. However, the sulfide sources (such as hydrogen sulfide gas, sodium sulfide, and thiourea) are very expensive on a "per pound of metal removed" basis, and can produce a noxious hydrogen sulfide gas emission from the mixing tanks. The chemistry of precipitation is discussed in more detail later.

2 CHEMICAL METALLURGY WASTES

Concentrates from the mines or impure metal products from metal recyclers must undergo further chemical processing steps to produce the metals for the manufacturing sector. Metallurgists classify these *chemical metallurgy* processes by the principal methods used to break the chemical bonds of the metals with the other metals or the mineral components. *Pyrometallurgy* is the field in which high-temperature processing (often with the metal in a molten state) is used to produce the metal ingots and sheets that will be used to manufacture various final consumer products. *Hydrometallurgy* is the process by which purified metals are won from aqueous solutions, either by electrowinning the metal from concentrated leach solutions or by electrorefining metal anodes to produce pure cathode sheets or plates. Finally, electrometallurgy is the field in which electric current is used to either melt or plate metals. For the sake of this chapter, let us consider that electrometallurgy could actually be classified into the other two areas. That is to say, electric current in small amounts can be used to recover metal from aqueous solutions (hence it is related to hydrometallurgy) or in large amounts can be used to resistively heat and melt metal scrap (hence it is related to pyrometallurgy).

2.1 Pyrometallurgical Processing Wastes

Nearly every metal produced for the manufacturing sector at one point has to be processed pyrometallurgically. Iron ore concentrates are first reduced to pig iron (high carbon iron), which must be converted to steel (low-carbon iron) in an open-hearth furnace, basic oxygen furnace, or steel processing plant. Gold recovered from processed ores is melted and poured into ingots at a mine. These impure bars of gold will have various other metals (most notably silver, copper, or iron) in the bars that must be separated pyrometallurgically to make 24-karat-gold bullion. Aluminum is either remelted and processed to remove alloying elements or is plated as pure metal in molten salt electrolytic processes. Suffice it to say that even the melting of scrap to make ingots is a high-temperature process that all metals must endure. These processes are the source of several different wastes. Solid wastes of pyrometallurgical processes, the predominant waste material, include cooled slag and dust. But gaseous wastes are also a part of the processes, as well. Combustion products and gaseous reaction products must also be considered when understanding all of the processes that lead to metal production.

Steel production has changed over the past century. It is an excellent example of how the landscape has changed in the metal manufacturing sector and the volumes of waste has been reduced. After the initial discovery of blast furnaces the late 1880s, integrated steel plants were constructed to convert iron oxide ores into pig iron in the blast furnace. The molten pig was then sent to open hearth furnaces or poured into ladles so that the carbon in the metal (up to 6% by weight in pig iron) could be reduced to produce steel (carbon content of less than 2%). *Integrated steel plants* are self-contained, including coke ovens to produce the coke for the blast furnace from bituminous coal, auxiliary operations to collect the coke oven gas, heat from waste gases, air preheaters, and so on. The large tonnages of iron ore and coal required that they be built near the Great Lakes or other large waterways where large ships can deliver the materials cheaply. However, as the demand for steel began to push these behemoth operations to their capacity, a push to recycling more spent metal has led to today's *minimill* operations. These operations were not restricted geographically to large bodies of water. Instead of blast furnaces, they used electric arc furnaces to melt the scrap. The molten scrap could be treated in ladles or basic oxygen furnaces (BOFs) to get the proper chemistry in the final metal. More and more of these operations are appearing as the need for recycling has increased.

Although blast furnaces and integrated operations are still important, the advent of the minimill has improved the steel manufacturing business. They are cleaner operations and produce much less solid or gaseous wastes than the previous processes. They produce steel for local consumers from local steel scrap, so transportation costs are minimal. Although this is only an iron and steel example, similar trends are happening with other metals as well. Conventional metal production from ores will continue to be the major source of most new metals, but recycling is catching on, and the benefits of reduced waste volumes is only one of the reasons. Conventional methods typically produce more wastes than the new recycling processes. For this reason we will focus primarily on the conventional processes, knowing that their overall impact has been reduced by incorporating more recycle.

Solid Wastes

Slags. Slags are defined as the waste products produced in the smelting or refining of molten metal products. Slags are produced in all metal pyrometallurgical processes. Although they are considered wastes, they are an important part of the metal-production process. Some slags are the result of additives with the feed, while some are the result of refining processes to clean the molten metal. The chemistry of the molten slags is varied to improve the metal properties during production. Slags are composed of fluxes that help to lower the melting temperature of gangue minerals that may be present in the final mineral concentrates feeding a smelter. Also, basic slags (i.e., slags that contain excess calcium, sodium, or iron oxides) help to reduce the concentration of detrimental elements (such as phosphorus, sulfur, and silicon) in pig iron and steels. The molten metal flows through the slag in droplets to improve the removal of these elements as the metal is collected at the bottom of the hearth. Inasmuch, slags can retain entrained amounts of the product metal of value. Because of this, much of the slag is recycled back into the furnace to reuse their chemical and thermal properties. For the most part, nonferrous metal slags are considered worthless and have little reuse; however, ferrous slags have many uses.

Given the huge steel market, the vast majority of the slags produced come from iron or steel mills. Iron and steel slag is considered a valuable commodity, worth as much as U.S. $340 million in 2005. Approximately 60 percent of this commodity is from blast furnaces, and the remainder is from steel manufacturing plants that use basic oxygen furnaces (BOF) and electric arc furnaces (EAF). Pig iron is the product of blast furnace processing of iron ores. For the production of 1 ton of pig iron, a charge will consist of about 1.7 tons of iron oxide ore or other iron-bearing materials, 0.5 to 0.65 tons of coke, and 0.25 ton of limestone or dolomite, which is added to make the slag. The blast furnace will require from 1.8 to 2.0 tons of air to convert the coke into CO for the reduction of the iron oxides to elemental iron. Along with the 1 ton of pig iron produced, the blast furnace will produce 0.2 to 0.4 tons of slag, 0.05 tons of flue dust and 2.5 to 3.5 tons of blast furnace gas (a combination of CO and CO_2). Given that there is from 35 to 45 million tons of pig iron produced in the United States annually, there is as much as 7 to 18 million tons of slag produced from this process annually.

Over the years, much has been done to make the slag more reusable. Regular cooling pits for the slag can be used to produce a slag that can be readily mixed with asphalt or cement to produce aggregates in paving asphalt and concrete. Some of the ground granular blast furnace slag (GGBFS) can be ground fine and used as feed material to make Portland cement in concrete mixtures. Air-cooled slag is used as concrete aggregate. Rapid quenching of the molten slag can produce a *popcorn slag*, which is an excellent aggregate for cement. The porous structure of the slag will securely bond with the cement matrix and produce a stronger concrete. Considering that over 1 billion tonnes of sand and gravel are produced each year in the United States, and the average production of slag is just 40 million, the construction industry has easily swallowed up the steel and iron slag production. Only about 1 million tons of slag ever end up in landfills in the United States each year.

Casting Sands. The production of metal products requires that the molten metal be cast into near-net shapes. The steel industry uses casting sand because it is inexpensive, is readily available, efficiently absorbs heat at a predictable rate, and is easy to use. The sand is blended with water, carbon, and other admixture compounds to improve the compaction of the sand, the heat capacity, and other properties. The sand is pounded around a mold, the mold is removed and the molten metal is poured into the sand molds. Unfortunately, the sand immediately next to the cast part is typically not reusable. Carbonization, fusion, and other factors make this material unusable and must be disposed of in landfills. Many of the companies that cast parts are implementing new programs to separate the *burned* material from undamaged sand to recycle the sand for more castings and reduce the amount shipped to landfills.

Dust. Particulates are a problem at many stages of the metal production process. The amount of air or oxygen that is passed through various pyrometallurgical processes invariable will kick up the fine particles in the material. Bag houses have been designed to collect even the finest particles for disposal or recycling of vented particles (typically metal oxide particles). Various processes have been designed to treat the broad spectrum of dusts produced in the metals industry.

Fluo-solid smelters use the suspension of fine particles as a means of more efficiently processing the particles. Blast furnace feed must be agglomerated to a more stable large size to prevent the immediate expulsion of fines with the vent gases. In an electric arc furnace, the dust generated is composed primarily of zinc oxide with small amounts of iron oxide and other base metal oxides. The electric arc furnace is fed recycled steel, direct reduced iron, or pig iron. It uses electric current to resistively melt the charge. But keep in mind that zinc has a boiling point ($900°C$) that is below the fusion point for steel, so any zinc galvanizing on the metal scrap will immediately be volatilized and react with the oxygen in air to produce a fine zinc oxide powder that is collected in the bag house filters. Several specialty companies are investigating methods to salvage the zinc value from these byproducts.

Gaseous Wastes

Sulfur Dioxide.   Smelting of sulfide minerals invariably leads to the production of sulfur dioxide at some point in the process. For instance, as mentioned previously, in the world today, only 25 percent of the copper produced comes from the more environmentally benign hydrometallurgical process (i.e., solvent extraction and electrowinning). An important aspect is that the aqueous processing of copper minerals obviates the need for high-temperature smelting of copper. However, this implies that 75 percent of the world's copper comes from a sulfide operation that first concentrates the copper using flotation or some other beneficiation process (tailings is sequestered behind a tailings impoundment), matte smelting of the copper-iron-sulfur concentrates in fluo-solids smelters, or reverberatory furnaces. The matte is converted to *blister copper* by injecting air and oxygen to first produce an FeO slag, and then SO_2 vapor from the melt. The Clean Air Act of 1977 has dramatically affected this practice. Long gone are the days of just sending the SO_2 up the stack. Today, the increase in copper produced hydrometallurgically has increased sulfuric acid demand, and given a natural path for the SO_2 to be disposed. Nearly every smelter in the industrialized world has inserted a gas scrubber to collect the SO_2 and convert it to sulfuric acid solution (usually by the contact process). This acid availability has directly influenced the market of sulfuric acid, and works in a symbiotic relationship with hydrometallurgical operations. Nearly every mine with a smelting operation has a leaching branch that uses the sulfuric acid produced to enhance metal production.

From a stoichiometric standpoint, nearly 0.35 tons of SO_2 are produced for every ton of copper generated. Nearly 0.65 tons of SO_2 are generated in the production of 1 ton of lead from galena concentrates. That means in 2005, approximately 400,000 tons of SO_2 were produced from the lead and copper production in the United States. This SO_2 made 613,000 tons of sulfuric acid, or about 1.5 percent of the total sulfuric acid consumed in the United States. Almost all of that acid was consumed by the hydrometallurgical portion of the metal mining industry.

Carbon Dioxide.   As the debate of the effect of greenhouse gases rages on, the simple fact remains that carbon dioxide production is one of the known side reactions of most metal-production operations. Carbon is an effective metal reductant. Coke is used to produce pig iron from iron oxide ores and lead from sulfide ores in blast furnaces, carbon electrodes are used to produce aluminum from bauxite leaching products, and coal is used in the reduction of zinc oxide in retorting furnaces. All told, the resulting product of metal reduction is the oxidation of carbon to carbon dioxide. It is important to keep in mind that the production of carbon dioxide has been reduced dramatically since the start of the Industrial Revolution of the late nineteenth-century. This is best exemplified by the history of steel making in the world.

In the late 1800s, the invention of the blast furnace greatly affected the production of pig iron, and consequently steel manufacture. Many historians credit this single invention with the beginning of the Industrial Revolution. Steel production prior to this unit process used as much as 2 to 3 tons of coke (fixed carbon from coal) for every ton of hot metal pig iron produced from iron oxide ores. Modern steel production has been far more efficient, with current production estimates ranging from 0.5 to 0.65 tons of coke per ton of iron produced. Add to this the recent trend away from blast furnace production from ores. Recycling of steel products (such as shredded automobiles, recycled appliances, construction materials, and other end-of-life cycled products) has lead to the direct production of nearly half of all steel from recycle through minimills. The electric-arc furnace (the workhorse of minimill recycling operations) does not require coke for the production of molten iron. Instead, the major waste product is small amounts of slag and EAF bag-house dust, which is primarily composed of the zinc oxide powder that is evolved from scrap that was galvanized. So in a way, the improved blast furnace operations and the advent of EAF recycling of steel have meant a dramatic drop in the carbon dioxide production of steel. Prior to 1880, steel production meant the production of as much as 11 tons of CO_2 per ton of iron, to the current rate of less than 1 ton per ton of steel consumed. The U.S. steel industry leads this charge, but other countries (namely Argentina, Brazil, Japan, and other major steel-producing nations) are also following suit by implementing better operational strategies in their steel production.

Lead is also produced in a blast furnace using coke as the primary reductant. Much like the iron blast furnaces, the off-gases from the smelting of lead oxides will primarily contain CO and CO_2. This gas is recycled back to the blast furnace to use the reducing power of the CO and the heating value otherwise lost. This helps to reduce the total amount of carbon that would be needed to produce lead. In many smelters, galena (the natural mineral PbS) is the feedstock. The galena is roasted to lead oxide, so some of the vapor waste from these operations will contain SO_2. As mentioned previously, this is collected and reacted to form sulfuric acid.

Primary aluminum production starts with the purified alumina (Al_2O_3) that is produced when bauxite is leached with caustic soda (NaOH). The pure Al_2O_3 is dissolved in molten cyrolite (Na_3AlF_6) held at $900°C$ to about 10%, by weight. Carbon electrodes are then put into the melt, and as the current is passed into the molten salt solution, the Al_2O_3 is reduced to Al liquid and O_2 gas. The oxygen from alumina reacts with the anodes to form CO_2. The molten aluminum sinks to the bottom of the melt, and is tapped off and cast in pure aluminum ingots. As the Al_2O_3 becomes depleted in the molten salts, the density of the solution raises until the aluminum cannot sink. At this point, the operators notice a large spike in the applied voltage and more alumina is charged and the process continues.

Zinc production follows a similar path to the iron and lead processes in which carbon monoxide from carbon is the reducing agent. Carbon (coke or charcoal) is blended with zinc oxide and fed to a retort. The boiling point of pure zinc is low ($\sim 900°C$), so as the carbon is burned to form CO, the direct reduction of the zinc oxide by CO will result in the direct volatilization of the zinc. Cooling the vapor will condense the zinc, which can then be cast into ingots or alloyed with other metals.

Table 1 shows the relative production of CO_2 per tonne of metal produced. As shown, the largest producer of CO_2 is indeed the iron and steel industry, but it should be noted that is primarily because of the high consumption of iron and steel. Lighter metals, such as aluminum, will likely be used as a replacement for steel in hybrid automobiles and in low-density building materials. So there could be an increase in CO_2 gas production as more of the lighter density metals are consumed to replace steel.

For the most part, there are few techniques in place that can be employed to remove and collect the CO_2 from metal production. Most efforts in CO_2 reduction have been in improved smelting processes and lower carbon consumption. However, monoethanolamine (MEA) and diethanolamine (DEA) in organic solutions

Table 1 The Approximate Amount of CO_2 Produced per Tonne of Metal

Metal	CO_2 per Tonne of Metal, Tonnes	Average Metal Production from Ore in U.S., Tonnes per Year	Annual CO_2 Produced per Tonne of Metal, Tonnes/Per Year
Pig iron	1.50	40,000,000	60,000,000
Lead	0.21	440,000	92,400
Aluminum	1.30	2,500,000	3,250,000
Zinc	0.67	760,000	509,200

are used to scrub CO_2 from vent gases, but its expense precludes widespread adoption by the metals industry.

2.2   Hydrometallurgical Processing Wastes

As mentioned previously, most of the wastes from hydrometallurgical operations are simply treated and then sequestered in tailings impoundments. For those operations that do not have tailings impoundments, more complete process scenarios have had to be designed. For nearly every operation, pretreatment wastewaters, rinse waters, or wash waters must include neutralization and metal removal before the solutions can be discharged, impounded, or reused. The most widespread method for removing dissolved metals from solution is to precipitate them as solids for separation.

Precipitation Overview

For the majority of metal-bearing solutions, simple chemical precipitation is the most effective means of removing a substantial amount of the dissolved metals. Hydroxide precipitation is a simple, inexpensive means of reducing most of the metal ions from solution, and neutralizing the pH. Sulfide precipitation using sulfide donors, such as hydrogen sulfide, sodium sulfide, thiourea and thioacetamide, is another means of reducing soluble metals from waste streams. The precipitation process is generally the same whether the final product is metal hydroxide precipitates or metal sulfide precipitates. The final size of precipitates is a function of the following factors:

- Initial metal ion concentration
- Rate at which precipitants are added, or chemically generated
- Precipitant concentration (i.e., the concentration of OH^- or $S^=$)
- Aging time

For most waste streams, the metal ion concentration is set by the process and is not variable. However, the concentration and rate of adding the precipitant can be varied. For instance, if H_2S is used as a precipitant, the actual sulfide ion concentration is defined by a complex series of equilibria:

$$H_2S(g) \rightleftharpoons H_2S(aq)$$

$$H_2S(aq) \rightleftharpoons H^+ + HS^-$$

$$HS^- \rightleftharpoons H^+ + S^=$$

As shown, hydrogen sulfide gas bubbled into water will produce an acidic pH. Consequently, the concentration of sulfide ion ($S^=$) will be very low for even a saturate hydrogen sulfide solution. The predominant species of sulfur in solution is aqueous H_2S or HS^-. As the metals react with the sulfide species, the equilibrium between the H_2S, HS^- and $S^=$ is shifted even more. This means that the net amount of reacting species with the metal ions is quite low, but will be continuously replenished as the precipitation process proceeds. Likewise, with such sulfide donors as thiourea and thioacetamide, the production of soluble sulfide species is directly dependent on the consumption of the $S^=$ by precipitation.

Keep in mind that all precipitation events are due to the fact that the concentrations of the soluble species (Me^{+2} and $S^=$, for example) have exceeded the solubility of the solid species. For the sake of discussion, let's look at a solution containing 1 g/L copper as cupric ions (Cu^{+2}). The solubility product for $Cu(OH)_2$ is defined as the equilibrium constant for the dissolution of copper hydroxide to form the individual ions of copper and hydroxide. Chemically, it is defined by the following equilibrium:

$$Cu(OH)_2(s) \rightleftharpoons Cu^{+2}(aq) + 2OH^-(aq)$$

The overall equilibrium constant of the reaction is defined as the solubility product

$$K_{SP} = \frac{a_{Cu^{+2}} \cdot (a_{OH^-})^2}{a_{Cu(OH)_2}},$$

where a_i is defined as the activity of species i.

For aqueous systems, a unit activity is expected for the solid species (i.e., we assume that the chemical reactivity of a solid in water is unchanging as long as there is solid in equilibrium with the solution). Also, for dilute concentrations, we assume that the activities are equal to the concentrations of the species. With these assumptions, we can reduce the solubility product constant equation to

$$K_{SP} = [Cu^{+2}] \cdot [OH^-]^2$$

(It is relatively simple to remember that the solubility product is the product of the soluble species.) For the solid precipitate to form, the product of the copper ion concentration and the hydroxide ion concentration squared must exceed the numerical value of K_{SP}. Thermodynamic data can provide the needed values of the free energies of the ionic species. Then the value of K_{SP} can be determined directly (the K_{SP} for $Cu(OH)_2$ is 1.58×10^{-20}). Keep in mind that the equilibrium hydroxide ion concentration is defined by the pH of the water solution. So for any metal ion concentration, we can predict the maximum pH before the precipitate will be formed.

Monhemius has gathered the K_{SP} values for several metal ion concentrations. He prepared an excellent nomograph that allows us to predict the final metal ion concentration in any water solution at a specific pH (i.e., for a specific hydroxide ion concentration.) Figure 3 shows the diagram from Monhemius. As you can see, based on this chart, if the pH is 4 or less, no copper hydroxide precipitate will form for our 1 g per liter copper solution. However, if we increase the pH of the solution to 6, the copper ion concentration will drop to the equilibrium concentration of 10^{-4} M (6 mg/L). Recall that these are equilibrium concentrations; that is, the final concentrations will be achieved only after a sufficiently long period of time has passed.

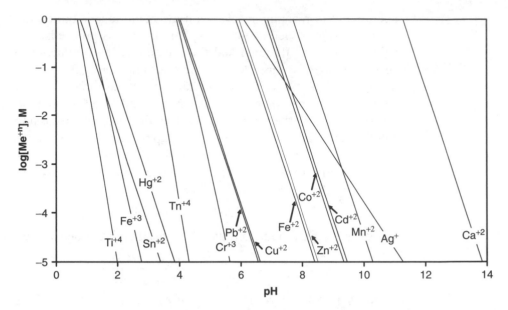

Figure 3 (Adapted from Ref. 10.)

This brings us to consider the actual precipitation reaction and the path at which it occurs. The copper ions must come in contact with two hydroxide species to form the copper hydroxide precipitate. But to initiate this transition from aqueous species to solid compound, the ions must congregate to form nuclei for the precipitation. Most experts consider a nucleus to consist of at least three molecules. Once these nucleates are formed, the rest of the ions will simply build onto these until the equilibrium solution concentrations are achieved. So we can think of the precipitation process as consisting of two principal steps—nucleation and growth. Nucleation is the slowest step in the process, and will only occur after saturation has been exceeded. It can be likened to an activation energy that must be overcome before a reaction will occur. Prior to forming nuclei, we can actually have a supersaturated solution that remains stable for some time. But once nuclei form, the solution rapidly changes to the final equilibrium concentration as precipitates form. One way to lower the nucleation energy is to seed the process with starter nuclei. Particles recycled from a precipitation process, or particles with similar crystallinity, can be used to initiate precipitation processes without having to achieve supersaturation first.

Here's where the concentration of metal ions becomes so critical. If the metal concentration is initially low, as nucleates form, there will be little more metal ions in solution to grow the precipitates. Therefore, the precipitation results in a relatively few colloidal (i.e., ultrafine) precipitates (also known as *sols*). If the metal ion concentration is higher, then there will be plenty of extra metal ions to grow the precipitates larger from the initial colloidal nuclei. However, at too high of a concentration, the solution becomes too dense for metal ion migration. Under these conditions, a tremendous number of nuclei are formed by the metal ions in close proximity. The shear number of nuclei consumes most of the metal ions, and again we have colloidal precipitates (gels). However, under these circumstances, an aging process can be utilized to grow larger precipitates at the expense of smaller ones. The natural tendency for precipitates will be to proceed to the lowest-energy state possible. A slurry comprised of a large number of colloidal particles will have a very large surface energy. The natural tendency will be for this slurry to reduce the net energy of the system by reducing the surface energy of the system. Particles coagulate and reform into larger particles, which will inherently have lower surface energies. Some smaller precipitates will actually be redissolved and reprecipitated onto larger particles (Smulchowski's theory). The aging time is a very big factor in the size that the precipitates will ultimately achieve. The key is to allow a very long time to achieve this new equilibrium. Typically, aging can be done is less than 30 minutes; however, longer times will result in larger solid particles.

Why do we want larger precipitates? Keep in mind that the whole reason for this process is to reduce the metal ion concentration in waste streams. Solid particles must be extracted from the waste streams before safe discharge or reuse of the water. Larger particles are much easier to remove by decantation processes (such as thickeners) or filtration. Colloidal particles are simply too difficult to remove from water.

Now let's look at the copper solution in a sulfide precipitation scenario. The K_{SP} for CuS is as follows:

$$CuS(s) \rightleftharpoons Cu^{+2} + S^{=} \quad _{SP} = 1.26 \times 10^{-36}$$

Notice that the equilibrium constant is an extremely small number compared to the hydroxide K_{SP}. This shows that for the same copper ion concentration, an immeasurable amount of sulfide ions is needed to achieve saturation. In other words, it will take extremely low sulfide concentration to precipitate more of the copper than just a pH change. As we have discussed previously, the sulfide ion concentration is a function of the solution pH. That is, the sulfide ion concentration will be considerably lower at low pH (where metal ions are more stable), and become higher as the pH is raised. So like the hydroxide precipitation process, the pH can be an important variable.

Again, Monhemius has simplified the sulfide precipitation process with the diagram reprinted in Figure 4. Notice that the hydrogen sulfide equilibrium changes with the pH. Also notice that metals such as copper and mercury form very stable sulfide precipitates at even negative pH values. Simply, this means that these sulfide species are very stable and easy to form at any pH. The iron and nickel sulfides are less stable at low pH, but are very stable at high pH.

Comparing the sulfide and hydroxide precipitation processes always tends to favor the sulfide precipitation process. First, the hydroxide ions are in large

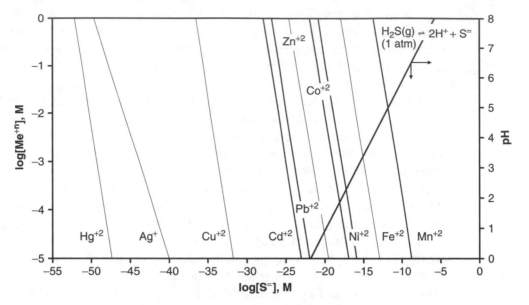

Figure 4 (Adapted from Ref. 10.)

concentrations immediately when a base is added. However, as we have seen, the sulfide ions that react with the metals must first pass through a series of equilibria to react. This simple pathway difference is one of the reasons sulfide precipitates are typically larger and more easily separated from the water. Also consider that hydroxide particles are simply isomers of the metal oxides that have waters of hydration incorporated in their matrix. For instance, the copper hydroxide precipitate $(Cu(OH)_2)$ of our previous example is actually a hydrated copper oxide; that is, it is actually $CuO \cdot H_2O$. Metal sulfides are crystalline at the point of formation, so the growth is more structured and thermodynamically favored. Aging of sulfide precipitates will result in much larger particles than for hydroxide precipitates.

Spent Electrolytes for Electrorefining or Electrowinning

In both the electrorefining (ER) and electrowinning (EW) processes, the primary product is a pure cathode sheet. Although similar processes, ER and EW differ only in the source of the metal ions that are reduced at the cathode. In electrorefining, impure, solid metal anodes are suspended in the electrolytic cell and oxidized to metal ions by the current. The metal ions are, in turn, reduced at the cathode. The net result is that metal ions are formed at the anode and removed at the cathode. Nickel, copper, and other metals can be cast in the impure form as anodes, which serve as the metal ion source during electrorefining.

In electrowinning, concentrated electrolytes from solvent extraction, ion exchange eluents, or carbon stripping are sent through an electrowinning cell to plate the metal on cathodes. The oxidation reaction at the anode varies, but typically involves the oxidation of water to form oxygen or the oxidation of ferrous ions to ferric ions. Copper and gold are the most common metals recovered by the direct plating of metal from leach solutions or concentrated electrolytes.

Despite the difference in source metals, both produce a concentrated electrolytic solution that must be recycled or treated and disposed. The refining of anodes by the electrolytic process is specifically designed to recover only one metal (copper, nickel, etc.). Consequently, all metals with a lower oxidation potential than the metal of interest will also be oxidized at the anode, but won't be reduced at the cathode. In copper refining, nickel and iron contamination in the anodes will dissolve into the electrolyte. However, because their oxidation potentials are below the copper oxidation potential, they will not plate on the cathode. Instead, they remain in the electrolyte solution and slowly build up in concentration. Arsenic is another common compound found in anodes that also dissolves and remains in the electrolyte. Periodically, these solutions have to be treated to reduce the level of contamination.

This is typically accomplished by first directly electrowinning all of the valuable metal from the electrolytes. Keep in mind that the primary metal of interest for these processes will be the metal that has the highest oxidation (and reduction) potential. So they must be eliminated before any of the other metals can be removed. Then the solutions are distilled to remove the excess water and form the precipitate of the other metals as sulfate salts or chloride salts (depending on the acid used).

As an example, a copper electrorefining plant will typically use electrolyte that contains 50–55 g/L copper with a 160 g/L sulfuric acid concentration. The copper concentration will vary only slightly as copper plates on the cathodes as an equal amount of copper is oxidized from the anodes. As the process continues, other metal values will likely increase. Arsenic levels can approach 10 g/L (as arsenate ions) and nickel concentrates can build to 20–40 g/L just by recycling the solutions. A portion of the solution is diverted to electrowinning cells where the copper concentration is reduced by plating it on cathodes. Oxygen is evolved at the anode and the copper concentration is reduced as much as possible. Then the solutions are distilled to 60 percent sulfuric acid. At this concentration, nickel sulfate will be precipitated and can be recovered by filtration and sold. The arsenic levels are not reduced much by this process, but can be reduced by adding ferric salts or other precipitating agents. The resulting solution (known as 'black acid') can now be recycled or sold to other industries interested in soluble metals in the solution. The solids are typically sold to other metal-processing companies. Occasionally, the solids are sent to landfills or tailings impoundments for sequestration.

Similar problems occur with the constant reuse of solutions used in electrowinning. Typically, leach solutions do not contain just the one metal of interest, but rather have any number of contaminating metals that were present in the ore.

Concentrating methods, such as solvent extraction for copper and carbon absorption for gold, will selectively strip just the metal of interest. Loaded solvents and carbon columns are stripped of their metal value in a concentrated solution that can now be electrowon to recover the metal value. The spent electrolytic solutions stripped of the copper or gold can be recycled back to recover more metal value from the solvents or carbon. Some contaminant metals may also pass through this process and continually build up in the electrolyte. For these operations, the solutions can be diluted into the original leach solutions as make up lixiviant, or a process similar to the one just described can be used to strip contaminant metals from the electrolytes.

Pickling Liquors

Rolling mills produce two of the more important forms in which steel is sold—rods and sheets. Rods are the source materials for steel wire, construction materials (i.e., rebar and beams), steel coils, and so on. Automobile manufacturers require large shipments of steel sheets wound in large rolls. The rolling mills process hot metal to make these products, which can lead to oxidation of the metal. Ingots of the metal are sent through successively narrower rollers until the desired thickness (or gauge) of the metal is achieved. Successive rollers will produce rods of various shapes and diameters, or sheets that vary from plate thicknesses to thin foils.

But one of the problems with continuously flattening or shaping metal is that it has to be hot. This means that metals such as steel and aluminum can oxidize rapidly. For aluminum this isn't a problem because the oxidation layer is thin and actually protects the aluminum from further oxidation. But in steel, oxidation dulls the finish and can affect the final properties of the steel. For instance, even the slightest surface oxidation can dramatically affect the drawing processes for making steel wire. *Pickling liquors* are concentrated acid solutions (hydrochloric or sulfuric acid, primarily) that are used to remove the oxidation products, such as iron oxides or scale, that can form on hot steel products. The pickling process is an important part of the final processing of steel products for market. However, the process is tricky to operate properly. Too aggressive of an acid, and the metal will be dissolved or corroded; not aggressive enough and the scale will not be removed. Several factors can influence this process:

- Acid concentration
- Temperature of the pickling bath
- Time that the steel remains in the bath
- Percent of ferrous ions in the solution
- Presence of inhibitors that slow or prevent attack on the steel surfaces
- Agitation of the bath

These solutions can contain as much as 60 percent (by volume) hydrochloric or sulfuric acid, depending on the shape of the steel and type of steel (high carbon or low carbon) to be cleaned. The solutions are used until either the acid content is too low (Baumé gravity scale), or the iron content is too high (titration technique). By the time it is ready to be discarded, pickling solutions can have very high iron concentrations (60 g/L or 0.5 lb/gal total dissolve iron). These solutions pose one of the most difficult waste disposal problems for the steel industry. There are a few other industries that can use the solutions (mainly for the acid content), but the volume of the solutions can literally flood the market.

Some of the end uses for this material are dictated by the species of iron in the solution. In the oxidized state, the ferric can be used as a flocculant in sewage treatment plants, phosphate precipitant, and pigment for paint, among other uses. Once the solutions are neutralized to a pH of about 5, the solutions can be oxidized with air (slowly) or hydrogen peroxide to create the ferric ions that are needed for these applications. In the reduced state (ferrous), the solutions can be neutralized and used as an iron supplement in livestock feed.

3 CONCLUSIONS

The wastes of the metals industry vary with the sector of the industry. Mining creates more wastes (primarily solids) than any of the other sectors. Tailings and waste rock are the two largest volumes of waste that metal production generates. While no attempt has been made to reduce this necessary waste, strict planning, federal and state regulations, and impoundment criteria have been adopted to lessen the impact of these wastes.

Chemical metallurgy (namely pyrometallurgical processing and hydrometallurgical processing) also generate large amounts of solid, liquid, and gaseous wastes, but the volumes of these wastes are generally less than the mining wastes. Also, these wastes typically have a better chance at being reused (such as iron and steel slags), or better treatment processes are available to more effectively treat them. A general reduction in the production of these wastes has been on-going, primarily due to the increased amount of recycling in recent years. The recycling trend has been market driven, to some extent, because of the increased worldwide demand for metals. The demand has exceeded mining production capacity, which is slow to develop new ore bodies with a minimal environmental impact. Energy conservation and environmental conscience are also factors in the voluntary trend of the metals industry toward more recycling.

REFERENCES

1. A. W. Adamson, *Physical Chemistry of Surfaces*, 5th ed., John Wiley, New York, 1990.
2. N. R. Ballor, C. C. Nesbitt, and D. R. Lueking, "Recovery of Scrap Iron Metal Value Using Biogenerated Ferric Iron," *Biotechnology and Bioengineering*, **93**(6), 1089–1094 (April 20, 2006).

3. R. Crowell, C. C. Nesbitt, J. L. Hendrix, and J. H. Nelson, "Reactions and Possible Reactor Design for the Precipitation of Metals Sulfides Using Thioacetamide," in *Hydrometallurgical Reactor Design and Kinetics*, R. G. Bautista, R. M. Weisley and G. W. Warren, eds., TMS-AIME, pp. 421–439, 1986.

4. H. M. Freeman, *Standard Handbook of Hazardous Waste Treatment and Disposal*, McGraw-Hill, New York, 1988.

5. J. D. Gilchrist, *Extraction Metallurgy*, 3d ed., Pergamon Press, Oxford, 1989.

6. International Network for Acid Prevention Web site, http://www.inap.com.au/.

7. D. Lueking and C. C. Nesbitt, "Ferric Iron Biogeneration: An Extreme Approach," *SIM News: The Official News Magazine of the Society for Industrial Microbiology*, **54**(1), 4–9 (January/February 2004).

8. H. E. McGannon (ed.), *The Making, Shaping and Treating of Steel*, 9th ed., United States Steel Corporation, Pittsburgh, PA, 1971.

9. J. Marsden and I. House, *The Chemistry of Gold Extraction*, Ellis Horwood, New York, 1992.

10. A. J. Monhemius, "Precipitation Diagrams for Metal Hydroxides, Sulphides, Arsenates and Phosphates," *Transactions of the Institute of Mining and Metals*, **86**, pp. C202-C206, 1977.

11. C. C. Nesbitt, J. L. Hendrix, and J. H. Nelson, "Use of Thiourea for Precipitation of Heavy Metals in Metallurgical Operation Effluents," in *Extraction Metallurgy '85*, The Institute of Mining and Metallurgy, London England, pp. 355–375, 1985.

12. J. Newton, *An Introduction to Metallurgy*, 2nd ed., John Wiley, New York, 1952.

13. D. J. Shaw, *An Introduction to Colloid and Surface Chemistry*, Butterworth-Heinneman, Oxford, 1980.

14. V. L. Snoeyink and D. Jenkins, *Water Chemistry*, John Wiley, New York, 1980.

15. W. Stumm and J. J. Morgan, *Aquatic Chemistry: An Introduction Emphasizing Chemical Equilibria in Natural Waters*, 2d ed., John Wiley, New York, 1981.

16. J. L. Uhrie, J. I. Drever, P. J. S. Colberg, and C. C. Nesbitt, "In Situ Immobilization of Heavy Metals Associated with Uranium Leach Mines by Bacterial Sulfate Reduction," *Hydrometallurgy*, **43** 231–239 (1996).

17. United States Environmental Protection Agency, *Identification and Description of Mineral Processing Sectors and Waste Streams*, 1991. Available at http://www.epa.gov/epaoswer/other/mining/minedock/id.htm.

18. United States Geological Survey, *Mineral Commodity Summaries*, U.S. Department of the Interior, Gale A. Norton, Secretary, U.S. Government Printing Office. Washington, 2006. Available at http://minerals.usgs.gov/minerals/pubs/mcs/.

19. United States Geological Survey, *Mineral Commodity Summaries*, U.S. Department of the Interior, Gale A. Norton, Secretary, U.S. Government Printing Office, Washington, 2005. Available at http://minerals.usgs.gov/minerals/pubs/mcs/.

20. N. L. Weiss (ed.), *SME Mineral Processing Handbook*, vols. 1 and 2, Society of Mining Engineers of the American Institute of Mining, Metallurgical, and Petroleum Engineers, Inc., New York, 1985.

21. B. A. Wills, *Mineral Processing Technology: An Introduction to the Practical Aspects of Ore Treatment and Mineral Recovery*, 5th ed., Pergamon Press, Oxford, 1992.

CHAPTER 3

LIFE-CYCLE EVALUATION OF CHEMICAL PROCESSING PLANTS

David Brennan
Department of Chemical Engineering, Monash University, Australia

1	LIFE CYCLES OF CHEMICAL PROCESSING PLANTS	59
2	ECONOMIC FEATURES OF PLANT LIFE CYCLES	61
3	ENVIRONMENTAL FEATURES OF PLANT LIFE CYCLES	63
3.1	The Concept of Cleaner Production	64
3.2	Product Life Cycles	65
4	WASTE-MINIMIZATION STRATEGIES	67
4.1	Waste Minimization in Reactors	67
4.2	Waste Minimization in Separation Equipment	69
4.3	Waste Minimization in Utility Systems	70
4.4	Waste Minimization in Operations and Maintenance	71
5	LIFE-CYCLE ASSESSMENT METHODOLOGY	72
5.1	Basic Steps in Life-cycle Assessment	74
5.2	Linking Environmental Impact Reduction with Costs	76
5.3	Continuing Development and Application of LCA Methodology	78
6	ECONOMICS OF CLEANER PRODUCTION	78
6.1	Costs of Managing Waste	78
6.2	Environmental Externalities	79
6.3	Economic Incentive Instruments	80
6.4	Eco-efficiency	80
7	SUSTAINABILITY	81
7.1	Sustainability Indicators	82
7.2	Temporal Characteristics of Sustainability	83
8	INDUSTRIAL ECOLOGY	83
8.1	Australian Example of Industrial Ecology	84
8.2	Extension of Industrial Ecology to Materials and Energy Recycling	85
9	REGULATORY DRIVERS	86
10	CONCLUDING REMARKS	86

1 LIFE CYCLES OF CHEMICAL PROCESSING PLANTS

Chemical processing plants add value to raw materials through the successful engineering of chemical and physical changes. The product of a chemical plant may be a consumer product, but is often an intermediate for further processing in a separate downstream plant. For example, ethylene produced by the steam

59

cracking of ethane becomes a raw material for making products such as polyethylene, ethylene oxide, and styrene in separate plants using distinct technologies.

The concept of *life cycle* is widely and usefully applied to both chemical processing plants and to chemical products. We first consider chemical processing plants, where the life cycle proceeds through six phases:

1. *Planning*, where a perceived business opportunity leads to the development phase of a project. In this phase, markets for products are evaluated, processing technology is selected and refined, suitable raw materials are sourced, and a plant site is selected. These activities lead collectively to the definition of a project for the design and construction of the plant. Planning must be subjected to a range of criteria assessments governing the economic, safety, and environmental viability of the project. The criteria assessments are made by the project team in consultation with government and the community.

2. *Design*, which progresses through several stages commencing with process design and leading to equipment and plant design. The plant includes foundations, structures, piping, electrics, and instrumentation. There is a flow of design information between project participants as the design proceeds from process conception through to the hardware that constitutes the physical plant. A degree of design detail is necessary to enable a plant cost estimate of sufficient accuracy to support an investment decision.

3. *Construction*, where the site is prepared, and equipment and plant hardware are delivered, installed, and assembled ready for operation.

4. *Commissioning*, where the plant is tested and production capacity and performance targets verified. If difficulties arise in meeting targets, some plant modifications might be required and time delays might be encountered.

5. *Operation*, where raw materials are converted to products at the desired production rates meeting required performance criteria. Performance criteria relate not only to production rates, but also to raw materials and utilities efficiencies, reliability and online time, and safety and environmental performance. In the operational phase, cash generated from sales revenue less plant operating costs provides the financial return on capital invested. Operating costs include raw materials, utilities, personnel, and a range of charges associated with owning, insuring, and maintaining capital assets. Operating costs also include costs associated with the generation, treatment, and disposal of any wastes.

6. *Decommissioning and dismantling*, when the plant is retired from service having been judged as no longer viable as an operating concern.

Figure 1 shows some key stages of the life cycle of a chemical processing plant.[1]

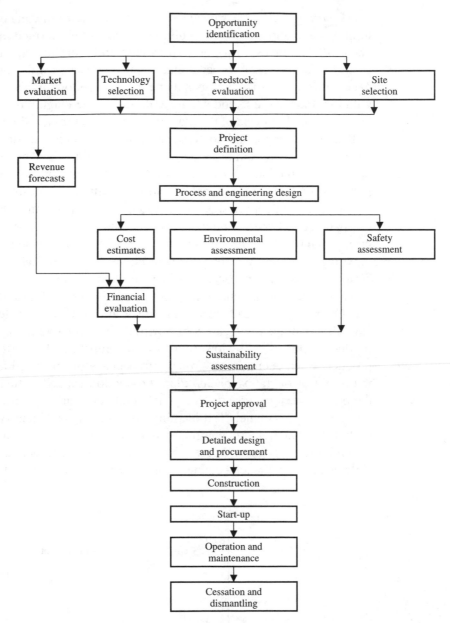

Figure 1 Key stages of the life cycle of a chemical plant (adapted from Brennan, 1998.)

2 ECONOMIC FEATURES OF PLANT LIFE CYCLES

The life cycle is important from an *economic* viewpoint, as major sums must be invested in the planning, design, and construction of the plant prior to the operational phase. Investment includes fixed capital costs of the processing plant (termed *inside battery limits investment*), storage facilities for raw materials and products, utilities generation plant, and buildings (collectively termed *outside battery limits* or *offsites* investment), as well as land and working capital. *Working capital* accounts for the cost of process material inventories as well as the credit extended to customers. Additional costs may be incurred following plant closure associated with plant dismantling and site rehabilitation. The magnitude and pattern of cash flow over the life cycle of the plant determine the profitability of the plant.

Planning, design, and contracting personnel exercise distinct roles, skills, and influence prior to commissioning, while operations and maintenance personnel exercise the predominant roles during the operating life of the plant. The experience of operations personnel should be an input into design. Designers should clearly communicate design intent to operations personnel. It is important that the balance between capital investment and operating cost that is established during planning and design minimizes the overall life-cycle cost of the plant and maximizes its life-cycle benefits (see Figure 2). Good dialogue and cooperation between designers and operators helps to achieve this.

There are many potential influences on the pattern of cash flow during project life. Operating costs can change due to variations in prices of feedstocks, utilities, and labor. Competing technologies with their distinctive life cycles can influence product selling prices by lowering costs in competing plants. Real selling prices of products often decline over project life because of improved technology embodied in newer operating plants. Plant production capacity often increases beyond the original design capacity due to deliberate capacity expansions and to learning by operating personnel. The learning process helps to identify capacity margins within the plant; these margins can be exploited, often with minimal capital expenditure, to achieve increased capacity. Figure 2 illustrates some of the possible changes in financial and technological performance over the life of a project.

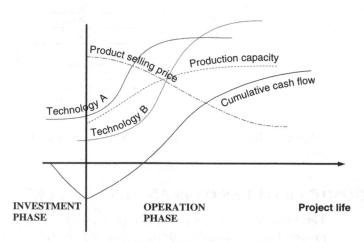

Figure 2   Cumulative cash flow and other performance profiles over project life.

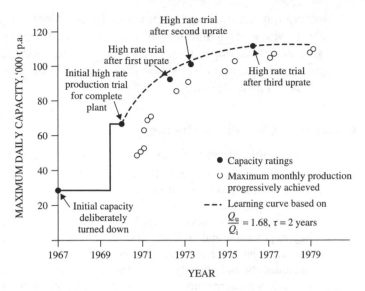

Figure 3   Capacity history of an ethylene plant. (From Ref. 1.)

The projected operating life of a process plant is difficult to predict; although 10 to 15 years is often assumed in conventional project evaluations, operating life is often longer. Factors limiting operating life are often economic rather than mere physical deterioration, and include feedstock availability, operating cost structure, market demand, product selling price, and technology competitiveness.

Figure 3 shows the production capacity profile for an Australian ethylene plant.[1] The plant based on steam cracking of naphtha was expanded in three stages following its initial construction. Capacity growth can be seen to follow a learning curve that is consistent with a model proposed by Malloy:[2]

$$(Q_t - Q_i)/(Q_u - Q_i) = 1 - \exp(-t/\tau)$$

where Q_t = production capacity at year t
   Q_i = initial production capacity
   Q_u = ultimate production capacity
   τ = time constant in years

Productivity gains were made over plant operating life in raw materials consumption, operating personnel, and capital; energy consumption productivity was maintained constant despite production increase. The plant was replaced at the end of its life by a larger-capacity plant processing ethane.

3   ENVIRONMENTAL FEATURES OF PLANT LIFE CYCLES

The life cycle of a chemical processing plant is also important from an environmental standpoint. Impacts can occur during construction (e.g., from site excavation and noise), but are mainly prevalent during the longer operational phase when raw materials and energy are consumed and wastes emitted to air, water, or land. Impacts can also occur during plant retirement through dismantling, removal, and disposal of materials, and can often require soil remediation and land rehabilitation.

3.1 The Concept of Cleaner Production

In designing, operating, and evaluating chemical plants and their products, the concept of cleaner production has a key role in improving environmental performance. An early definition of cleaner production states:

> Cleaner production involves applying an integrated, preventive, environmental strategy to processes and products to increase efficiency and reduce risks to humans and the environment. For processes, this implies conserving raw materials and reducing quantity and toxicity of wastes before they leave the process. For products, it implies reducing impacts along the entire product life cycle, from raw material extraction through to product disposal. Cleaner production involves applying know-how, improving technology and changing management attitudes and operating practices.[3]

Some key points arise from this definition.

Integration
Cleaner production initiatives must be integrated

- With business objectives including needs of employees and the community
- With technical, safety and operational considerations
- Over the complete spectrum of industrial activity from planning, design, construction, through to operation

Prevention
Cleaner production emphasizes elimination or reduction of waste at source. This approach is vastly preferred to *end-of pipe* treatment of terminal waste streams from a process. Elimination of waste at source not only averts the need for subsequent treatment, but also avoids consuming material and energy resources in waste generation, separation, treatment, or disposal.

Product Life Cycle
Cleaner production is applied to the entire life cycle of a product. The life cycle extends from extraction of basic raw materials, through processing stages, product assembly, packaging and distribution, to use and final disposal. The life cycle incorporates any recycle loops.

Global Applicability

Although not explicit in the UNEP definition, the involvement of the United Nations parallels an international acceptance of the need for cleaner production. International participation in regional and global programs devoted to cleaner production is active.

Terms closely related to cleaner production are *pollution prevention* and *waste minimization*, which emphasize the preventive aspect of cleaner production. Closely associated with cleaner production is the *hierarchy of waste treatment strategies,* enunciated in the U.S. 1990 Pollution Prevention Act. Environmental agencies in most countries have adopted this hierarchical approach in their policies. Waste treatment strategies, in order of preference, are

1. Source reduction
2. Recycling and reuse
3. Waste treatment to render the waste less hazardous
4. Secure disposal

Source reduction is the most preferred strategy and disposal is the least preferred.

3.2 Product Life Cycles

A key aspect of the cleaner production philosophy is its application to the product life cycle. Figure 4 shows the life cycle for an aluminum beverage can. For this product, environmental impacts are associated with the following:

- Mining bauxite ore
- Refining bauxite ore to produce alumina
- Smelting alumina to produce aluminum
- Fabricating the aluminum can
- Using the aluminum can
- Recycling aluminum from a used can
- Ultimate disposal of a used can

Further impacts may be associated with transport at various stages of the life cycle.

Impacts from the various stages of the aluminum life cycle are many and varied, but include consumption of large amounts of energy at the mining, refining, and, particularly, the smelting stage. However aluminum is a light, strong metal, which enables energy efficiency in its use. Further, most aluminum products can be readily and efficiently recycled both at the fabrication stage and after use. Benefits of recycling include avoiding impacts upstream of the point where recycled aluminum is reintroduced, including reduced consumption of bauxite. Thus, aluminum can be considered energy intensive in its production, but energy efficient in its use. The life of a product in use is also important in determining its

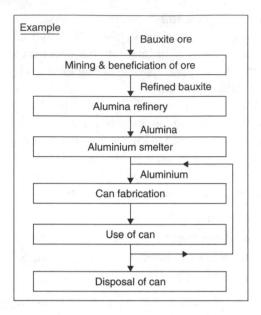

Figure 4   Diagram of the product life cycle for an aluminium beverage can.

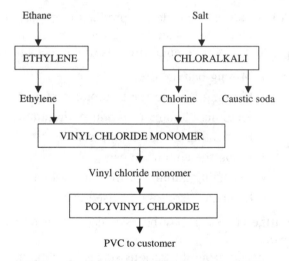

Figure 5   Cradle-to-gate life cycle for PVC.

life-cycle impacts; for example, aluminum has a longer life in building products and engine components than it does in beverage containers.

Figure 4 is an example of a *cradle-to-grave* life cycle. Figure 5 shows a further example of a product life cycle for polyvinyl chloride (PVC) manufacture, taken from the chemical industry and this time represented as *cradle-to-gate*, terminating at the stage where PVC product leaves the gate of the PVC factory for conversion by customers to fabricated products. Each processing stage of the life cycle is conducted in a separate chemical processing plant. Waste minimization in each plant (chlorine, ethylene, vinyl chloride monomer, and PVC) is important in minimizing waste generated in the production of the product PVC.

4   WASTE-MINIMIZATION STRATEGIES

Waste minimization in processes can be approached in a hierarchical order similar to that adopted in process design, starting with reactors and working through to utility systems. Thus waste is progressively identified and minimized in

- Reactor(s)
- Separation and recycle systems
- Heat exchanger network(s)
- Utility systems

4.1   Waste Minimization in Reactors

Waste minimization in a reactor requires a critical approach to both the reaction chemistry and reactor design. Many chemical synthesis routes are inherently prone to generation of waste. In the worst cases, there may be a complicated sequence of reaction steps, or environmentally hazardous chemicals may participate in the chemical synthesis path. Thus, there has emerged an awareness of the importance of clean chemistry leading to major research initiatives devoted to exploring fundamentally new process routes. In some cases, stoichiometric quantities of waste are generated as byproducts of the main chemical reaction. Such byproducts may be useful if saleable; however, in the event of market failure they become environmental and economic burdens. In other cases, waste is generated through side reactions. Often the reaction chemistry, even in mature industrial processes, is complex and not completely understood. Nevertheless, an understanding of reaction chemistry is fundamental to identifying and minimizing waste.

Reactor performance is generally characterized by

- *Conversion*, which measures the conversion of feedstock to product in one pass through the reactor. Conversion is usually quantified as the ratio of feedstock consumed in the reactor to feedstock fed to the reactor.
- *Selectivity*, which measures the extent to which the desired product is produced relative to undesired products of side reactions. Selectivity is usually quantified by the ratio of desired product produced to feedstock consumed in the reactor.
- *Yield*, which quantifies the mass of desired product relative to mass of reactor feed.

Secondary reactions can occur in parallel or series with the desired primary chemical reaction leading to reduced selectivity and yield. Key approaches to minimizing secondary reactions are as follows:

- Good temperature control of the reaction within the reactor
- Good mixing of the reactants fed to the reactor
- Rapid quenching of the reaction products immediately following reaction, either by indirect generation of steam or direct quenching in a liquid, often water

Many key industrial reactions operate at elevated temperatures. For example, formaldehyde, hydrogen cyanide, and ethylene production all occur at elevated temperatures and involve secondary reactions. In each case, reaction time is minimized and a waste heat boiler is used for rapid cooling of reaction products.

For catalytic reactors, catalyst performance is often key to minimizing waste, and much research effort is devoted to developing new catalysts. Improved catalysts can improve conversion, yield, and selectivity, but can also reduce required operating temperature and pressure requirements for reactors, thus reducing energy consumption. Catalyst degradation can cause deterioration in conversion, selectivity, or yield. Shortening of catalyst life has cost implications in terms of catalyst replacement and production loss through increased plant downtime. In some cases, catalyst degradation leads to fine particles accumulating in the catalyst bed, causing increased pressure drop across the reactor. The viability of regeneration or disposal of spent catalyst is also an important waste consideration.

Feedstock impurities are often a source of reactor waste. In different cases, impurities can cause these problems:

- They can poison a catalyst, reducing its effectiveness or life.
- They can react with feedstocks to generate wastes.
- They can be unreacted with feedstocks, but contaminate the product or require separation from the product downstream of the reactor.

The viability of feedstock purification should be evaluated. Basic raw materials such as crude oil and minerals inevitably contain impurities. In most cases, it is best to remove impurities as early as possible in the processing chain.

Some industrial reactions have been enabled by agent materials that do not participate directly in the stoichiometry of the reaction. An example of an agent material is mercury in chlorine-caustic soda production. Mercury can contaminate a number of effluent and product streams, as well as contaminating soil or groundwater if spillage occurs. Replacement technologies are being developed in many cases to avoid using such agent materials. In the case of chlorine manufacture, membrane cells have been developed and have largely displaced mercury cells.

Waste arising from limited feedstock conversion can be minimized in the overall process by recycling unreacted feedstock from the reactor effluent back to the reactor feed. Benefits of such recycling include reduced raw material consumption and reduced waste disposal or emissions. The burdens of recycling such streams include the need for separation of feedstock from reaction products and any necessary compression or pumping of recycled streams. These benefits and burdens have both economic and environmental consequences, which must be identified and evaluated.

4.2 Waste Minimization in Separation Equipment

Separation equipment has a diverse range of functions:

- Separation of distinct phases
  - Liquid from solids—settling, filtration, centrifugation, drying
  - Solids from gases—cyclones, bag filters
- Separation of components from a gas stream, for example by gas absorption
- Separation of components from a liquid stream, for example by distillation

Different separation steps typically have inherent sources of waste generation. These sources must first be identified prior to waste minimization. In considering distillation, gas absorption and drying as examples of separation steps, we can identify characteristic waste sources.

In distillation, wastes can occur through

- Accumulation of noncondensables in the overhead condenser
- Accumulation of nonvolatiles in the reboiler
- Fouling of trays and heat transfer surfaces

In gas absorption, wastes can occur through

- Contamination of effluent gas due to incomplete gas absorption or solvent entrainment
- Contamination of recirculating solvent by dissolved species or corrosion products

In drying, waste can occur through

- Product quality deterioration due to thermal effects or particle degradation
- Solids entrainment in effluent gas
- Fouling of heat transfer surfaces due to thermal effects

Such sources of waste must be systematically identified and avoided by appropriate design and operation strategies. For example, in distillation, selection of operating temperature and pressure in the distillation column can sometimes minimize fouling; in gas absorption, careful attention to liquid distributor and packing support plate can improve mass transfer; in drying, limiting gas velocities can often minimize degradation of solid particles.

Apart from the sources of waste inherent in different types of separation, energy consumption is common to all separation equipment. Energy may be consumed directly as a utility or indirectly through pressure loss from equipment contributing to compression or pumping energy. Energy or utility consumption implies a spectrum of impacts through resource use and waste emissions. When aggregated over all stages of a product life cycle, these impacts represent a substantial environmental burden.

4.3   Waste Minimization in Utility Systems

Wastes are generated within utilities production systems. Since utilities are consumed within chemical processing plants, these plants and their products contribute indirectly to utility waste generation. Utility consumption in chemical processing plants arises mainly from the following:

- Heating by steam, hot oil, other heat transfer fluids, or indirect heating from hot gases
- Cooling by cooling water and refrigeration
- Electricity for driving agitators, pumps, compressors, and solid conveyors

Utility waste can be minimized both by reducing consumption of utilities within chemical plants and by improving the design of utility systems. Environmental impacts result from the generation and distribution of utilities. Even though the utility generation may occur remote from the site of the process plant, the environmental burdens from utility generation and distribution may be attributed to the demand caused by the utility user. Common examples of remote utility generation and distribution include electricity and fossil fuels, as illustrated in Figure 6.

Much of utility waste results from a combustion of fuel, whether for direct heating use, or indirect use in steam or electricity generation. Fuel combustion both depletes a resource and emits wastes; wastes include carbon dioxide, sulfur dioxide, the oxides of nitrogen, and ash from unburned components of the fuel. The use of water in utility production results in water depletion and emission of aqueous wastes, often as purges necessary to avoid concentration of impurities.

Local temperature excursions arise where the latent heat of steam condensation is released to the environment through the cooling water return streams from steam condensers.

Many utility-generation systems have a spectrum of impacts. For example, recirculated cooling water from cooling towers involves power consumption in air fans and cooling water pumps, raw water consumption as make-up, and a purge to avoid excessive concentration of contaminants. Refrigeration systems involve

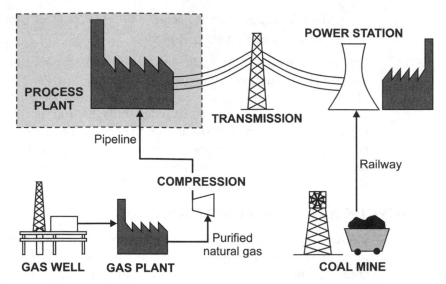

Figure 6 Possible wider systems for utilities supply to process plants.

power consumed in refrigerant compression and cooling water use for refrigerant condensation, as well as impacts from fugitive emissions of refrigerants to air.

Energy efficiencies and environmental impacts in electricity generation depend on the technology employed and the fuel adopted. Impacts also occur in fuel extraction and supply, and in electricity transmission to the consumer. Electricity generation efficiency may be defined as the ratio of electrical energy exported from power station to fuel energy consumed. Typical efficiency ranges, reflecting the effects of design and operation are:

Steam turbine—black coal	30–37%
Steam turbine—brown coal	23–32%
Open cycle gas turbine	17–37%
Combined cycle gas turbine	44–57%
Integrated gasification combined cycle—black coal	40–52%

4.4 Waste Minimization in Operations and Maintenance

Some emissions from an operating plant can be identified from analysis of a detailed process flowsheet that defines all process streams and utility consumptions. However, these data are based on steady-state design considerations. Additional emissions can result from the following:

- Abnormal operation
- Start-up and shut-down
- Maintenance
- Cleaning and purging

- Grade changes, made to achieve different product properties
- Defects in equipment and piping

One common source of waste in operations is fugitive emissions from leakage of valves, pump and compressor seals, and pipe flanges. Fugitive emissions of hydrocarbons are a major source of emissions in petroleum refineries, for example. Abnormal operation includes spillages from accidents, including those incurred in transport and storage of materials. Since risk of containment loss is always present in transport, avoidance of transport through appropriate site location and integration of manufacturing is an important strategic consideration.

5 LIFE-CYCLE ASSESSMENT METHODOLOGY

Life-cycle assessment (LCA) is a method of providing a *quantitative* assessment of the environmental impact of a product over its *entire life cycle*. Raw materials consumptions, energy (or utilities) consumptions comprising *inputs,* and emissions to air, land, and water comprising *outputs* are accounted for. The life cycle encompasses all stages of product manufacture, use, and disposal. Figure 7 shows the basic concept of LCA, with inputs and outputs at each stage of the product life cycle. Byproducts are grouped with emissions as outputs in the diagram as they are distinct from the main product processing and use chain. Market demand for byproducts dictates whether byproducts can be usefully sold into other product chains or must be disposed of.

LCA has been most widely applied to products. One objective has been to establish which product is the most environmentally friendly for a particular function. Another objective has been to determine which part of a product's life cycle incurs the greatest environmental burden as a precursor to sourcing alternative raw materials, developing improved processing technology, or improving product design. Product functions can be compared related to those objectives:

- Beverage containers (e.g., a comparison of glass bottles and aluminium cans)
- Alternative adhesives
- Alternative washing machines

LCA can also be applied to competing materials:

- Steel versus aluminium in a range of applications
- Plastic versus steel components for motor cars
- Alternative construction materials for buildings

LCA can also be applied to chemical processes, with a view to process comparison and selection, process development or process synthesis. LCA can likewise

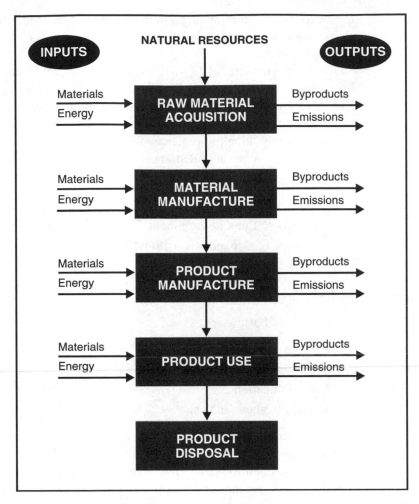

Figure 7 Diagram showing inputs and outputs at various stages of a product's life cycle.

be applied to the design of utility systems. Application of LCA to processes may be done in various contexts:

- At the research and development stage for an emerging process
- At the process design stage of a project
- For competitive analysis in business review

A chemical processing plant can be seen as a module of the extended life cycle for a product. For example, both alumina refining and aluminium smelting are separate and significant processes in the life cycle of the product aluminium. As such, LCA is a promising tool for quantitative evaluation of the cleanliness of a process, or for ranking alternative processes according to their environmental merits. LCA also enables identification of weak environmental links in a processing chain.

Although elements of LCA thinking emerged in 1960s, LCA became established as a methodology in the early 1990s. The methodology is still evolving, and there are distinctions in approaches, terminology, and definitions used. Not all aspects of methodology have been universally agreed. However, industries are increasingly adopting practices which incorporate LCA principles.

Some of the more accepted approaches to LCA methodology follow:

- Centre for Environmental Studies (CML), Leiden University[4]
- Society of Environmental Toxicology and Chemistry (SETAC)
- United Nations Environment Program (UNEP)
- International standards—ISO 14040, 14041, 14042, 14043, 14047–49

5.1 Basic Steps in Life-cycle Assessment

LCA has four basic steps:

1. Goal definition
2. Inventory analysis
3. Classification
4. Improvement analysis

In *goal definition*, the scope and purpose of the LCA is defined. The functional unit and system boundaries are also established. The *functional unit* is the reference point to which environmental impacts are attributed. The choice of functional unit depends on the application of the LCA. For a process it could be a tonne of product or feedstock, or a tonne of impurity removed. For a commodity product the functional unit could be a tonne of product, but for specialty chemicals and most consumer products performance properties are more important and the choice can become complex. For example, performance of a paint would be related to its surface coverage and durability; hence a possible functional unit for the paint might be the quantity required to cover a square meter of surface over a time span of 20 years.

An important early step is the definition of *system boundaries* within which impacts are accounted for. A good example of the issues in system boundary definition is the case of utility systems. The process plant in Figure 6 may consume electricity, but this consumption implies impacts at the power station through fuel combustion and water use. For a coal-fired power station, this, in turn, implies impacts at the coal mine. If the plant consumes natural gas for heating, impacts in exploration, extraction, treatment, and compression of the natural gas are implied. The environmental impacts incurred in such utility systems have a strong influence on the *aggregated impact* over the product life cycle, since impacts at each step of the life cycle invariably involve energy consumption, which, in turn, implies impacts from utility generation.

Inventory analysis involves the identification and quantification of all material and energy resources consumed, and wastes emitted. Note that *inventory* in LCA refers to the raw materials and energy inputs and emission outputs, not a quantity of stored chemical as the term denotes in safety analysis. For inventory data to be valid, the process technology, design, and operation of the plant must be considered and understood, and mass and energy balance requirements satisfied. Operational performance data are an essential input since deviation from design conditions often occurs during plant operation, and fugitive emissions, leaks and spillages must be accounted for. Published inventory data, whether in commercial software or literature, often fails to provide a clear account of the system boundary adopted and the basis for the data.

In *classification,* inventory data are grouped according to categories of specific environmental effects caused by particular molecular species. These effects include acidification, global warming, nutrification, photochemical smog, ozone depletion and toxicity as well as resource depletion. The inventory data are weighted using equivalency factors prescribed for the various molecular species and aggregated within their respective impact category to provide an impact score. The score for each impact category is frequently expressed in terms of an equivalent tonnage of a molecular species; for example global warming potential is quantified in tonnes equivalent of carbon dioxide, photochemical smog potential in tonnes equivalent of ethylene, and so on.

Often a *normalization* step follows directly from the classification step. In the normalization step, the score for a given impact category is divided by the total score for that impact category attributed to a specific country, region or the world. Normalization allows a perspective of the relative contributions to environmental damage categories for that country. Figure 8 illustrates this relativity for an LCA study on hydrotreating of diesel where energy depletion and global warming are shown to be the predominant normalized impacts.[5] In this case the impact score for global warming expressed in tonnes equivalent of CO_2 per bbl of straight run gas oil (SRGO) feed, is divided by the global warming score for Australia (in tonnes equivalent of CO_2) for a calendar year. Thus, the units for normalized global warming impact are yrs/bbl SRGO.

The final step in LCA, *improvement analysis*, seeks to identify where in the life cycle (or subsystem under study) the major environmental impacts occur and how they might be reduced. For example, in the hydrotreating study, the hydrotreater unit is energy intensive due to the high temperature and pressure required for the reactor. Improved catalysts might lower energy consumption by permitting less severe reactor conditions. Burgess and Brennan provide details of system boundary definition and data for the inventory, classification, and normalization steps for the diesel hydrotreating study.[5]

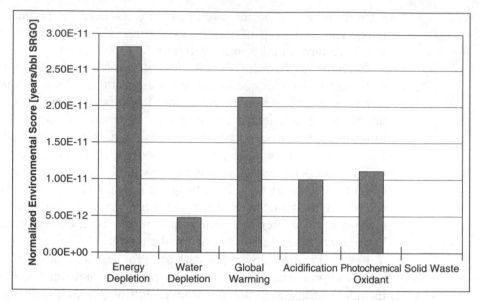

Figure 8 Normalized Environmental Scores using Australian data for an LCA study on hydrotreating of diesel. (From Ref. 5.)

A further step, *evaluation*, which was established in earlier expressions of LCA methodology, aims to weight the impact scores according to their relative environmental importance. For example, it might be judged that global warming and associated climate change represent a greater environmental threat than nutrification. The evaluation step, however, has been subject to much debate and relies on subjective judgment to a much greater extent than the other LCA steps. The judgment may be influenced not only by local environmental concerns but also by socioeconomic and political factors for a particular nation or region.

5.2 Linking Environmental Impact Reduction with Costs

In some LCA studies, normalized scores have been summed to provide an environmental impact score for a process or product, implying a relative equal weighting of impact scores. This approach has not had universal agreement and is not part of accepted LCA methodology. However, it can be a useful means of linking initiatives to reduce environmental burdens with the corresponding costs incurred. For example, Figures 9 and 10 shows the relative magnitudes of capital costs and internal operating costs and environmental impact scores for an extended diesel hydrotreating system. The hydrotreater is identified as the major contributor to both cost and environmental burden, compared with the hydrogen plant (using steam reforming of natural gas), the amine unit for recovering hydrogen sulfide from the hydrotreater product streams, and the sulfur recovery plant for converting hydrogen sulfide to sulfur.

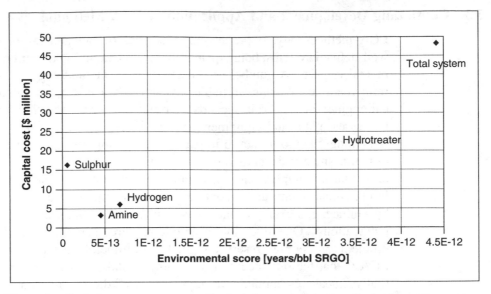

Figure 9   Capital costs versus normalized environmental impact for hydrotreating system. (From Ref. 5.)

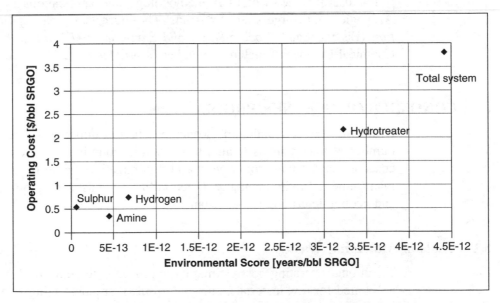

Figure 10   Operating costs versus normalized environmental impact for hydrotreating system. (From Ref. 5.)

5.3   Continuing Development and Application of LCA Methodology

LCA methodology, if properly applied, has the advantages of being based on well-defined systems, being quantitative, and examining the full range of environmental impacts. A number of aspects of LCA methodology are still undergoing refinement. Establishing reliable inventory data is closely aligned with chemical engineering, since it depends on process system definition, understanding of processes, plant and equipment, and verification of mass and energy balances. Assigning inventory data to impacts and deriving reliable impact scores is more difficult, since it depends on a knowledge of the science of environmental systems that is complex and not completely understood. Equivalency factors used in classification are under continuous review; equivalency factor data published by Guinee, for example, indicates numerous revisions of data in earlier CML publications.[4] Quantification of abiotic resource depletion is difficult because of problems in defining reserves of fuels and minerals. Nevertheless, a strength of LCA is that the full spectrum of environmental impacts is considered and quantitative measures applied. An appreciation of the nature, causes, and effects of the various types of environmental damage can be obtained from environmental engineering texts.[6] Approaches such as environmental impact assessment and environmental risk assessment can be used to supplement LCA.

For useful reviews of LCA methodology and applications to processes, see Azapagic[7] and Burgess and Brennan.[5] The journals *Journal of Cleaner Production* (Elsevier) and *Process Safety and Environmental Protection* (Institution of Chemical Engineers) include a number of articles on LCA-based studies.

6   ECONOMICS OF CLEANER PRODUCTION

Evaluating the economics of cleaner production initiatives is complicated by a number of uncertainties related to realistic accounting for internal and external costs associated with pollution, waste minimization, waste treatment, and waste management. Environmentally driven initiatives must satisfy both environmental and economic criteria to be sustainable.

6.1   Costs of Managing Waste

Traditional operating cost estimating approaches for industrial production have accounted for raw materials, energy (utilities), any packaging required for products, personnel, and capital-related costs. Since raw materials and utilities costs are derived from consumptions of raw materials and utilities, there is an important link here between economic and environmental assessments. Traditional operating cost estimating methods may omit certain costs associated with waste management or environmental control. In order to account for these costs, the U.S. EPA has developed a classification to encourage a *total cost assessment*:

Tier 0   Traditionally recognized costs associated with production

Tier 1   Hidden costs of monitoring waste, applying for permits, and related administration costs

Tier 2   Liability costs of penalties and fines arising from regulatory noncompliance and personal injury and property damage settlements, future liabilities

Tier 3   Opportunity cost related to consumer responses, employee health and safety, employee relations, and corporate image related to improved environmental performance

An important cost consideration in cleaner production initiatives is the cost of any necessary technology. Many environmental problems require novel technical solutions based on research and development, involving cost, time, and technical risk. Uncertainties in estimating Tier 1,2, and 3 costs, and the costs of technology development, invite a probabilistic approach to estimation of costs, revenues, and related profitability. This approach was adopted by Moilanen and Martin, for example, using *expected monetary values* or risk-weighted cash flows.[8]

6.2 Environmental Externalities

The major proportion of costs arising from environmental damage is borne by the natural environment and the wider community. Since these costs fall outside the accounting framework of the polluter, they are often called external costs, or *externalities*. Examples derived from air pollution include costs of

- Degradation of architectural buildings, structures, landscapes
- Deterioration of human health (e.g., due to respiratory problems)

Examples derived from water pollution include costs of

- Loss of marine life and, hence, related food sources
- Loss of recreational value of a river or lake

Examples derived from soil pollution include costs of

- Eventual remediation of soil
- Loss of biodiversity

An externality of current interest and debate relates to the potential implications of climate change derived from global warming, which can include costs of extreme climate disasters such as hurricanes, floods and bushfires, and impacts on soil productivity in agricultural production.

Estimating external costs is uncertain and difficult. Various proposals have been suggested for estimating externalities. *Contingent valuation techniques* use a survey technique to assess people's willingness to pay for environmental amenities. Although there are difficulties and disagreements in estimating external costs due to pollution, it is clear that these costs can be major and rarely borne directly by those who inflict them. An example of externalities estimation is a study on environmental costs of electricity.[9]

There may be little economic incentive for the polluter to reduce pollution. Emission charges, disposal fees, and environmental penalties (e.g., fines) may be too low. Market prices of raw materials and utilities seldom account for external costs; low prices of inputs reduce the incentive to consume less. The situation can be worsened when governments subsidize raw materials or utilities prices.

6.3 Economic Incentive Instruments

Economic instruments have been used at times by governments in order to encourage greater environmental responsibility by individuals or companies:

- Emission charges related to quantity and quality of pollutant and the damage done.
- User charges for treatment of discharges, related to cost of collection, disposal and treatment.
- Tradable or marketable permits, which enable pollution (or resource consumption) control to be concentrated amongst those who can do it economically without increasing total emissions. Trading can be internal within a company for different company sites, as well as between companies. Tradable market permits have been used for SO_2 emissions in the United States in the 1990s and were introduced in Europe for CO_2 emissions in 2005.
- Deposit refund systems involving a refundable deposit paid on potentially polluting products.

Economic policy instruments should be economically efficient, equitable, cost effective to administer, and successful in achieving improved environmental performance. Economic policy instruments and externality valuation are further discussed by Turner, Pearce, and Bateman.[10]

6.4 Eco-efficiency

The concept of *eco-efficiency* has been developed primarily by business for business needs, through the World Business Council for Sustainable Development (WBCSD). *Eco* encompasses both *eco*nomic and *eco*logical criteria. Important aspects of eco-efficiency include

- Resource productivity (or 'doing more with less')
- Creating additional value for customers while reducing environmental impacts
- Drawing on other business approaches such as cleaner production and total quality management

One example of eco-efficiency is making products with consistent properties but incorporating less material. This has also been termed *dematerialization*. An example is the 375 ml glass bottle (stubby), which in Australia weighed 260 gm in 1986 but less than 160 gm in 2000.

7 SUSTAINABILITY

Sustainability and sustainable development are now major priorities for government and industry policy and the scientific profession.[11] The Brundtland report defined sustainable development as meeting "the needs of the present without compromising the ability of future generations to meet their own needs."[12] Most appraisals of sustainability refer back to this definition. Although there is lack of unanimity about the exact meaning of sustainable development and its implications for industry, most agree that it is concerned with satisfying social, environmental, and economic goals. The *triple bottom line* reflects acknowledgment of the need to meet threshold standards in each of these criteria; the extent to which the threshold standards are reached is commonly referred to as *sustainability*.

Figure 11 shows a simplified view of social, environmental, and economic systems. The economic system draws on materials and energy resources from the environment, and on labor and intellectual capital from society. The economic system generates a diversity of goods and services for society's benefit, but at

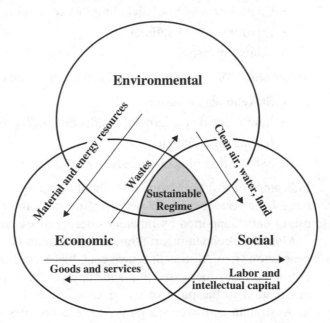

Figure 11 Simplified view of environmental, economic, and social systems—their interactions and sustainable regime.

the expense of impacts on the environment. The environment provides clean air, water, and land, as well as the diversity of flora and fauna, to society; these provisions are essential to the development of the social system.

7.1 Sustainability Indicators

Sustainable development implies both spatial and temporal properties, since the criteria of sustainability must be met locally and globally, and for both present and future generations.

Sustainability is important in the context of industry sectors, products, process technologies, and individual process plants. In assessing performance achievement in sustainability, there is a need for suitable *sustainability indicators* to measure progress in environmental, economic, and social performance.

Work is proceeding to develop suitable scenarios and indicators for assessing sustainability.[13] Many of the *environmental indicators* proposed are based on LCA methodology and broadly correspond to those used at the classification stage.

Economic indicators include

- Value added
- Contribution to Gross Domestic Product (GDP)
- Capital expenditure, including that on environmental protection
- Environmental liabilities
- Ethical investments

Social indicators include those related to ethics and human welfare, for example

- Stakeholder inclusion
- International standards of conduct regarding business dealings, child labor
- Income distribution
- Satisfaction of social needs, including work

Indicators of safety performance should also be included here, particularly for more hazardous industries. Accident statistics and reported case studies provide useful data, supported by inherent safety indices and risk assessments.

May and Brennan report a range of sustainability indicators for electric power generation from Australian brown coal, black coal and natural gas.[14] System definition includes fuel extraction and supply, electricity generation, and electricity distribution to the point of use. Existing and developing technologies, as well as Australian and overseas power generation sites are considered. The study is an interesting case example of applying indicators for sustainability assessments where many criteria must be considered.

7.2 Temporal Characteristics of Sustainability

Since changes occur over time in markets, costs, prices, technologies, company policy, and government policy, a desirable characteristic of a sustainable venture is robustness to such changes. The extent of robustness can be explored through sensitivity and scenario analyses.

Since sustainability is concerned with intergenerational equity, evaluation of technologies, products, and processes should consider extended time periods, desirably those encompassing at least two generations. These horizons exceed those traditionally used in project evaluation.

Such considerations increase the time and cost of evaluation and conflict with commercial opportunism in seizing perceived market opportunities, but are valuable in assessing and understanding risk and are essential to sustainability assessment.

8 INDUSTRIAL ECOLOGY

Extension of system boundary beyond the geographic boundary of a process plant was discussed earlier in the context of encompassing utility systems. System boundaries can also be extended beyond what has been termed the *micro-scale* (process plant), through the *meso-scale* (refinery, petrochemical, or metallurgical complex) to the *macro-scale* (interaction of processing plants from different industry sectors).[15] This extension to the macro-scale (depicted in Figure 12) is an important aspect of what is increasingly being termed *industrial ecology*.

Industrial systems have local, regional, and global dimensions, and interact with natural ecosystems on these scales. A need is perceived for transition from current industry structures of predominantly linear flow of materials and energy, to a structure with greater recycle of materials and energy, such as occurs with living organisms. Natural cycles for carbon, oxygen, or nitrogen, which are essentially closed, are contrasted with industrial cycles, which are predominantly open. Currently, recycle of process streams to minimize raw material consumption occurs in process technology at the micro-scale, while exchange of utility streams and some process streams is common at the meso-scale. Some industrial complexes are increasingly transferring process and utility streams across company and industry sector boundaries at the macro-level.[15,16]

Exchange of materials and energy flows between industries demands a knowledge of the total industry sector, and is important under these conditions:

- Where a concentration of industry exists, opening up opportunity for exchange of material and energy streams. Examples include separation of hydrogen from hydrocarbons in hydrogen-rich gas streams from petroleum refineries to supply different user plants or external markets; and exchange of hydrocarbons between petroleum refineries and petrochemical plants to maximize use of specific hydrocarbons.

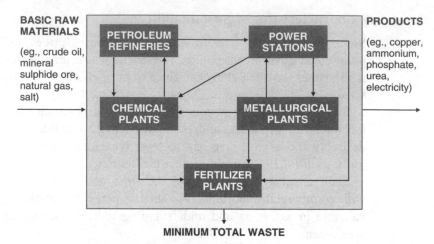

BASIC RAW MATERIALS

(eg., crude oil, mineral sulphide ore, natural gas, salt)

PRODUCTS

(eg., copper, ammonium, phosphate, urea, electricity)

PETROLEUM REFINERIES

POWER STATIONS

CHEMICAL PLANTS

METALLURGICAL PLANTS

FERTILIZER PLANTS

MINIMUM TOTAL WASTE

Figure 12 Materials and energy flows between distinct industries to minimise waste.

- In the planning of new industries to ensure a degree of integration to achieve waste minimization.
- In planning resource-based industries in remote regions, where transport of low-value byproducts can be costly. An important case in Australia is sulfur dioxide emissions capture from metallurgical smelters to make sulfuric acid. Sulfuric acid can be used for phosphate fertilizer manufacture, but depends on availability of phosphate rock deposit, natural gas supply (for ammonia production), and infrastructure for transporting the fertilizer to markets.
- Where transport of products is hazardous because of the safety or environmental properties of the cargo—for example, in the bulk transport of ammonia or vinyl chloride monomer.

Risks exist in industrial ecology initiatives in relation to sustainability of markets, individual plants, technologies, and raw materials supplies, as well as to the reliability of participating plants and their management. Such risks must be evaluated as part of establishing industrial partnerships.

8.1 Australian Example of Industrial Ecology

Figure 13 shows an Australian example of industrial ecology. Sulfur dioxide is emitted from the metallurgical smelting of sulfide minerals at Mt. Isa in a remote area of Queensland. For many years, large quantities of SO_2 had been emitted to the atmosphere. The development of phosphate rock mining at Phosphate Hill some 150 km away provided an opportunity to capture SO_2 and convert it to sulfuric acid. The sulfuric acid could be used to digest the phosphate rock to make phosphoric acid. Ammonia manufacture from natural gas was possible, since natural gas from the Cooper Basin field in central

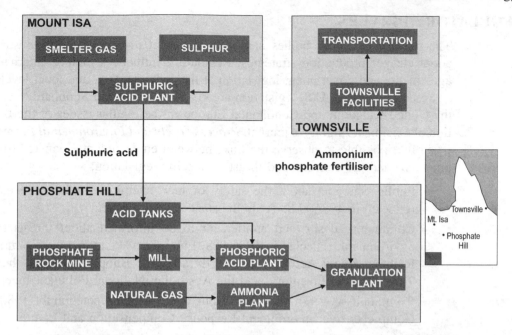

Figure 13 Queensland fertilizer project— an example of industrial ecology.

Australia was available from a pipeline. Ammonia could be reacted with phosphoric acid to make ammonium phosphate fertilizer, which could be transported by rail to Townsville on the Queensland coast to supply Australian and export markets.

8.2 Extension of Industrial Ecology to Materials and Energy Recycling

The concept of industrial ecology can be extended beyond manufacturing frameworks to the flow of materials or energy, and the recycling of material and energy streams through an entire regional or national economy.[15,16] A distinction is commonly made between closed-loop and open-loop recycling. *Closed-loop recycling* involves the return of material to the life-cycle chain from which it originated, as occurs with aluminium in Figure 4. Closed-loop recycling occurs in many individual chemical processes where unreacted feed leaving a reactor is separated from a product and recycled back to the reactor feed. *Open-loop recycling* involves recovery of material from one product life-cycle chain and feeding it into a different, often unrelated product life-cycle chain. Examples include hydrogen recovery from catalytic reformer unit off gases from petroleum refineries for external sales, but also materials reuse initiatives such as recovery of PVC from PVC bottles to make drainage pipes, and recovery of waste polyurethane foam for use as carpet underlay.

9 REGULATORY DRIVERS

The role of regulatory bodies has been important in encouraging industries to adopt cleaner production strategies. Regulatory influences include international agreements and government legislation at the federal, state, and local levels.

Some accounts of U.S. legislation are given by Allen and Shonnard,[17] and for the United Kingdom and Continental Europe in the regular issues of the Institution of Chemical Engineers journal, *Process Safety and Environmental Protection*. It is often possible to observe the links between environmental damage (past or potential) and the contents and thrust of specific legislation:

- Public concerns about the safety of new chemical products developed, reflected in 1976 U.S. Toxic Substances Control Act
- Community desire and entitlement to be informed about threats to the environment, reflected in 1986 U.S. Emergency Planning and Community Right to Know Act (enacted shortly after the Bhopal incident), the U.S. Toxic Release Inventory, and the Australian National Pollutant Inventory
- Community concern about hazardous waste sites, reflected in the 1980 U.S. Comprehensive Environment Response Compensation and Liability Act
- Government desire for cleaner production initiatives, reflected in the 1990 U.S. pollution Prevention Act

The regulatory process, particularly through *command and control policies*, has limitations; for example:

- Damage (often irreversible or difficult to reverse) usually precedes legislation by a significant time interval.
- Legislation is costly and often of limited effectiveness.

These limitations have led to the acceptance by government and industry of the need for voluntary approaches to cleaner production principles by industry and the community.

10 CONCLUDING REMARKS

Chemical processing plants have extended life cycles encompassing planning, design, construction, operational, and retirement phases. There is an associated cash-flow profile for the life of the plant incorporating investment costs and profitable operation. Chemical processing plants embody technologies that themselves have life cycles and are subject to competition from newer and improved technologies.

The manufacture of a chemical product within a chemical processing plant is one stage of a product life cycle that includes the entire manufacturing and product assembly chain, as well as the use and disposal phases of the product. It is important to minimize the waste in the entire product life cycle, implying waste minimization in each chemical processing plant contributing to the product life cycle. Waste generation occurs in chemical plants within reaction and separation processes, and through consumption of energy and related utilities. Waste minimization is important in both process and utility generating plants. Waste can be generated through abnormal as well as normal operation, and cognizance must be taken of fugitive emissions and wastes derived from maintenance, cleaning, storage and transport.

The quantitative assessment of environmental impacts can be made using life-cycle assessment (LCA) methodology, which accounts for both inputs and emissions. LCA can be used to identify the major environmental impact categories and the sources of those impacts within a chemical processing plant. LCA can also be used to identify the major contributions to environmental impact within a product's life cycle. Impact scores derived from LCA can be used along with economic assessment scores and social indicators to provide indicators of overall sustainability of processes and products. Economic assessments are often limited through failure to account for all internal costs and especially the external costs associated with waste.

Opportunities exist to minimize waste through the planning and structure of industrial development, where surplus process or utility streams from one plant or industrial facility can usefully be utilized by separate plants or facilities. This exchange can often be across distinct industry sectors.

REFERENCES

1. D. Brennan, "Process Industry Economics. An International Perspective," Institution of Chemical Engineers, 1998.
2. J. B. Malloy, "Risk analysis of chemical plants," *Chemical Engineering Progress*, **67**(10), 68–77 (1971).
3. United Nations Environment Program (UNEP), Proceedings of Asia Pacific Cleaner Production Conference, Melbourne, Victoria, Australia, 1992.
4. J. B. Guinee et al. (eds.), *Handbook on Life Cycle Assessment*. Kluwer Academic Press, 2002.
5. A. A. Burgess and D. J. Brennan, "Desulfurisation of Gas Oil. A Case Study in Environmental and Economic Assessment." *Journal of Cleaner Production* **9**(5), 465–472 (2001).
6. G. M. Masters, "Introduction to Environmental Engineering and Science," 2nd ed., Prentice Hall, Englewood Chffs, NJ, 1998.
7. A. Azapagic, "Life-cycle Assessment and Its Application to Process Selection, Design and Optimization," *Chemical Engineering Journal*, **73**, 1–21 (1999).
8. T. Moilanen and C. Martin, "Financial Evaluation of Environmental Investments," IChemE, 1996.
9. U.S. Office of Technology Assessment, *Studies of Environmental Costs of Electricity*, OTA-BP-ETI_134, 1994.

10. R. K. Turner, D. Pearce, and I. Bateman, *Environmental Economics. An Elementary Introduction*, Harvester Wheatsheaf, 1994.

11. A. Azapagic, S. Perdan, and R. Clift, *Sustainable Development in Practice*, John Wiley, New York, 2004.

12. World Commission on Environmental Development (WCED), *Our Common Future*, Oxford University Press, 1987.

13. A. Azapagic and S. Perdan, "Indicators of Sustainable Development. A General Framework," Transactions of the Institution of Chemical Engineers. Part B, 243–261, July 2000.

14. J. R. May and D. J. Brennan, "Sustainability Assessment of Australian Electricity Generation," Transactions of the Institution of Chemical Engineers, Vol. 81 Part B, 131–142, 2006.

15. B. R. Allenby and D. J Richards (eds.), *The Greening of Industrial Ecosystems*, USA National Academy of Engineering, 1994.

16. D. J. Richards (ed.), *The Industrial Green Game. Implications for Environmental Design and Management*, USA National Academy of Engineering, 1997.

17. D. T. Allen and D. R. Shonnard, *Green Engineering. Environmentally Conscious Design of Chemical Processes*, Prentice Hall, Upper Saddle River, NJ, 2000.

CHAPTER 4

WASTE REDUCTION FOR CHEMICAL PLANT OPERATIONS

Cheng Seong Khor, Chandra Mouli R. Madhuranthakam, and Ali Elkamel
Department of Chemical Engineering, University of Waterloo, Waterloo, Ontario

1 INTRODUCTION   90
   1.1 Waste Management Hierarchy   90
   1.2 Definition of Waste Reduction
        and Pollution Prevention Used in
        this Chapter   92

2 DEVELOPMENT OF
   POLLUTION PREVENTION
   PROGRAMMES   93
   2.1 Identifying Pollution Prevention
        Opportunities   93
   2.2 An Integrated Methodology for a
        Successful Pollution Prevention
        Program within Corporate
        Management Structures,
        Strategies, and Practices   102

3 ECONOMICS OF
   POLLUTION
   PREVENTION   105
   3.1 Frameworks for Assessing and
        Quantifying Environmental
        Costs   105

3.2 The AIChE–CWRT Total Cost
     Assessment (TCA) Method
     for Evaluating Environmental
     Costs   105
3.3 DuPont's "10-Step Method of
     Engineering Evaluations
     for Pollution Prevention"   107

4 SURVEY OF TOOLS,
   TECHNOLOGIES, AND
   BEST PRACTICES FOR
   POLLUTION
   PREVENTION   108
   4.1 Pollution Prevention at the
        Macroscale   108
   4.2 Pollution Prevention at the
        Mesoscale   110
   4.3 Pollution Prevention at the
        Microscale   120

5 CONCLUDING REMARKS   121

The paradigm of green engineering that advocates environmentally conscious practices and operations has emerged as the perennial driving force of the industrial sector of the global economy. This chapter aims at providing an integrated treatment of objectives and methods for identifying and implementing waste reduction and pollution prevention (P2) initiatives in the chemical industry. We consolidate a set of tools, techniques, and best practices that would ensure the successful development of an economically sustainable pollution prevention program with a multiscale survey and categorization of these available approaches at the levels of macro-, meso-, and microscale.

89

1 INTRODUCTION

This chapter provides a survey treatment on the subject of *waste reduction* programs and practices in chemical plant operations, extended to include the wider scope defined by the term *pollution prevention*. (A discussion on the definition of the most commonly encountered terms in the literature of the subject follows.) In the context of current developments and state-of-the-art industrial practices and academic research, waste reduction and pollution prevention can be broadly and comprehensively defined as practices that reduce or eliminate the creation of pollutants and wastes, at the source. The basic paradigm of the strategies variably known as *waste reduction, waste minimization, pollution prevention*, or *cleaner production* is that avoiding the generation of wastes or pollutants can often be both more cost-effective and better for the environment, compared to the traditional end-of-pipe abatement solutions of controlling or disposing of pollutants only after they have been formed. This is typically accomplished via a four-pronged approach:[1]

1. Increase efficiency and long-term sustainability in the utilization of raw materials, energy, water, or other resources (e.g., chemical synthesis routes with minimum or no byproducts that consume sustainable or renewable sources)[2]
2. Substitute hazardous substances with less harmful alternatives
3. Eliminate toxic substances from production/manufacturing processes
4. Protect natural resources by means of conservation efforts

1.1 Waste Management Hierarchy

Elements of the waste management hierarchy can be placed in the following order of preference:[3,4]

1. Source reduction
2. In-process recycling (i.e., within a given production process)
3. On-site recycling (i.e., out of the process but still within the same production site)
4. Off-site recycling (i.e., within different production sites)
5. Waste treatment to reduce hazards imposed by the waste
6. Safe environmental disposal
7. Direct disposal into the environment

The waste management hierarchy introduces a myriad of terms that require definitions due to subtle differences between them. To precisely define these terms is an arduous task that becomes even more complex when regulatory definitions are added to an originally operational or technical definition. Table 1 provides generally accepted definitions for some of the most commonly encountered terms in the literature of this field.

Table 1 General Definition for Commonly Encountered Terms in the Pollution Prevention Literature

Term	Generally-accepted definition(s) (and related remarks)
Source reduction	Process or product modifications that reduce waste.[3]
Waste reduction (WAR)/waste minimization	Any activity that reduces or eliminates the generation of wastes and hazardous pollutants at the source, e.g.: (i) changes in industrial processes in the form of operating practices, technologies, inputs, and products; and (ii) recycling (in-process, in-line, closed-loop) that reuse and recycle wastes for the original purpose or others, such as for materials recovery or for energy production.[5]
	Remarks:
	• The term *waste reduction* and the related term *waste minimization* project broader meanings than *source reduction*. Waste reduction and waste minimization generally incorporate both source reduction and on-site recycling (i.e., the first two or three elements of the waste management hierarchy.)[3] For the purpose of this article, the definition offered here will/would be adopted.
	• While the terms *waste reduction* are more expansive than *source reduction* in certain ways, they are more restrictive in the sense that while source reduction can be employed in the context of gaseous, liquid, or solid wastes, the term *waste reduction* is generally used to refer to solid and liquid waste management activities.[3]
Pollution prevention (P2)	"A multimedia concept that reduces or eliminates pollutant discharges to air, water, or land and includes the development of more environmentally acceptable products, changes in processes and practices, source reduction, beneficial use, and environmentally sound recycling."[3]
	Remark: It is generally acknowledged that P2 encompasses the first two or three elements of the waste management hierarchy for all wastes and emissions.[3]
Cleaner production (CP)	• "Continuous application of an integrated preventive environmental strategy to processes and products to reduce risk to humans and the environment."[6]
	• "Maximum feasible reduction of all wastes generated at production sites."[7]

1.2 Definition of Waste Reduction and Pollution Prevention Used in this Chapter

We are inclined to adopt a broader definition of waste reduction that includes gaseous wastes as well, which will cover the definition of pollution prevention, hence, encompassing the first four elements of the waste management hierarchy (i.e., source reduction and in-process, on-site, and off-site recycling), as depicted in Figure 1. Thus, the terms *waste reduction* and *pollution prevention* will be used interchangeably throughout this chapter to essentially denote the same concepts. This, we hope, will enable us to capture a holistic view and coverage of the field, by enabling the examination of more alternatives for reducing industrial impacts on the environment and for eliminating adverse health effects that are due to chemical plant manufacturing activities. This will also provide a true depiction of the wide-ranging extent of this tremendously exciting and meaningful area for both academic research and industrial practices. On a lighter note, we will also extensively use the shorthand term for pollution prevention, P2, in this chapter in keeping with the P2 spirit of conserving ink and paper.[8]

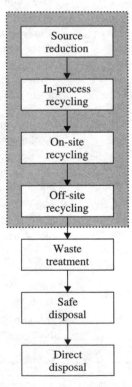

Figure 1   The waste management hierarchy with shaded boxes indicates elements that are included in the definition of the terms *waste reduction* and *pollution prevention* that are used interchangeably throughout this chapter. (Adapted from Refs. 3 and 4.)

Additionally, the term *process* implies *product*, for instance, as in *product/process engineering*. For the sake of habitual convenience, the term *process* will be more frequently encountered throughout the chapter, but when a clear distinction between the two is warranted, an explanation will be provided.

The extensive breadth that spans the coverage of P2 has entailed a sheer amount of development and coverage of new knowledge and information in both industrial practices and academic research.[3,9,10] To the best of our knowledge, there has been only one comprehensive review article on the subject—that is, the 1995 survey by Freeman and co-workers of the U.S. Environmental Protection Agency (U.S. EPA)'s Pollution Prevention Research Branch in the Risk Reduction Engineering Laboratory.[10] The review focuses on P2 in the industrial sector, which is by far the sector in which major initiatives of P2 have been undertaken and will continue to be so.

2 DEVELOPMENT OF POLLUTION PREVENTION PROGRAMMES

2.1 Identifying Pollution Prevention Opportunities

The life cycle of chemical manufacturing is the key toward understanding the driving forces that contribute to initiatives for P2 and the associated risk reduction. Allen and Shonnard provide a unified representation of the typical life cycle of chemical products and processes, as shown in Figure 2.[9,11] In general, chemical processes evolve through life-cycle phases that begins with research and development (R&D) activities, which include conceptual design (also known as product/process synthesis), followed by product/process engineering, plant operations, and eventually, decommissioning.[12] In relation to this, Mulholland and Dyer of DuPont suggest a continuum as shown in Figure 3 that depicts the relative scale of opportunities for implementing P2 along the life cycle of a product/process, corresponding to the magnitude of benefits expected out of such initiatives.[13] It is clearly shown that the number of technology options available for reducing environmental impact is greatest early on in the life cycle, and then the options drastically decrease. In contrast, costs associated with resolving an environmental problem typically increase exponentially as the process matures and the scale of equipment becomes larger. Thus, there is considerable incentive, apart from being more effective, to consider a P2 implementation earlier (rather than later) in the life cycle of a product/process.[12,13]

Indeed, opportunities for cost-effective solutions to P2 problems are most prevalent at the R&D and process engineering stages for new units and plants. After the basic process chemistry and basic system design philosophy are in place, significantly fewer opportunities exist, comparatively, to undertake P2-related cost-effective decisions in the later stages of the project (i.e., during detailed engineering, construction and start-up, operations and maintenance, or the dismantling phases). It is noted as well that this decision on the extent to

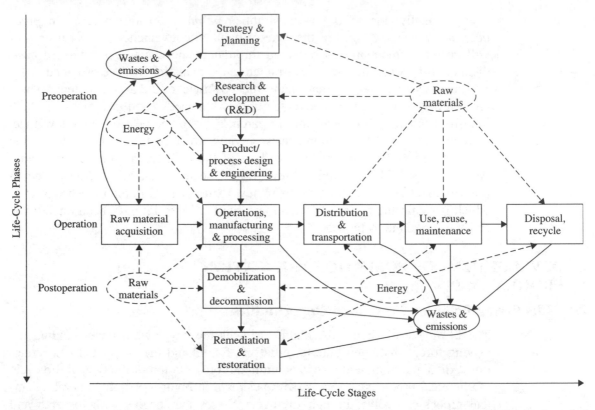

Figure 2   Product life cycles, including raw material extraction, material processing, distribution and transportation, use, and disposal, are depicted along the horizontal axis. Process life cycles (or process development cycle), including strategy and planning, R&D, process engineering, process operations, and decommissioning steps, are illustrated along the vertical axis. In both product and process life cycles, raw materials and energy are consumed at each stage of the life cycle while wastes and emissions are generated. (Adapted from Refs. 9 and 11.)

which a P2 initiative should be implemented will also largely depend on the interplay of factors that primarily consist of corporate and business environmental goals, economics, regulatory requirements, and process maturity or product life.[14]

Figure 4 provides an overview of the elements of a complete typical chemical process for consideration in targeting P2 opportunities. Although the focus for identifying P2 opportunities is necessarily on the process, other factors, such as off-specification (off-spec) product, maintenance wastes, leaks, and spills can also be significant causes of pollution. Table 2 identifies numerous examples of potential sources of pollution and waste in chemical process plants. Once these have been identified, P2 solutions can be subsequently evaluated based on the general guidelines provided in Table 3.[14]

We have highlighted potential opportunities for P2 on a general basis. Next, we categorize the discussion according to P2 opportunities that potentially arise in the

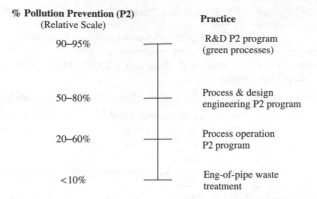

Figure 3   The P2 methodology continuum. (Adapted from Ref. 13.)

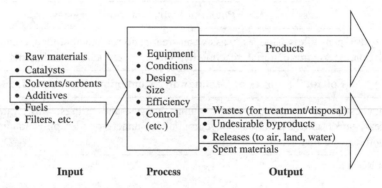

Figure 4   Process elements for consideration of P2 initiatives. (Adapted from Ref. 14.)

three most prominent phases in product/process life cycle: R&D, product/process design and engineering, and manufacturing or process operations.

Pollution Prevention during the Research and Development (R&D) Stage

As emphasized earlier, the greatest opportunity for cost-effective P2 at the source is present at the R&D stage. R&D programs typically progress through three distinct phases:

1. Conceptual design to determine economic feasibility
2. Bench-scale laboratory development studies and testing to validate the process chemistry
3. Pilot-plant-scale development and testing to address engineering issues for process scale-up

Conceptual Design.   During the conceptual design phase, the reaction pathway synthesis is undertaken; the inherent process safety is incorporated; the general

Table 2 Potential Sources of Pollution and Wastes

Potential Sources of Air Emissions

- Point-source emission: stack, vent (e.g., laboratory hood, reactor, storage tank vent), material loading and unloading operations, and others
- Fugitive emissions: pumps, valves, flanges, mechanical seals, relief devices, tanks, and others
- Secondary emissions: wastewater treatment unit, cooling tower, process sewer, sump, spill and leak areas, and others

Potential Sources of Liquid (Organic or Aqueous) Wastes

Equipment wash solvent and water, laboratory samples, surplus chemicals, product washes and purifications, seal flushes, scrubber blowdown, cooling water, steam jets and vacuum pumps, leaks, spills, spend or used solvents, housekeeping (pad washdown), waste oils and lubricants from maintenance, and others

Potential Sources of Solid Wastes

Spent catalysts, spent fillers, sludges, wastewater treatment biological sludge, contaminated soil, old equipment and installation, packaging material, reaction byproducts, spent carbon and resins, drying aids, and others

Potential Sources of Groundwater Contamination Due to Leaks or Spills

Unlined ditches; process trenches; sumps; pumps, valves, and fittings; wastewater treatment ponds; product storage areas; tanks and tank farms; aboveground and underground piping; loading and unloading areas and racks, manufacturing maintenance facilities

Note: From Ref. 14.

environmental impacts are investigated; and the potential waste streams are identified and studied.[13] This culminates with the formulation of relevant P2 concepts that influence the determination of the following key process parameters: (1) the operating methodology (e.g., batch vs. continuous or an operating mode between these two extremes); (2) the process conditions (e.g., temperatures and pressures); (3) the production procedures; (4) the selection, integration, and design of processing equipment; and (5) the process control schemes. Naturally, the potential for P2 in subsequent stages (particularly the preliminary product/process engineering phase) are contingent upon these decisions.

Table 4, "Twelve Principles of Green Chemistry,"[15] serves to provide a road map for chemists and chemical engineers to implement the use of sound, responsible chemistry for P2, or *green chemistry*, during the conceptual design of chemical products/processes in order to reduce or eliminate the use and generation of hazardous substances.[16]

Additionally, in borrowing ideas from the industrial ecology concept, Allen and Shonnard advocate the relatively new approach of identifying uses for chemical byproducts in industries that extend further beyond chemical manufacturing.[11] Acknowledging that the principle of identifying productive uses for byproducts

Table 3 Guidelines for Identifying P2 Solutions/opportunities

Inventory Management and Operations

- Improvement in material purchasing, receiving, storage, and handling practices, which minimize materials that may exceed their shelf life, be left over or not needed, or have the potential for accidental release
- Installation of the right equipment and maintaining strict preventive maintenance programs to prevent leaks, spills, or accidental releases
- Effective and efficient production scheduling by reducing the number of production changes to reduce equipment cleaning and to minimize wastes generated from production transitions or turnovers/changeovers

Equipment

- Installation of equipment that produces less waste
- Redesign of equipment or production lines to produce less waste and to enhance or permit recovery or recycling options
- Improvement in operating efficiency of equipment
- Elimination of sources of leaks and spills

Process/Product

- Process changes that optimize reactions and raw material use to reduce waste generation or releases
- Input material changes: material purification and material substitution that allow a change in the process to reduce waste or eliminate a hazardous constituent used in the product or during the manufacturing phase
- Product substitution: developing a new product that is less hazardous in use or in ultimate disposal compared to the original
- Additional automation to increase the reliability of the operations and to reduce the occurrence of process upsets and production of off-specification material

Recycling

- Participation in waste exchanges
- Installation of closed-loop systems for in-process recycle
- Recycling on-site at other process units or off-site for reuse
- Finding new uses for previously unwanted byproducts
- Segregation of wastes by type to allow for recovery
- Reclamation: processing of waste for resource recovery

Note: From Refs. 5, 6, and 14.

has long been a mainstay in the chemicals industry, they promote the idea that chemical engineers should take on design tasks such as managing the heat integration between a power plant and a petroleum refinery, or integrating the water usage between a semiconductor fabrication plant and a commodity chemical manufacturing facility. Indeed, such design tasks are still at the conceptual design

Table 4   Twelve Principles of Green Chemistry

---

1. *Prevent waste*: Design chemical syntheses to prevent waste, leaving no waste to treat or clean up.

2. *Synthesize safer-chemicals and products*: Design chemical products to be fully effective functionally, yet have little or no toxicity.

3. *Design less hazardous chemical syntheses*: Design syntheses that use and generate substances that pose little or no toxicity to human health and the environment.

4. *Use renewable feedstocks*: Use raw materials and feedstocks that are renewable rather than depleting whenever technically and economically feasible. Renewable feedstocks are often synthesized from agricultural products or are the wastes of other processes; depleting feedstocks are derived from fossil fuels (petroleum, natural gas, or coal) or are mined.

5. *Use catalysts, not stoichiometric reagents*: Minimize waste by using catalysts that are highly selective for the desired reactions. Only small amounts of catalysts are required for carrying out a single reaction multiple times, and furthermore, they can be regenerated (up to a certain number of times). This is in contrast to the typically excess amount of stoichiometric reagents that are required and that only work once (i.e., consumed in every reaction and cannot be regenerated).

6. *Avoid chemical derivatives*: avoid using unnecessary derivatization in the form of blocking or protecting–deprotecting groups or any temporary modification of physical or chemical processes. Derivatives consume additional reagents, thus, generating waste.

7. *Maximize atom economy*: Design syntheses so that the final product contains the maximum proportion of the starting materials. There should be few, if any, wasted atoms.

8. *Use safer solvents and reaction conditions*: Avoid using solvents, separation agents, or other auxiliary chemicals whenever possible. Use innocuous chemicals if their usage is necessary.

9. *Increase energy efficiency*: recognize the environmental and economic impacts of energy requirements and minimize them by conducting chemical reactions at ambient temperature and pressure whenever possible.

10. *Design chemicals and products to degrade after use*: Design chemical products to break down to innocuous degraded substances at the end of their function or life cycle so that they do not persist and accumulate in the environment.

11. *Analyze in real-time to prevent pollution*: Develop the tools available to perform real-time analysis of in-process monitoring and control during syntheses to minimize or eliminate the formation of byproducts, particularly hazardous substances.

12. *Minimize the potential for accidents*: Design chemicals and their forms (solid, liquid, or gas) as used in a chemical process to minimize the potential for chemical accidents including explosions, fires, and releases to the environment.

Note: Adapted from Ref. 16.

stage of our profession. Hence, these perspectives must be broadened if we desire to make these design undertakings as common in the near future as heat integration for processes. To embark on this form of P2 initiative, we ought to begin integrating process design tools and methodologies from our traditional domain of chemical manufacturing with industries as diverse as polymer processing, semiconductor manufacturing, and pulp and paper processing.

Table 5 P2 Data to Be Collected for Bench-scale Laboratory Development Studies and Testing

Unit Operation	Data Required
General	• Determine corrosion rates of potential construction materials. • Screen for catalytic effects of potential materials, corrosion products, and feed impurities.
Reactors	• Determine reaction stoichiometry. • Determine equilibrium yield. • Measure catalyst selectivity, activity, and lifetime. • Identify and characterize reaction byproducts. • Determine kinetics of major side reactions. • Determine effects of recycle.
Distillation and other separation processes	• Obtain vapor pressure data for products and intermediates. • Obtain vapor–liquid equilibrium (VLE) data for potential entrainers, diluents, and trace compounds. • Determine loading capacity and regenerative properties of absorbents.

Note: Adapted from Ref. 17.

Bench-Scale Laboratory Development Studies and Testing.   During laboratory studies, the reaction chemistry is verified and validated; the waste streams are characterized; the process variables are tested; the P2 options are identified; the data are collected for the subsequent pilot-plant testing and process engineering; and the potential impact of environmental regulations are investigated.[13] Table 5 proposes relevant P2-related data to be gathered according to unit operations.[17]

Pilot-Plant-Scale Development and Testing.   During pilot-plant studies, laboratory results are verified and validated; process chemistry is established; key process variables are tested; equipment design is evaluated; and waste characteristics are defined.[13] Butner highlights the following issues to be thoroughly explored in this phase to assist the direction of P2 initiatives during the ensuing stage of process engineering:[17]

  • *Effect(s) of reactor mixing and feed distribution on formation of byproducts.* Pilot testing is often the first opportunity for conducting the process reactions in equipment that approximates subsequent scale-up to commercial reactor implementation. Therefore, it is highly imperative to meticulously inspect and analyze for unwanted byproducts in the slate of products obtained, which could form as a result of residence time distributions, due to poor mixing or feed distribution, hot spots on catalysts, impurities (that are not present in laboratory-grade chemicals), or catalytic effects of construction materials.

- *Fouling rates in heat exchangers.* Pilot-plant studies should be utilized to thoroughly investigate fouling due to process wastes that originate from (1) cleaning waste generated by routine heat exchanger maintenance, and (2) products of thermal decomposition formed by increased wall temperatures as a result of reductions in heat-exchange coefficients.

- *Sedimentation rates and product stability.* Pilot-scale tests provide a good chance to identify formation of sedimentation or sludge in process vessels, which indicate product stability during storage in the vessels.

- *Corrosion studies.* It is equally essential to monitor corrosion rates at this stage, although ideally, these analyses should have been initiated during the previous phase of laboratory studies.

Pollution Prevention during the Product/Process Engineering Stage

Although there are limited opportunities for major product/process changes at this stage, there still exist many opportunities for incorporating P2 into a process. In fact, as the emphasis shifts from conceptual to operational aspects of a project, it may be easier to implement P2 initiatives since we can more assuredly evaluate uncertainties that arise in the suggestions for P2. Table 6 offers such suggestions that are particularly suitable for the preliminary engineering phase of a project.[17]

As a project progresses into the detailed design (also known as mechanical design or production design) stage, source reduction opportunities practically diminish chiefly because the process and preliminary plant design have become fixed, and the project on the whole becomes schedule-driven. The focus at this stage departs from the process per se to the design of the associated equipment and facility; thus, requiring P2 emphasis on protecting groundwater from spills (see Table 2) and minimizing (or even eliminating) fugitive emissions.[13]

Pollution Prevention during the Process Operations Stage

Experience documented across all industries has shown that processes that minimize waste generation at the source are the most economically optimal, especially for existing operating plants. Perhaps, the greatest success stories of P2 during process operations stage have been the application of process integration tools for heat integration via the pinch analysis or pinch technology method[18] and

Table 6   Strategies for P2 during Product/process Engineering

Activity	Strategy
Materials selection	• Consider costs of waste disposal when selecting maximum allowable corrosion rates. • Consider foul-resistant materials (e.g., Teflon) on heat exchange surfaces and vessels requiring frequent cleaning. • Consider glass- or polymer-lined vessels when frequent cleaning is required.
Cost estimation	• Incorporate "hidden" waste costs (e.g., regulatory tracking and reporting costs) in cost equations. • In absence of detailed waste costs, consider *penalty function* for releases based on environmental objectives. (A detailed treatment is provided in Section 3 on P2 economics.)
Piping design	• Recover waste streams separately. • Minimize length of piping runs (because this reduces material inventory). • Minimize valves and flanges (e.g., by reducing valve counts and by using welded fittings). • Route drains, vents, and relief lines to recovery or treatment. • Specify bellow-seal or zero-emission valves.
Instrumentation design	• Select in-line process analyzers to reduce sampling wastes. • Use closed-loop (purge style) sampling ports. • Install preventive maintenance monitoring equipment (e.g., vibration monitors, torque sensors) to optimize maintenance schedules. • Install instrument on heat exchangers that permit real-time monitoring of fouling and leakage. • Consider advanced control strategies such as model-based control (MPC).

Note: Adapted from Ref. 17.

the emerging principle of mass integration.[19] These have been documented by Rossiter for heat integration[20] and Dunn and El-Halwagi[21] for mass integration. The concepts of mass and heat integration will be revisited in this chapter under the section on quantitative methods for P2 at the mesoscale. Table 7, provides further examples of possible operational changes for reducing waste generation.[8]

Table 7 Examples of Operational Changes to Reduce Waste Generation

- Reduce raw material and product loss due to leaks, spills, drag-out, fugitive emissions, tank breathing, and off-specification (off-spec) process solution.
- Minimize number of shutdowns by designing for high availability or installing more equipment (or standby equipment).
- Inspect parts before they are processed to reduce number of rejects.
- Allow for enough intermediate storage to provide flexibility for reprocessing off-spec materials.
- Consolidate types of equipment or chemicals to reduce quantity and variety of waste.
- Improve cleaning procedures to reduce generation of dilute mixed waste with methods such as using dry cleanup techniques, using mechanical wall wipers or squeegees, and using "pigs" or compressed gas to clean pipes and increasing drain time.
- Segregate wastes to increase recoverability.
- Optimize operational parameters (such as temperature, pressure, reaction time, concentration, and chemicals) to reduce byproduct or waste generation.
- Evaluate the need for each operational step and eliminate steps that are unnecessary.
- Collect spilled or leaked material for reuse.
- Design a continuous process for flexible operation such as high turndown rate rather than shutdown.
- Increase level of automation or better management of process to assist in reducing wastes due to poorly operated process.
- Study the potential for converting the operating mode from batch to continuous since batch processes are always at unsteady-state condition by nature and are therefore, difficult to maintain at optimum condition.

Note: Adapted from Refs. 8 and 22.

2.2 An Integrated Methodology for a Successful Pollution Prevention Program within Corporate Management Structures, Strategies, and Practices

Incorporating environmental management, particularly P2, as a part of a company's existing setup, with the hope of institutionalizing it, is a demanding task. P2 initiatives typically cut across the traditional management functional boundaries of R&D and product development, manufacturing, engineering, business management, customer and supplier relationships, and risk management. Many companies encounter difficulty in maintaining the momentum of P2 efforts, managing only to undertake portions of a P2 project, often due to a lack of serious commitment to P2 programs. Unless P2 is integrated into a company's business practices, early successes with P2 projects do not ensure long-term sustainability of the program. To overcome such a situation, Case et al. and Bishop propose general procedures for establishing and maintaining a sustainable P2 program that is applicable to the reduction of all wastes (for all medium, quantity, or toxicity), as summarized in Table 8.[23,4]

Therefore, after identifying the potential P2 initiatives, an effective work process or methodology is crucial in ensuring the successful implementation of

Table 8 General Procedure for Establishing and Maintaining a Successful P2 Program

1. Obtain support from top management in the form of a corporate policy statement of support.
2. Initialize the program by starting to incorporate small changes throughout the company, training staff in P2, and developing a written P2 plan that describes the P2 team makeup, authority, and responsibility.
3. Describe how all the groups in the company (production, engineering, maintenance, laboratory, marketing, shipping, retailing, and other functions) will cooperate to reduce waste production and energy consumption.
4. Devise a plan for publicizing and gaining companywide support for the P2 program.
5. Develop a plan for communicating the successes and failures of P2 programs within the company.
6. Review and describe in detail, the processes within the facility to determine the raw materials used and the sources of waste generation, particularly the processes that produce, use, or release hazardous or toxic materials, by clearly defining the amounts and types of substances, materials, and products under consideration; this will provide a baseline inventory to be used to set goals and evaluate progress.
7. Identify potential P2 opportunities for the facility by considering the treatment, disposal, and recycling facilities and transporters that are currently used.
8. Determine the cost of current waste generation and develop a (preliminary) system of proportional waste management (in terms of the ensuing pollution control and waste disposal measures) charges for the departments that generate waste.
9. Describe and evaluate currently ongoing and past P2 activities at the facility.
10. Present criteria for prioritizing candidate facilities, processes, and streams for P2 projects.
11. Select the best P2 options for the company and implement these choices.
12. Evaluate the implemented P2 program on a companywide basis as well as evaluating specific P2 projects and activities.
13. Maintain and sustain the P2 program for continued growth and benefits to the company; reevaluate the program as economic situations change and/or process equipment requires upgrading.

Note: Adapted from Ref. 4 and 23.

the plans. DuPont propose a methodology, displayed in its adapted form in Figure 5, that provides a solid foundation for a practical P2 implementation.[13] It incorporates many of the features initially recommended by the Pollution Prevention Assessment (PPA) framework of the U.S. EPA, which extensively draws upon DuPont's recognized leadership and long experiences in implementing P2 projects.[7] Focusing on the implementation phase, the tasks that are outlined hold the key for turning the selected P2 options by a core assessment team into projects that will actually accomplish the targeted waste reductions. It is noteworthy that many P2 methodologies unnecessarily emphasize the development of an overall corporate program, failing to focus on the process operations level where waste reductions actually take place. Mulholland and Dyer highlight the following vital attributes for consideration in implementing an effective P2 program, which are often overlooked by many P2 implementation methodologies:[13]

1. The core activities vying for people's attention in the process operations stage, namely routine production, quality and safety, maintenance,

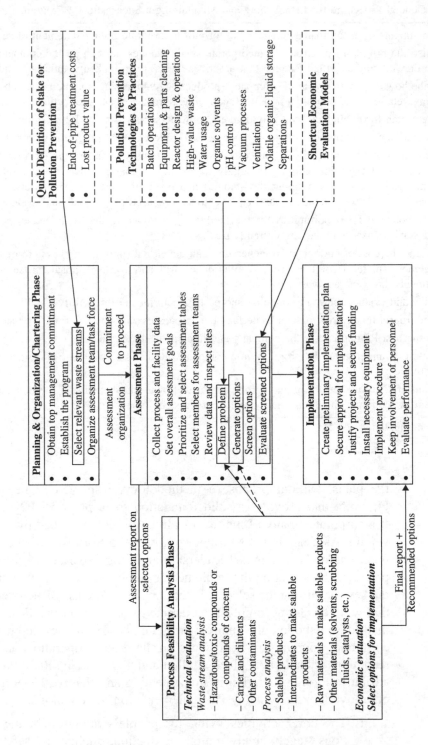

Figure 5 An integrated methodology leading to the successful implementation of a P2 program. (Adapted from Refs. 4, 7, and 13.)

problem troubleshooting, and turnaround activities comprising start-ups and shutdowns

2. The limited capital, time, and people resources available to execute P2 initiatives

3. Elimination of time-consuming and unnecessarily rigorous analytical methods in performing waste assessment activities that do not commensurate with their benefits

3 ECONOMICS OF POLLUTION PREVENTION

3.1 Frameworks for Assessing and Quantifying Environmental Costs

The traditional approach to chemical process design has been to first engineer the process, and only then to subsequently engineer the treatment and disposal of waste streams. However, the exponential escalation of waste management costs resulting from regulatory and societal pressures to eliminate emissions to the environment has led to capital investment and operating costs for waste management becoming a prominent fraction of the total cost of any chemical plant operation. Thus, to assess the economic viability for the implementation of a P2 program, significant efforts have been undertaken to evaluate environmental costs and benefits in assisting business decision making of determining the optimal minimum economic option, which accounts for both the process and the waste management components. Here, we present two such methods:

1. The Total Cost Assessment (TCA) method, proposed by the American Institute of Chemical Engineers' Center for Waste Reduction Technologies (AIChE–CWRT) is perhaps the most relevant work for the design of chemical processes.[9,11]

2. DuPont's "10-Step Method of Engineering Evaluations of Pollution Prevention" outlines the process for choosing the best approach for solving an engineering problem.[13]

Allen and Rosselot have also reported on the environmental costs estimation methodology proposed by the U.S. EPA.[3]

3.2 The AIChE–CWRT Total Cost Assessment (TCA) Method for Evaluating Environmental Costs

The AIChE–CWRT TCA method identifies five tiers for estimating environmental costs as listed in Table 9, beginning with the most tangible costs in Tier I and gradually extending to the least quantifiable costs in Tier V.

Tier I types of costs, with examples listed in Table 10, are effectively quantified in traditional economic analyses using conventional accounting systems. However, these systems, which focus on Tier I costs, often charge some types

Table 9   The AIChE–CWRT Total Cost Assessment (TCA) Method

Tier	Type of Cost
I	Costs normally captured by engineering economic evaluations
II	Administrative and regulatory environmental costs not normally assigned to individual projects
III	Liability costs
IV	Costs and benefits, *internal* to a company, associated with improved environmental performance
V	Costs and benefits, *external* to a company, associated with improved environmental performance

Note: Adapted from Ref. 11.

Table 10   Costs Traditionally Evaluated during Financial Analyses of Projects (Tier I costs)

• Capital cost of equipment	• Materials
• Labor	• Supplies
• Utilities	• Structures
• Salvage value	

Note: Adapted from Ref. 9.

of environmental costs to overhead. Thus, this belies a complete estimation of project costs because it results in the possibility of these Tier II costs (see Table 11 for examples) being *hidden* and often difficult to distinguish from general overhead expenditures. Note also that these hidden costs are actually borne by facilities, irrespective of whether they are quantified by the facilities or are assigned to project or product lines.

The less tangible set of liability costs is designated as Tier III and includes compliance obligations; remediation obligations; obligations to compensate private parties due to personal injury, property damage, and economic loss; fines and penalties; punitive damages; and natural resource damages. The final component of Tier IV and V costs, often referred to as image or relationship costs, arises in interactions with customers, investors, insurers, suppliers, lenders, employees, regulators, and communities; they are probably the most difficult to quantify. In many cases, the Tier III to V costs are incurred by unplanned events (e.g., incidents that result in civil fines and remediation costs). Therefore, we evaluate the expected value of these costs by estimating the probability and possible timing that an event will occur, with its associated cost. In the case of events that will take place in future years (e.g., costs of complying with anticipated future regulations), knowledge of the timing of event is crucial for a more precise estimation of the expected costs' present value.[9,11]

Table 11 Environmental Costs that Are Often Charged to Overhead

- Treatment costs
  - Off-site waste management charges
  - Waste treatment equipment
  - Waste treatment operating expenses
- Handling costs
  - Taking samples
  - Sampling stormwater
- Regulatory tracking/monitoring and paperwork reporting and notification (for compliance in order to obtain permit requirements)
  - Filing for permits
  - Completing sample reporting forms
  - Conducting waste and emission inventories
  - Completing hazardous waste manifests
  - Inspecting hazardous waste storage areas and keeping logs
  - Preparing and updating emergency response plans
  - Preparing chemical usage reports
  - Reporting on P2 plans and activities
- Disposal costs
- Liability costs

Note: Adapted from Refs. 3, 9, and 17.

3.3 DuPont's "10-Step Method of Engineering Evaluations for Pollution Prevention"

The DuPont's 10-Step Method (see Figure 6) is a generic framework that outlines the necessary steps for identifying, evaluating, and choosing the best alternative(s) for a specific engineering problem. Successful applications throughout DuPont have been reported for new process development programs, as well as environmental-related programs, testifying to its suitability for evaluating the environmental costs and benefits associated with a P2 program.[13]

As depicted in Figure 5,[13] economics feature prominently in two key areas: (1) the planning and organization or chartering phase and (2) the assessment phase. In the former, we are interested in quickly defining the incentives for undertaking P2 initiatives. These incentives are largely in financial terms (e.g., prevented investment in end-of-pipe treatments, lower manufacturing costs, prevented environmental fines and penalties, improved yield, and increased plant operating time). However, they can also be less financially quantifiable when dealing with benefits derived from improved public image of the company, environmentally friendly products/processes, and fewer permitting barriers to business growth. Quantifying the financial incentives will especially help to rank the process waste streams for source reduction during the waste-stream selection step,

---

The DuPont's 10-Step Method of Engineering Evaluations of Pollution Prevention: A Road Map to the Best Solution

Define Problems and Goals
1. Define the problem.
2. Set the goals.

Identify the Alternatives
3. Identify all possible alternative solutions.

Define the Alternatives
For each alternative:
4. Develop flowsheets and facility scopes.
5. Define development and operating requirements.

Evaluate the Alternatives
For each alternative:
6. Estimate the new investment required.
7. Estimate the change in cash costs and revenues.
8. Estimate the net present value (NPV).
9. Determine all 'noneconomic' considerations.

Then
10. Choose the best alternative.

Each step builds on previous steps, so don't skim over the early steps.

Iterative Evaluations
Give insight on what's important, the detail required, potential problems and risks, and how to formulate better alternatives.

Figure 6 The DuPont's "10-Step Method of Engineering Evaluations for Pollution Prevention." (From Ref. 13.)

besides securing support for the program from business leadership. As described by Figure 5, there are two aspects to evaluate the incentives obtained from P2 initiatives: (1) the cost associated with new end-of-pipe treatment for the selected waste streams and (2) the lost product value that goes into the waste leaving the process, either in terms of loss of raw materials alone or loss that includes the manufacturing cost.

4 SURVEY OF TOOLS, TECHNOLOGIES, AND BEST PRACTICES FOR POLLUTION PREVENTION

For the rest of the chapter, we present a more specific survey treatment of tools, technologies, and best practices for undertaking P2 by adopting a multiscale analysis of categorizing these tools into macro-, meso-, and microscale.[3]

4.1 Pollution Prevention at the Macroscale

Opportunities for P2 at the macroscale can be identified from three distinct perspectives.[3,26] First, waste-generation audits identify flow rates and compositions of materials in the industrial economy, which are potential P2 targets, from natural resource extraction to consumer product disposal. Second, industrial ecology studies examine the uses and the wastes associated with a particular material. Third, life-cycle analyses (LCA) assess the environmental impacts due to the life cycles of individual products/processes by determining waste generation rates, energy consumption, and raw material usage.

Industrial Ecology

Audits of waste generation and management via inventory of wastes represent a first step in performing macroscale studies of P2. The next step is to integrate waste-generation data with production data, recognizing that the huge amount of wastes generated annually ought not to be ignored as a potential source of reuse as industrially valuable materials. On the whole, the conversion of wastes from something undesirable into something of value eliminates the need for the release of these substances into the environment. This reduces the need for raw material extraction. In line with this, studies in the field of industrial ecology or also known as industrial metabolism examine the material efficiency of large-scale industrial systems, searching for ways that improve their efficiency. By integrating models of the uses of selected materials with data on waste generation for the same materials, it may be possible to identify targets of opportunity for recycling between industrial sectors. Studies can also be conducted to determine whether the contaminants in wastes are potentially recoverable, and what factors promote and hinder the "mining" of wastes. Finally, the mix of technologies used to produce goods can be investigated to determine the most environmentally benign combination of technologies. In summary, studies of industrial ecology attempts to present macroscale views of the integration of feedstock and waste data.[3]

Materials Integration across Industrial Sectors. Substantial environmental and economic benefits are possibly derived when the waste from one facility can be used directly as feedstock in another facility: (1) the facility producing the waste is spared the burden of treatment or disposal; (2) the facility accepting the waste can reduce the costs of feedstock materials; and (3) the environmental burdens associated with disposal and raw material extraction are reduced. At present, arrangements for materials integration between facilities are made more on an ad hoc basis than from careful planning with a thorough understanding of what types of feedstocks and wastes lend themselves to assist toward integration. However, the direct utilization of wastes is bound to be more systematically exploited as data on waste generation and management improve and the whole field of industrial ecology matures.[3]

Recovery of Valuable Materials in Waste. Factors affecting the potential for extraction and recovery of valuable contaminants in waste streams depend on: (1) the concentration of the contaminants and (2) their market value. Over and above these two factors that indicate whether a waste can be economically mined, the existence of a recycling infrastructure and the technologies currently in place influence recycling rates as well.[3]

Life-cycle Assessment (LCA)

Life-cycle assessment (LCA) is an increasingly popular macroscale approach of analytical framework tool for environmental auditing that identifies opportunities for P2 by assessing the environmental impacts associated with a product or a material. It considers an individual product or a material and traces the flows of energy, raw materials, and waste streams that were required to create, use, and dispose of the product.[3]

LCA integrates environmental impacts over the entire life cycle "from cradle to grave," as it is often dubbed as. Thus, it is conducted by companies wishing to improve the environmental characteristics of their products and by companies or industrial groups intending to promote their product as environmentally superior to a competitor's product. Additionally, it is also used as a tool in product design; for corporate strategic planning and environmental planning in order to reduce possible environmental risk and thereby attain competitive advantage in the marketplace;[12] and for public policy making that involves government initiatives in environmental labeling, acquisition, procurement, and legislation.[3,25]

The traditional main focus of chemical product and process designers on product life-cycle stages from raw material extraction up to manufacturing is increasingly changing to one that considers how products will be recycled. In other words, product designers need to consider the way customers use their products, while process designers ought to avoid contamination at the sites where their processes are executed. Rosselot and Allen perceive this as design engineers having to become stewards for their products/processes throughout their life cycles.[25]

We will not delve further into this subject on LCA, as it is treated elsewhere in this book as a separate dedicated chapter.

4.2   Pollution Prevention at the Mesoscale

Macroscale studies of P2 are valuable for identification of P2 opportunities, typically beginning with waste audits and emission inventories. Subsequently, the building blocks of chemical processes (i.e., the unit operations) can be examined for designs and practices that contribute toward reduction of wastes and emissions. Systematic tools are then employed for optimizing the integration of the unit operations with enhanced environmental performance in process flowsheets by exploiting suitable methods for assessing the ensuing economic performance of the optimal flowsheets.[3]

Pollution Prevention for Unit Operations

As emphasized in the preceding section, it is desirable to consider the environmental ramifications of each unit operation early in the process rather than postponing until the flowsheet has been completed. An environmentally benign design is bound to be more profitable, as it will incur lower waste treatment and environmental compliance costs while converting a higher percentage of raw materials into saleable products. Allen and Shonnard stress the importance of the considerations outlined in Table 12 for identifying P2 opportunities for unit operations in the design of chemical processes.[9] Table 13 (adapted from Ref. 3,

Table 12 Important Considerations for Identifying P2 Opportunities for Unit Operations

1. Select raw material(s)—Nelson advocates the following suggestions to minimize human health impact and environmental damage due to any possible emissions:
   - Eliminate impurities in the feed
   - Use substitute chemicals that are less hazardous
   - Utilize waste materials from other processes.
2. Carefully evaluate in-process waste generation mechanisms to direct the process designer toward environmentally sound material choices and other P2 options.
3. Optimize operating conditions for each unit to achieve maximum reactor conversion and separation efficiencies.
4. Minimize losses from cleaning operations:
   - Use mechanical cleaning methods to reduce solvent losses from both parts and equipment cleaning operations
   - Consider continuous processing when batch cleaning wastes are likely to be significant (e.g., with highly viscous, water-insoluble, or adherent materials).
5. Adopt best material storage and transfer technologies to minimize releases of materials to the environment.
6. Minimize energy consumption and the release of utility-related emissions.
7. Enforce safe working conditions.

Note: Adapted from Ref. 9.

17) provides an overview of current P2-inclined operating practices and process modifications, with indication of research needs for the necessary technology aid development.[3,17]

Flowsheet Analysis for Pollution Prevention

Flowsheet analysis examines a process flowsheet in terms of the following: (1) the performance of the individual unit operations with improved environmental performance that make up the flowsheet, and (2) the level to which the process streams are sequenced and networked or integrated for enhanced environmental performance. The analysis of the degree of integration can be executed via the process integration of mass and energy since chemical processes and chemical plant operations essentially involve a strong interaction between mass and energy, with the overall objective of converting and processing mass via the use of energy to drive reactions, effect separations, and steer pumps and compressors.[26]

This section describes a set of methods for synthesizing or restructuring (retrofitting) process flowsheets that can be generally categorized into qualitative and quantitative methods. We begin with tools that are largely qualitative in concept and implementation, focusing on the following three approaches:

1. Material flow analysis, in which mass balances are used to identify P2 opportunities

Table 13   Process Modifications of Unit Operations for P2

Unit Operation	Changes in Operating Practices	Currently Feasible Modifications	Process Modifications Requiring Further Research and Technology Development
Reactors	Improve/Increase selectivity through: • Better mixing of reactants • Elimination of hot and cold spots • Improved feed distribution (avoids short-circuiting) • Careful control of catalyst regeneration	• Modify catalysts to enhance selectivity or to prevent catalyst deactivation and attrition. • Provide separate reactors for recycle streams to permit optimization of conversions (when warranted by conditions).	Consider: • New catalyst technologies • Changes in process chemistry • Integration of reaction and separation units
Separators	Reduce waste from boilers.	• Improve separation efficiencies. • Insulate the column. • Change tray configuration, or consider high-efficiency packing rather than conventional tray-type columns (this reduces pressure drop and decreases reboiler temperatures). • Avoid adsorptive separations when adsorbent beds cannot be readily generated. • Consider low-temperature distillation columns when dealing with thermally labile process streams.	Consider new separation devices that efficiently separate dilute species.
Heat exchangers (HEX)	• Use anti-foulants. • Use innovative cleaning devices for HEX tubes.	• Use staged HEXs and adiabatic expanders to reduce HEX temperatures and reduce sludge formation. • Implement heat integration. • Execute online cleaning.	Consider scraped-wall exchangers and evaporators with viscous materials to avoid thermal degradation of product.

Table 13 (*continued*)

Unit Operation	Changes in Operating Practices	Currently Feasible Modifications	Process Modifications Requiring Further Research and Technology Development
Vessels (e.g., storage tanks)	• Use mixers to reduce sludge formation. • Add emulsifiers to dissolve tank bottoms (when the bottoms and the emulsifiers do not interfere with downstream processing), which is formed as wastes during storage. • Maintain seals and paint in good condition. • Reduce losses during loading and unloading. • Minimize wasted container material.	• Ensure easy (but safe) access to simplify cleaning. • Design to ensure complete draining. • Use floating-roof tanks. • Use high-pressure tanks. • Use insulated tanks. • Use variable-vapor-space tanks.	Consider process-specific changes to eliminate need for storage of hazardous intermediates.

Note: Adapted from Refs. 3 and 17.

2. Frameworks for examining flowsheets that account for modifications to existing processes in operations (i.e., design retrofitting)
3. The Douglas hierarchical design decision procedures for analyzing flowsheets, particularly in the design of new processes[27,28]

The second set of more quantitative and rigorous flowsheet analysis tools consists of energy (heat) integration and mass integration that have been shown to contribute significantly to overall energy and mass efficiency. Two major approaches are available to implement these tools:

1. The pinch analysis or pinch technology method introduced by Linnhoff and co-workers[18]
2. The mathematical programming or optimization techniques[29]

In this chapter, emphasis is placed on the former approach of pinch technology. It is noteworthy that the mass integration concept incorporates the use of both approaches. Also, a notable work that combines the three approaches of hierarchical design via process simulation using the Aspen Plus™ commercial simulator, pinch technology, and mathematical programming is that of Dantus and High.[30,31]

Qualitative Methods.

MATERIAL FLOW ANALYSIS.   The simplest analytical tool for waste minimization is the mass balance. Although most process engineers are comfortable with the concept of mass balances, a number of new issues arise when mass balances are applied to P2 and waste reduction. Two key issues are (1) defining the system boundaries and (2) determining the level of detail required in the mass balances and waste stream measurements. The former is a key aspect in determining whether an adopted strategy actually accomplishes the desired P2, thus contributing to its viability. After the system boundaries have been identified and selected, the second major decision concerns whether detailed mass balance techniques would actually yield significant information about the waste.[3]

FRAMEWORKS FOR EXAMINING FLOWSHEETS OF EXISTING PROCESSES FOR POLLUTION PREVENTION OPPORTUNITIES.   Hazard and Operability (HAZOP) studies identify potential hazards associated with each process stream via qualitative and sometimes quantitative evaluation by systematically considering possible deviations of a stream from normal operating conditions.[32] A parallel analysis for P2 would entail a series of systematic questions concerning each process stream (particularly the waste streams) and cluster of unit processes, with typical questions listed in Table 14. In addition, Allen and Rosselot suggest the following framework for evaluating an existing process flowsheet for P2 opportunities:[3]

1. Establish the system boundaries of the process.
2. Execute a waste stream audit using the material flow analysis, primarily for determining the composition of the waste stream(s) and the associated mechanism of waste formation.
3. Examine systematically the P2 options for each waste stream.
4. Examine systematically the P2 options for each unit operation.

Table 14   Typical Questions in P2 Assessment Analogous to HAZOP Analyses

- Would a change in process configuration or topology reduce waste generation?
- What changes in operating procedures might reduce wastes?
- Would changes in raw materials or process chemistry be effective in eliminating a waste stream?
- Would a change in process conditions (e.g., temperature, pressure, reactant concentrations) reduce waste generation?
- Would improvements in process control be effective?
- Would the waste be a raw material for another process?

Note: From Refs. 3 and 9.

5. Carry out an appraisal on the impact of each potential P2 program on the whole process or on the finished product.

HIERARCHICAL DESIGN PROCEDURES FOR ANALYZING FLOWSHEETS FOR THE DESIGN OF NEW PROCESSES. Chemical processes are often synthesized by employing the Douglas hierarchical design procedure with waste minimization (as summarized in Table 15).[27,28,33,34] This procedure provides opportunities to perform environmental assessments by identifying P2 options via a series of heuristic rules at each level in the design process. The hierarchy is organized in such a way that decisions that affect generation of wastes and emissions at each level limit the decisions in the subsequent levels below it. Accordingly, computer-aided analysis tools providing the integrated capabilities indicated in Figure 7 will therefore be necessary to efficiently link chemical process and product design with critical environmental processes in a larger systems analysis. At the earliest stage of design, choices are made among competing alternatives for reaction pathways, raw materials, solvents, and energy sources. Next, technological choices are made regarding equipment for carrying out these reaction and separation steps. A flowsheet containing connections between these unit operations is then generated, and the material and energy requirements are computed (based on reasonable estimates) using commercial process simulation software. Opportunities for energy and mass integration are then evaluated in efforts to reduce energy consumption and waste generation. In the final process design step, the flowsheet is optimized to determine process operating parameters that maximize profit (or minimize production and operating costs, accordingly) and minimize environmental impacts.[35]

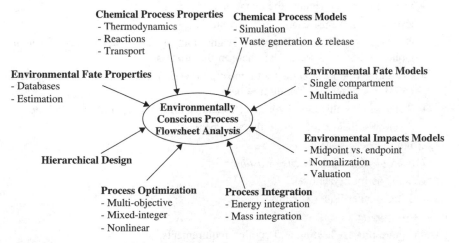

Figure 7 Integrated methods and capabilities for environmentally conscious design and assessment for waste reduction and P2. (Adapted from Ref. 35.)

Table 15 Levels of Douglas's Hierarchical Procedure for Process Synthesis with Waste Minimization

Design Level	Waste Minimization/Reduction Problem and the Associated Strategy

1. Identify the input information of the type of problem: the material to be manufactured.

2. Specify the input–output structure of the flowsheet.

For problems caused by:

- Reaction chemistry: change the chemistry
- Air oxidation to NO_x: change to O_2 in recycle CO_2 oxidations
- Spent catalysts: regenerate the catalyst

3. Design the recycle structure of the flowsheet.

For problems caused by:

- Adding reactor diluents to shift the product distribution or to shift the equilibrium conversion: change the diluent
- Adding heat carriers: change the heat carrier
- Adding reactor solvents: change the solvent

4. Specify/stipulate the separation system.

4a. General structure: phase splits

4b. Vapor recovery system

For problems caused by:

- Absorber solvents: change the solvent
- Regeneration of adsorption beds: change the bed stripping agent
- Removing spent adsorbents: change to absorption or condensation
- Use of reactive absorbers to remove toxic materials

4c. Liquid recovery system

For problems caused by:

- Stripping agents: change the agent
- Extraction solvents: change the solvent
- Crystallizer (recycle and) purge streams (almost pure water): reuse (through mass integration techniques) the purged water elsewhere in the process
- Crystallizer purge streams (not almost pure water): remove the contaminants and recycle the water or look for a different separation system
- Reactive crystallization byproducts: identify different separation technique
- Spent absorbents: regenerate the adsorbent

4d. Solid recovery system

For problems caused by cake washing:

- Same as for crystallizer
- Filter mother liquor streams

5. Process energy and mass integration.

5a. Integrate process heating and cooling requirements.

5b. Identify process waste recycling and water reuse opportunities.

Table 15 (*continued*)

Design Level	Waste Minimization/Reduction Problem and the Associated Strategy

6. Evaluate opportunities.
   6a. Different decisions
   6b. Reactor configuration
   6c. Complex distillation columns
7. Address flexibility and control.
8. Address process safety

Note: Adapted from Refs. 9, 28, and 34.

Quantitative Methods.

PROCESS ENERGY INTEGRATION/HEAT INTEGRATION.   The core principle in process energy integration is to use the heat from streams that need to be cooled for heating streams that need their temperature/energy level or content raised. Heat transfer between streams in a process prevents pollution by reducing the nccd for fuels and for cooling rower operation. Heat integration is generally accomplished via the method of heat-exchanger network synthesis (HENS) pioneered by Linnhoff and co-workers.[18,36,37] In HENS, all of the heating and cooling requirements for a proccss are systematically examined to compute the extent to which streams that need to have their temperature raised can be heated by streams that need to be cooled.[18,27,36–38]

PROCESS MASS INTEGRATION.   Mass integration is a systematic methodology that provides fundamental understanding of the global flow of mass within a process and employs this understanding for identifying performance targets and for optimizing the generation and routing of species throughout the process. P2 denotes a mass-allocation objective that is at the heart of mass integration. In fact, mass integration is more general and more involved than energy integration. Due to the overriding mass objectives of most processes, mass integration can potentially provide much stronger impact on a process than energy integration. Since both integration branches are compatible, mass integration coupled with energy integration essentially provides a systematic framework for understanding the big picture of a process, for identifying performance targets, and for developing solutions to improve process efficiency, which automatically translates as a P2 initiative.[39]

Mass integration is properly defined as the use of materials that would otherwise be wasted, analogous to the notion of heat integration, which is the use of *heat* that would otherwise be wasted. Three tools are available for determining process configurations that result in mass integration:[9]

1. the source–sink mapping, which is the simplest and most visual and intuitive of the methods for identifying candidate streams for mass integration
2. strategy for determining optimum mixing, segregation, and recycling
3. the mass exchange network (MEN) synthesis, which is the mass integration analog for heat exchange network (HEN) synthesis that is developed by Vasilios Manousiouthakis and coworkers at the University of California at Los Angeles (UCLA).[19]

SOURCE–SINK MAPPING.   Source–sink mapping is used to determine whether waste streams can be used as feedstocks. The first step in creating a source–sink diagram is to identify the sources and sinks of the material for which integration is desired. As an example, for the problem of water utilization (treatment and reuse) in petroleum refineries, we need to identify wastewater streams as the *'sources'* of water and the processes that require water as the *'sinks'* of water. The flow rates of the sources and sinks must be known, with the latter able to accept a range of flow rates. Contaminants and their associated concentrations that are present in the source streams and that potentially pose significant problems for the sinks need to be identified, as well as the tolerances of each sink for these contaminants. Some processes may require very pure feed, rendering infeasibility to the waste streams that contain the feed material as well as some contaminants. Nevertheless, many processes are still able to make use of material that contains impurities, while some sinks possess widely flexible tolerances.[9]

STRATEGIES FOR OPTIMIZATION OF SEGREGATION, MIXING, AND RECYCLE OF STREAMS.   Processes with increased number of sources and sinks that demand more complex analyses, beyond the capability of source–sink mapping, require the employment of mathematical programming techniques, combined with process simulation packages for segregation, mixing, and recycle of streams. These involve linear and nonlinear optimization approaches.[9]

MASS EXCHANGE NETWORK (MEN) SYNTHESIS.   Unlike source–sink mapping and optimization of segregation, mixing, and recycle, MENs achieve mass integration by executing direct exchange of mass between streams through the systematic generation of a network of mass-exchangers, which preferentially transfer compounds that exist as pollutants in the streams where they are originally present to streams where they have a positive value (this will be explained later in the section). MEN synthesis can be adopted for any countercurrent, direct-contact mass-transfer operation such as absorption, desorption, or leaching. These concepts are illustrated by Figures 8 and 9.

Analogous to energy integration in which heat transfer is limited by energy balances and a positive driving force, the limits to mass transfer are enforced by

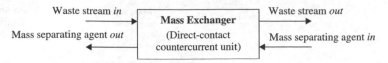

Figure 8 Schematic of a single mass exchanger for environmental process design. (From Ref. 21.)

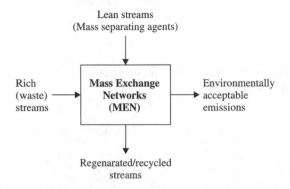

Figure 9 Mass exchange network (MEN) synthesis. (From Ref. 21.)

mass balance constraints and equilibrium constraints:

1. The total mass transferred by the rich stream (i.e., the stream from which a material is to be removed) must be equal to that received by the lean stream (i.e., the stream receiving the material).

2. Mass transfer is possible only if a positive driving force exists for all rich stream/lean stream matches.

The MEN concept has also been successfully extended to novel process design concepts of wastewater-minimization systems, waste-interception and allocation networks (WINs), heat-induced waste-minimization networks (HIWAMINs), and energy-induced waste-minimization networks (EIWAMINs). The latter two approaches serve to improve process economics by simultaneously addressing issues of waste reduction and energy conservation in process design.[40–45] Figures 10 and 11 display the schematic representations of the design methodologies of WIN and HIWAMIN together with EIWAMIN, respectively.

A commendable amount of work on the synthesis of mass exchange network for P2 is available in the literature for dephenolization of petroleum refinery wastewaters, in which phenol is a pollutant in the water effluent streams of the catalytic cracking units, desalter wash water, and spent sweetening waters, but in other streams, phenol is present as a valuable additive.[46,47] Other relevant works include that of Dunn and El-Halwagi[40] for reducing the emission of hydrogen

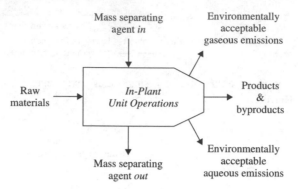

Figure 10 Schematic representation of waste interception and allocation networks (WINs). (From Ref. 21.)

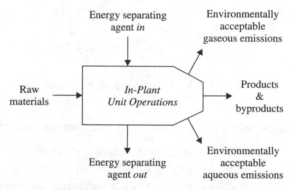

Figure 11 Schematic representation of heat-induced waste minimization networks (HIWAMINs) and energy-induced waste minimization networks (EIWAMINs). (From Ref. 21.)

sulfide from pulp and paper plants; the concept-oriented papers by Gupta and Manousiouthakis[48] and Papalexandri et al.;[49] and Hallale and Fraser[50] for water minimization. Due to the limited space here, the interested reader is referred to the review article by Dunn and El-Halwagi[21] and the comprehensive text by El-Halwagi[26] for a detailed exposition on the application of MENS at industrial scale and for academic research.

4.3 Pollution Prevention at the Microscale

Many of the macroscale and mesoscale P2 approaches rely on a molecular-level understanding of chemical and physical process at the microscale. For example, the synthesis of new catalysts to achieve higher yields with less wastes relies on a fundamental understanding of surface chemistry. There are a large number of methods available 'for undertaking P2 at the microscale via the molecular-level redesign of chemical products and processes. In line with the emergence of the power of computational tools that employ mathematical programming or optimization techniques, we mention here two methods that adopt such tools:

1. Systematic design of substitute materials for environmentally harmful chemicals such as solvents
2. Systematic analysis of reaction pathways

The former approach utilizes *group contribution theory*, in which molecules are systematically designed to incorporate a distribution of functional groups resulting in desired properties. It relies on an understanding of the ways in which the building blocks of a molecule (i.e., its functional groups) affect its properties. The objectives of the second approach of reaction pathway synthesis are to reduce both the use of toxic precursors and the formation of unwanted byproducts. In spite of current significant limitations confronting these techniques, they nonetheless hold much promise for further improvement into robust methods for the systematic identification and evaluation of desired synthesis routes.[3]

5   CONCLUDING REMARKS

This chapter strives to compile and consolidate a broad set of tools, techniques, and best practices for identifying and subsequently implementing pollution prevention (P2) programs, with a multiscale analysis of macro-, meso-, and micro-scale employed to systematically categorize these approaches. It is hoped that over time, a shift in focus could be expected and targeted, from mere compliance to a point in which environmentalism, as driven by the P2 paradigm, will be fully justified and integrated into corporate business culture, similar to the overriding emphasis, nowadays, that is placed on the health and safety elements of the ubiquitous concept of "health, safety, and environment" in chemical plant operations.

REFERENCES CITED

1. A. Elkamel, "Preface of Special Issue on Pollution Prevention and Waste Minimization via Simulation and Optimization," *International Journal of Environment and Pollution*, 2006.
2. M. P. Harold and B. A. Ogunnaike, "Process Engineering in the Evolving Chemical Industry," *AIChE Journal*, **46** (11), 2123–2127 (2000).
3. D. T. Allen and K. S. Rosselot, *Pollution Prevention for Chemical Engineers*, John Wiley, New York, 1997.
4. P. L. Bishop, *Pollution Prevention: Fundamentals and Practice*, Waveland Press, Long Grove, Illinois, 1999.
5. U.S. Environmental Protection Agency (U.S. EPA), *Waste Minimization Opportunity Assessment Manual*, EPA/625/7-88/003, Washington, DC, 1988, http://www.p2pays.org/ref/10/09929.pdf. Retrieved on June 25, 2006.

6. United Nations Environment Program (UNEP), *Cleaner Production Programme* (a pamphlet on UNEP's cleaner production program). UNEP, Geneva, 1996.

7. U.S Environmental Protection Agency (U.S. EPA), *Facility Pollution Prevention Guide*, EPA/600/R-92/088, Washington DC, 1992, www.p2pays.org/ref/01/00370.pdf or http://www.epa.state.oh.us/opp/tanbook/fppgbgn.html. Retrieved on June 25, 2006.

8. H. Freeman, T. Harten, J. Springer, P. Randall, M. A. Curran, and K. Stone, "Industrial Pollution Prevention: A Critical Review," *Journal of the Air and Waste Management Association*, **42**, 618–656 (1992).

9. D. T. Allen and D. R. Shonnard, *Green Engineering: Environmentally Conscious Design of Chemical Processes*, Prentice Hall, Upper Saddle River, NJ, 2002.

10. H. Freeman, *Industrial Pollution Prevention Handbook*. New York, McGraw-Hill, 1995.

11. D. T. Allen and D. R. Shonnard, "Green Engineering: Environmentally Conscious Design of Chemical Processes and Products," *AIChE Journal* **47** (9), 1906–1910 (2001).

12. C. J. Pereira, "Environmentally Friendly Process," *Chemical Engineering Science*, **54** 1959–1973 (1999).

13. K. L. Mulholland and J. A. Dyer, *Pollution Prevention: Methodology, Technologies, and Practices*, American Institute of Chemical Engineers, New York, 1999.

14. F. L. Moore, "Pollution Prevention in the Chemical Industry," in *Industrial Pollution Prevention Handbook*, Chapter 44, ed. Harry Freeman. McGraw-Hill, New York, 1995.

15. P. Anastas and J. Warner. *"Green Chemistry: Theory and Practice,"* Oxford University Press, New York, 1998.

16. U.S. Environmental Protection Agency (U.S. EPA), http://www.epa.gov/greenchemistry/principles.html. Accessed on June 15, 2006.

17. R. S. Butner, "Pollution Prevention in Process Development and Design," in *Industrial Pollution Prevention Handbook*, ed. Harry Freeman, Chapter 44, pp. 329–341, McGraw-Hill, New York, 1995.

18. B. Linnhoff, D. W. Townsend, D. Boland, G. F. Hewitt, B. E. A. Thomas, A. R. Guy, and R. H. Marsland, *User Guide on Process Integration for the Efficient Use of Energy*, The Institution of Chemical Engineers, Rugby, 1982.

19. M. M. El-Halwagi and V. Manousiouthakis, "Synthesis of Mass Exchange Networks," *AIChE Journal*, **35** (8), 1233–1244 (1989).

20. A. P. Rossiter, "Process Integration and Pollution Prevention," in *AIChE Symposium Series on Pollution Prevention via Process and Product Modifications* **90**(303), 12–22 (1994). Edited by Mahmoud M. El-Halwagi and Demetri P. Petrides. New York: American Institute of Chemical Engineers (AIChE).

21. R. F. Dunn, and M. M. El-Halwagi, "Process Integration Technology Review: Background and Applications in the Chemical Process Industry," *Journal of Chemical Technology and Biotechnology*, **78**, 1011–1021 (2003).

22. R. Smith, and E. Petela, "Waste Minimization in the Process Industries," *The Chemical Engineer*, **17**, 21–23 (1992).

23. L. Case, L. Mendicino, and D. Thomas. "Developing and Maintaining a Pollution Prevention Program," in *Industrial Pollution Prevention Handbook*, ed. Harry Freeman, Chapter 8, McGraw-Hill, New York, 1995.

24. D. T. Allen and K. S. Rosselot, "Pollution Prevention at the Macroscale: Flows of Wastes, Industrial Ecology and Life-cycle Analyses," *Waste Management*, **14** (3–4), 317–328 (1994).

25. Kirsten Sinclair Rosselot and David T. Allen. "Flowsheet Analysis for Pollution Prevention." In: *Green Engineering: Environmentally Conscious Design of Chemical Processes*. Edited by David T. Allen and David R. Shonnard. Upper Saddle River, NJ: Prentice Hall, 2002.

26. M. M. El-Halwagi, *Pollution Prevention through Process Integration*, Academic Press, San Diego, 1997.

27. J. M. Douglas, *Conceptual Design of Chemical Processes*, McGraw-Hill, New York, 1988.

28. J. M. Douglas, "Process Synthesis for Waste Minimization," *Industrial & Engineering Chemistry Research*, **31**, 238–243 (1992).

29. I. E. Grossmann and Z. Kravanja, "Mixed-integer Nonlinear Programming Techniques for Process Systems Engineering," *Computers & Chemical Engineering*, **19**, supplement: S189–S204 (1995).

30. Aspen Technology, Inc. (AspenTech), *Aspen Plus™*, http://www.aspentech.com. Accessed June 23, 2006.

31. M. M. Dantus and K. A. High, Economic evaluation for the retrofit of chemical processes through waste minimization and process integration. *Industrial & Engineering Chemistry Research*, **35** (1996): 4566–4578.

32. D. A. Crowl and J. F. Louvar, *Chemical Process Safety: Fundamentals with Applications*, Prentice Hall, Englewood Cliffs, NJ, 1990.

33. J. M. Douglas, "A Hierarchical Decision Procedure for Process Synthesis," *AIChE Journal*, **31**(3), 353–362 (1985).

34. J. M. Douglas, "Synthesis of Multistep Reaction Processes," *Proceedings of the Third International Conference of the Foundations of Computer-Aided Process Design (FOCAPD) at Snowmass Village, Colorado, USA on July 10–14, 1989*, ed. J. J. Siirola, I. E. Grossmann, and G. Stephanopoulos, pp. 79–103, Amsterdam, The Netherlands: CACHE–Elsevier, 1990.

35. J. R. Mihelcic, J. C. Crittenden, M. J. Small, D. R. Shonnard, D. R. Hokanson, Qiong Zhang, Hui Chen, S. A. Sorby, V. U. James, J. W. Sutherland, and J. L. Schnoor, "Sustainability Science and Engineering: The Emergence of a New Metadiscipline," *Environment Science & Technology*, **37**(23), 5314–5324 (2003).

36. B. Linnhoff and E. Hindmarsh, "The Pinch Design Method for Heat Exchanger Networks," *Chemical Engineering Science*, **38**, 745–763 (1983a).

37. B. Linnhoff, H. Dunford, and R. Smith, "Heat Integration of Distillation Columns into Overall Processes," *Chemical Engineering Science*, **38**, 1175–1188 (1983b).

38. Robin Smith, *Chemical Process Design and Integration*. New York: John Wiley & Sons, 2005.

39. M. M. El-Halwagi and H. D. Spriggs, "Educational Tools for Pollution Prevention through Process Integration," *Chemical Engineering Education* (Fall 1998), 246–249.

40. R. F. Dunn and M. M. El-Halwagi, "Optimal Recycle/reuse Policies for Minimizing the Wastes of Pulp and Paper Plants," *Journal of Environmental Science & Health Part A*, **A28** (1), 217–234 (1993).

41. Y. P. Wang and R. Smith, "Wastewater minimization," *Chemical Engineering Science*, **49**, 981–1006, (1994).

42. Y. L. Huang and T. F. Edgar, "Knowledge-based Design Approach for the Simultaneous Minimization of Waste Generation and Energy Consumption in a Petroleum Refinery,"

in *Waste Minimization through Process Design*, ed. A. P. Rossiter, McGraw-Hill, New York, 1995, pp. 181–196.

43. Y. L. Huang and L. T. Fan. "Intelligent Process Design and Control for In-plant Waste Minimization," in *Waste Minimization through Process Design*, ed. A. P. Rossiter, McGraw-Hill, New York, 1995, pp. 165–180.

44. M. M. El-Halwagi, A. A. Hamad, and G. W. Garrison, "Synthesis of Waste Interception and allocation networks," *AIChE Journal*, **42**, 3087–3101 (1996).

45. R. F. Dunn and B. K. Srinivas, "Synthesis of Heat-induced Waste Minimization Networks (HIWAMINs)," *Advances in Environmental Research*, 1, 275–301 (1997).

46. D. T. Allen, N. Bakshani, and K. S. Rosselot, *Pollution Prevention: Homework & Design Problems for Engineering Curricula*, American Institute of Chemical Engineers' (AIChE) Center for Waste Reduction Technologies (CWRT), New York, 1992.

47. M. M. El-Halwagi, A. M. El-Halwagi, and V. Manousiouthakis, "Optimal Design of Dephenolization Networks for Petroleum Refinery Wastes," *Transactions of the Institution of Chemical Engineers Part B*, **70** (B3), 131–139 (1992).

48. A. Gupta and V. Manousiouthakis, "Waste Reduction through Multicomponent Mass Exchange Network Synthesis," *Computers & Chemical Engineering*, **18**, supplement, S585–S590 (1994).

49. K. P. Papalexandri, E. N. Pistikopoulos, and C. A. Floudas, "Mass Exchange Networks for Waste Minimization: A Simultaneous Approach," *Chemical Engineering Research & Design, Part A: Transactions of the Institute of Chemical Engineers*, **72**(A3), 279–294 (1994).

50. N. Hallale and D. M. Fraser, "Capital Cost Targets for Mass Exchange Networks. A Special Case: Water Minimization," *Chemical Engineering Science*, **53** (2), 293–313 (1998).

CHAPTER 5

INDUSTRIAL WASTE AUDITING

C. Visvanathan
Environmental Engineering and Management Program, Asian Institute of Technology, Pathumthani, Thailand

1	OVERVIEW	125
2	WASTE-MINIMIZATION PROGRAMS	127
3	WASTE-MINIMIZATION CYCLE	129
4	WASTE AUDITING	130

4.1	Phase I: Preparatory Work for a Waste Audit	131
4.2	Phase II: Preassessment of Target Processes	139
4.3	Phase III: Assessment	142
4.4	Phase IV: Synthesis and Preliminary Analysis	148
5	CONCLUSION	152

1 OVERVIEW

In the pursuit of *sustainable production and consumption*—as the true value of natural resources and nonrenewable energy sources are being globally perceived—wastes can no longer be viewed as substances that are spendable. Research shows that wastes traditionally discharged into natural bodies as unwanted substances still possess some economical value. What is useless in one context can be useful in other. Importantly, pollution problems can be significantly reduced if wastes can be reused and recycled instead of being discharged to natural bodies. There is a radical shift in the perception of opportunities with industrial waste, and the current tone is to *conserve and cultivate* rather than *deploy and deplete*

In many countries, the manufacturing industry is one of the largest polluting sectors, and every year enormous effort and financial resources are spent worldwide to deal with industrial waste. Therefore, from an industry perspective, a global change in the environmental perception has a profound significance. Industrial processes, management, goals, and ethics are under pressure from a rising environmental awareness. There is ever-increasing demand to externalize the environmental cost of industrial activities. Gradually, a stage is set to internalize the environmental cost, not by marginalizing environmental concerns but, contrarily, by increased environmental stewardship of products and processes.

In response the global industry is embracing proactive methods that principally focus on energy conservation and waste minimization by the application of cleaner production techniques. These techniques are progressively conceived to meet the goals of environmental and economical sustainability of industries in a more dependable way. Use of traditional *end-of-pipe* treatment of waste alone is progressively becoming inadequate to satisfy the tightening requisites of modern environmental legislations. Industrial processes are also impacted by progressive phasing out or banning of several chemicals that have been regularly used in industries. These include ozone depleting substances (ODS), and persistent organic pollutants (POPs), for example. Several national governments have pledged in various global multilateral environmental agreements to gradually eliminate such substances.

Open or global market regimes are bringing additional complications on top of environmental needs by putting up a stiff pricing competition. Over and above, the questions of sustainability are becoming more pressing as several global institutional buyers are including environmental criteria (like ISO 14000 certification or others) in the procurement specification. This is driving industries to reevaluate their activities and associated costs, which also include waste treatment cost that covers roughly 15 to 30 percent of the total operational cost. The most assertive way is to reduce energy consumption and waste generation in the first place, which improves the overall process efficiency. Several case studies proved that such proactive approaches reduce the overall production cost and environmental liabilities. Moreover, being *greener* is also helping companies to market their products with institutional buyers and attract wider public attention.

With increasing popularity and attention to proactive methods, structured methodologies are being developed to systematically explore, analyze, and implement energy conservation and waste minimization or cleaner production programs in industries. In many industries, such techniques are now being used as one of the management tools to monitor and control process efficiency and environmental liabilities.

Traditionally, environmental impacts from industry are mainly assessed based on the type, characteristic, and volume of waste that it generates. However, recent analyses show that higher use of energy create significant environmental impacts when environmental issues related to energy production are taken into account (typical example is GHG emission). Therefore, energy use and waste generation in industries have been recognized as interlinked systems in the way that the more energy is used, the more pollution is produced, and the more waste generated, the more energy is required. As such, this chapter will refer to both waste and energy, beginning with waste minimization.

2 WASTE-MINIMIZATION PROGRAMS

All manufacturing processes will require raw materials and energy to produce a product (or an intermediary) and will generate waste in some form. Each manufacturing plant is unique in the type, characteristics, and quantity of waste generation. In other words, the manufacture of specific products creates particular waste quantity and quality. Thus, it is difficult to make generalizations regarding waste.

Since manufacturing is one of the largest single polluting sectors, several driving factors are compelling the manufacturing industries to change their outlook about waste management. Six major factors are described:

1. *Changing perception of industrial pollution:* There has been a tremendous rise in awareness about industrial pollution in general public and institutions. This has forced governments to take steps to control pollution from industries.

2. *Changing legislations:* Environmental laws and regulations for industries are tougher, and implementation is more rigorous. As a result, waste treatment technologies now require a stringent level of efficiency to meet the discharge standards.

3. *Changing waste treatment and discharge costs:* As a result of tightening legislation and discharge standards, waste treatment cost is continuously increasing.

4. *Changing availability and cost of raw material and energy:* Greater demand of raw material and energy has tightened supply, causing their price to rise. This will definitely affect the cost of the finished products.

5. *Changing traditional markets and trade barriers:* The concept of protected markets is giving way to more competitive global markets. This is forcing industry to increase efficiency and reduce raw material and energy consumption. Buyers in many developed countries are progressively incorporating environmental specification (ISO 14000 certification, eco-labeling, green productivity) in their procurement processes to screen companies.

6. *Changing international commitments:* More and more national and local governments and institutions are committing to international bilateral and multilateral agreements (like UN-mediated multilateral environmental agreements) to curb pollution by reducing and eliminating known harmful substances like ozone depleting substances and persistent organic pollutants. Many of these substances are heavily used in manufacturing. This pressures industries to change their manufacturing processes and design new products.

As a result of these driving factors, many industries are taking new perspectives and strategies in handling their waste management issues, and are trying to resolve them in a more sustainable way. The current trend of waste management is to balance proactive methods with traditional reactive methods. The concept of *reactive method* is about treating the waste once it is generated, also called *end-of-pipe treatment*, while the *proactive method* includes energy and waste minimization, waste recycling and reuse, and cleaner production that reduces end-of-pipe waste. In brief, the main advantages of energy conservation and waste minimization are as follows:

- Raw material consumption can be reduced, which in turn reduces the product cost
- Energy consumption can be reduced, thereby reducing specific energy required for the product
- Process efficiency can be improved, increasing product yield and quality
- Waste generation can be reduced, thereby reducing waste treatment and disposal cost
- Waste materials can be segregated, leading to containment of hazardous and toxic waste, which in turn can improve workers' health and safety.
- Byproducts can be recovered from waste. In addition to recycle and reuse of waste, waste heat recovery and waste exchange (with other industries) can generate additional income
- Increased environmental stewardship can lead to higher attention from institutional buyers and marketing of products.
- Investor confidence can be increased

There are also some known barriers to implementing waste minimization:

- Some waste minimization or cleaner production techniques may involve significant capital investment.
- There may be obvious risk involved in implementing new systems.
- There is a lack of proper manpower and expertise in appropriate technology.
- There is a lack of information and awareness, especially among small and medium scale industries.
- Often there is a hesitation to change traditional ways of doing things.

The response of an industry will largely depend on several factors:

- Nature of the industrial process
- Size and structure of the firm
- Technology and information available to the company
- Economics of prevention
- Attitude of the government to control industrial pollution through legislations, incentives, and penalties

3 WASTE-MINIMIZATION CYCLE

Typically, a development cycle of industrial waste minimization programs comprises six phases: inception, audit, analysis, design and development, implementation, and evaluation, as illustrated in Figure 1. The overall goal of waste reduction and cleaner production program is to critically investigate, evaluate, design and implement such environmentally benign processes and process improvements that would minimize consumption of resources and energy and reduce waste generation in order to reduce adverse environmental impacts and effect in overall economic benefits.

These six phases can be discussed in more detail:

1. *Inception phase:* This phase comprises setting up the goals, commitments, methodologies, task force, time frame, and budget for the project. Such goals should be quantifiable, measurable, achievable, and usable to measure the success or failure of any waste minimization or cleaner production program in real terms. Senior management plays a key role in setting up the project framework, resources, and the project team.

2. *Audit phase:* In the audit phase, the relevant factory processes include management processes and waste treatment processes. These processes are thoroughly investigated to obtain a complete balance sheet of the raw material and energy input and output, including waste. Over and above collecting and compiling all facts and figures, the audit exercise should recommend energy and waste minimization options to attain the desired goals of the program that could be carried to the detail analysis phase.

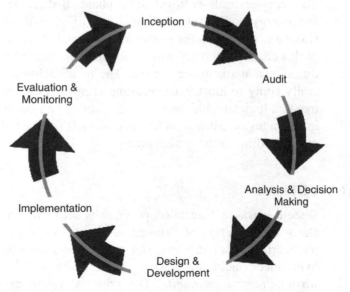

Figure 1 Typical waste minimization and cleaner production cycle.

If possible, preliminary technical and economical analysis may also be carried out to prioritize the options.

3. *Analysis phase:* The analysis phase starts with detailed analysis of the findings and recommendation from the audit phase to explore the various opportunities and risks associated with the various options. It also explores further possibilities. This phase ends with making decisions on which options to be pursued and which to be dropped.

4. *Design and development phase:* The design and development phase starts with setting up the framework of design, development, and implementation of selected options. This would require planning of all actions. It is a good idea to implement different waste minimization and cleaner production improvements in stages to reduce impacts from introducing new process and process modifications. At this stage, process changes and new processes are designed and procured, and all preparatory works for implementation are undertaken.

5. *Implementation phase:* In the implementation phase, processes and process modifications are installed and integrated to the existing system, commissioned, and put into operation. All operators and workers are trained for the changes in the process and new processes.

6. *Evaluation phase:* This phase continues after the changes or new systems are fully integrated into the normal production processes. In this phase, the actual results from the process modifications are monitored, evaluated, and compared to that originally conceived. The cost of such monitoring is normally included in routine quality assurance/quality control activities.

This chapter deals with the audit phase. It discusses different aspects of waste and energy audit systems. The scope of such discussions has been limited to waste and energy audits within a typical manufacturing industry. For the purpose of this chapter, industry would be treated in general, with specific examples from different manufacturing sectors. The methodologies described here would generally apply to most manufacturing sectors. Some modification to the described methodologies would be necessary, depending on specific activities undertaken in the industry. Such modifications are left for the industry professionals to work out according to the requirements.

4 WASTE AUDITING

Once a waste-minimization program is set to be undertaken, the physical work starts with a series of detailed surveys of ongoing activities inside the industry, starting from raw material entering the premises to finished products and byproducts (including wastes). These audits can be termed as *waste audits* or *waste-minimization audits*. The principal intent of such audits is to critically assess various inputs, processes and outputs to find methods and practices for minimizing waste and reduce the resource consumption in a more sustainable environmentally benign way. Traditionally, industrial waste audits do not include examination of the design of the product itself but investigate all activities of the production processes and opportunities of waste recycling/reuse, including waste treatment systems. Other terminologies may be used, such as pollution prevention audit, eco audit, or green audit which essentially focus on some of the common objectives of preventive approaches.

The phases of a typical waste audit process are illustrated in Figure 2. Note that an energy audit process can also be divided into similar phases. The rest of the chapter discusses each of these steps in detail with illustrations, examples, and workouts.

4.1 Phase I: Preparatory Work for a Waste Audit

Preparatory work for a waste audit consists of three main steps:

1. Getting the management and staff involved in the program
2. Forming an audit team and appointing a team leader
3. Planning the audit exercise

Once a waste minimization program is begun, it's time to provide the program with personnel, technical, and financial resources. The first step is to involve stakeholders in the program to get management and staff involved directly or indirectly in the program. This should be mainly done by one of the core management groups who would ultimately be responsible for managing the overall pollution prevention program. Normally, the production management or the environmental management group have the responsibility to execute such programs. Although all stakeholders can be involved in this process, more emphasis should be given on internal stakeholders like the management, supervisors, and workers in taking up the initiative.

Getting Management and Staff Involved
Commitment from different management groups is a decisive factor in the success of a pollution prevention program. Management representatives from all relevant departments should be involved in the program. In a typical manufacturing industry, such departments may include the following:

- Executive management (typically CEO or a deputy)
- Product development and design (if present)
- Production
- Procurement and inventory control
- Operation and maintenance
- Environmental
- Marketing

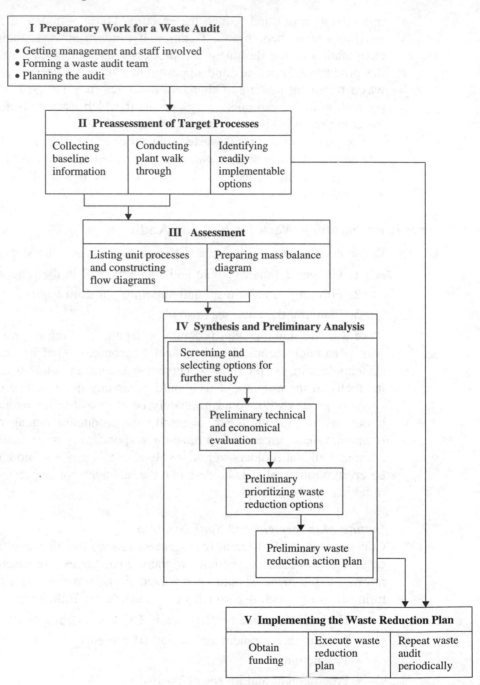

Figure 2 Steps of waste audit methodology.

- Finance
- Quality control

As the management staff would be supposedly aware of the environmental concerns and directly related to the welfare of the company, less effort would probably be required to motivate them for voluntary participation. However, it has to be confirmed that during the whole tenure of the program there is a high level of cooperation and support, even if such a program may cause short-term disruption to normal activities.

Involvement of the supervisors and workers is another key factor in the success of such programs. They have the hands-on involvement with each of the individual activities of the factory. In order to make the best use of their day-to-day experience with the machineries, processes, and a myriad of issues with the factory processes, audit exercise should be carried out with their full involvement and support. Typically, the barrier to such an involvement is fear on the part of certain supervisors and workers that a waste audit will expose inefficiencies that lead to job cuts. In order to overcome such a barrier, supervisors and workers should be assured of their job security, barring evidence of Fraud or sabotage.

The exercise can also be used as an opportunity to create more employee awareness about the environmental issues and energy and waste-minimization and cleaner-production program. Rewards in the form of bonuses, prizes, or acknowledgment would motivate employees to voluntarily participate in the program. There are several common ways of raising interest and motivation that could benefit the program (Box 1).

Box 1 Common Ways to Raise Interest in Waste and Energy Minimization Program

- Use posters or banners to inform the staff about the pollution scenario and the requirements, benefits, objectives, and goals of the upcoming waste minimization and cleaner production program.
- Publicize the upcoming audit exercise.
- Provide some prominent identification mark (like badges) to the members of the audit team and also provide them some special privileges like free access to any place or information, etc. during the audit program.
- Organize "Environment Week," with programs such as speeches, videos, skits, or tree planting.
- Offer cash prizes/monetary incentives for the staff members who come up with innovative idea leading to cleaner production.

- Offer incentives for the machine operators/staff for quality production with minimum resources.
- Hold inter-departmental competition for waste minimization.
- Share the financial gains due to waste/energy minimization programs.

Formation of an Audit Team

One of the key requirements of an audit exercise is to form a proper and balanced audit team that would be responsible for all subsequent audit works. The team is normally formed by the direction of the responsible management group, in discussion with senior management and all other management groups. If required, expert advice from external consultant may be sought at this stage.

Waste audit is an interdisciplinary activity. Therefore, the team should be formed with diversified expertise from representative groups or departments, which will have major contribution and interest in the program. Usually, an audit team is a subset of the project team but may include various other personnel whose contribution may be useful. For example, an audit team can be formed from these groups:

- Environmental managers
- Plant managers
- Process or operation and maintenance engineers
- Occupational health and safety officers
- Supervisors and operators
- Laboratory technicians

The team should appoint a team leader, who will lead the audit team and coordinate the activities of all other team members. The plant manager or the environmental manager can be prospective candidates for team leader. It is recommended that a balanced audit team should be made up of three to six persons, including the team leader. This recommendation is based on typical industry structures but may be varied. In practice, the selection of the team leader and composition of the audit team will depend on the nature of the processes in the industry, scope of audit, and the scale of the industry.

Each of the audit team members should be aware of the goals and objectives of the overall energy and waste minimization program and should be able to contribute to it. The team leader should delegate duties to each of the members, depending on strength and capabilities, and should monitor them throughout auditing phase. Members of the audit team should be relieved from their routine activities during the audit exercise.

The requirement of external consultants or experts in the team would depend on the objectives of the program and complexity of the industrial processes. If the goals are to explore simple and readily applicable waste minimization or cleaner production opportunities (e.g., reduction in general water and electricity consumption), plant engineers and supervisors may be sufficient for auditing purposes. However, if the requirement is complex, such as performance of cryogenic units or distillation columns, experts in these technologies may be required. In these cases, hiring an external consultant may be justified.

Planning the Audit Exercise

Before the actual audit exercise is undertaken, a substantial amount of planning work is to be done in order to carry out the audit within time and budget and least interference to normal activities of the plant. Such planning works would mainly include the following:

- Define the scope of the waste audit
- Develop an audit program
- Prepare specific workplan and checklists
- Develop uniform reporting procedures
- Inform all departments about upcoming audit programs

However, depending on the specific need of the program, additional works may be added to the list.

Defining Scope of the Waste Audit.   Determining the scope of an audit is one of the fundamental steps before taking up the actual auditing process. Scope should be defined according to the goals and objectives of the waste minimization program. Goals should be quantitative, realistic, and achievable. Compliance with a set of legislative requisites within a fixed time period could be the prime goal in some instances. In other instances, there could be the need to reduce (by a certain percentage) the operation of waste treatment cost by reducing the quantity and strength of some priority pollutants (e.g., chromium, copper, or organic pollutants like phenols, cyanides, etc.). In some other cases, the goal may be to reduce consumption of chemicals (by certain tonnes per unit product) by improving certain processes. Waste audits can also focus on specific objectives such as improving the general environmental management system in environmental management system audits, improving operational health and safety by reducing use of toxic chemicals in operational health and safety audits, or total environmental risk audits, and so forth. In every case, there should be some target sectors or processes, which have to be closely scrutinized. Identification of these target processes is crucial in defining the scope. The most common target processes of waste or energy audits are as follows:

- *Production processes:* All unit production processes in the plant need to be audited to check for process inputs, outputs, operating conditions and controls, and process efficiency.

- *Inventory processes:* Inventory processes like storage and handling of chemicals, machine parts, and other accessories need to be evaluated to check loss of material while handling or storage, loss of chemicals that have expired, and so on.
- *Housekeeping processes:* Housekeeping practices should be audited in order to check various losses during cleaning, arranging, and various types of maintenance practices. Better housekeeping practices can reduce leaks, spills, dragout, losses in rinsing, cleaning, and so on.
- *Waste treatment processes:* All waste treatment processes such as effluent treatment plants, gas treatment, and dust filters should be audited to determine treatment efficiency, operating conditions and controls, and so on.
- *Packaging processes:* Packaging sometimes requires specialized systems that can generate wastes. Therefore, such processes must also be audited if waste generation or energy consumption is high in such processes.

Unless the scope of the waste audit is positioned to achieve the overall goals of the program, many efforts can go waste. Table 1 shows how some of the common objectives are related to the target processes. Defining anything more than what is required or less may result in a poor outcome. For example, if the objective of a waste-minimization project is to reduce water consumption in a factory, only those processes/activities (target processes) that are using significant amount of water need to be checked. In this case, it is not particularly useful to audit inventory processes or final packaging systems that consume no water or very little water. However, when the goals are to explore opportunities of electricity reduction, probably all activities need to be audited, as there are practically no areas in a factory where electricity is not consumed. But in the first instance, it may be

Table 1 Audit of Target Processes as per Objectives of Waste-minimization Program

Main Objective	Target Processes
Legislative compliance	Probably all processes, with more emphasis on production and waste treatment processes.
Reduction of toxic and hazardous wastes	Production and inventory processes where hazardous and toxic chemicals are used. Special attention to be given on transport, handling, storage, and use of such chemicals, as well as treatment and disposal of the hazardous wastes.
Operational health and safety Improvement	Special attention to production processes, especially with high temperature, high voltage applications, etc. Attention to be given to handling, storage, and use of hazardous and toxic chemicals. Housekeeping processes need to be checked as well.
Compliance audit	Mainly the effluent treatment units only, with emphasis to check if they are meeting the required discharge standards.

better to audit those sections where large amounts of electricity are consumed. The scope can thus be much wider compared to water-reduction programs.

Defining a balanced scope is not generally a one-step procedure. Starting with preliminary scope, the audit team may modify the scope as the audit is undertaken and new information emerges, which may need extension of scope to certain target processes and elimination of certain other sectors that may not significantly contribute to the overall objective. The scope should also be subjected to debate by all team members of the program to arrive at a consensus.

Develop an Audit Program. Like all other projects, an audit exercise should stick to a program and plan to avoid any overrun. The audit team should develop a series of programs for auditing each of the target processes and timeframe for each of the activities. Generally, for a complete waste and energy reduction audit, the following time frames can be used for manufacturing plants:

- Small scale (less than 50 workers): 3 to 4 weeks
- Medium scale (50–100 workers): 4 to 6 weeks
- Large scale (more than 100 workers): 6 to 10 weeks

This timeframe is generalized; several factors like the number of steps in the production process, the degree of complexity, the level of automation, the quantity, and characteristics of different feedstock used should be considered for determining the timeframe. In the absence of any specific condition for the program, there are two ways of setting up an audit program:

1. Follow the material flow path though the industry.
2. Audit target processes according to the significance in terms of waste reduction.

Both systems have some merits. For example, following the material flow helps keep good track of several raw material as these are transformed into products and byproducts, which gives a complete account of material inputs and outputs. While auditing according to priority of target processes, allow a thorough investigation of major target processes that can identify more waste-minimization opportunities. Development of program is a dynamic activity and can be modified as more and more audit exercises are undertaken and new clues evolve.

Audit timing and frequency are important factors in planning audit exercises. Timing and frequency can be fixed, depending on the following:

- Type (continuous, batch, etc.) of the production process
- Number of parameters to be audited
- Scale of production
- Accuracy of audited information required

Auditing can be scheduled during peak production hours or uniformly throughout a complete production cycle, depending on the nature of the production process.

For example in a typical batch production process (like textile dyeing), quantity of wastewater discharged at the end of each cycle need to be audited, while in a continuous production process (like pulp processing in a pulp and paper industry), discharge of wastewater may be monitored at four-hour intervals. Care should be taken not to disrupt the normal factory processes.

It is recommended that the major target processes are audited more than once, typically three times, to obtain good representative results and to assure repeatability of the information, while less important sectors can be audited once. Quality and reliability of an audit exercise would very much depend on the frequency of audit on each process, as each time a process is studied, there is an increased chance of getting a new clue.

Prepare Specific Workplan and Checklists. Each team member should prepare his or her own specific workplan and checklists, depending on the nature of the activities to be audited. Preparing workplan and checklists would enable auditors to focus on key activities to be audited and collect all necessary information for those processes and activities. Some understanding of the processes would be required to prepare the workplan and checklists.

Develop Uniform Reporting Procedures. The audit report is the final product from the audit phase. All subsequent activities would very much depend on the audit report, and its importance in the program is substantial. It is therefore recommended that the audit report should be well planned and easily comprehensible, and should contain all the information that may be required during the subsequent phases. The contents of the report should be determined based on the objectives of the program. A typical table of contents for a report is given in Box 2. Depending on the requirement, some of the sections can be excluded. However, important sections should be retained in a good audit report.

Box 2   Table of Contents for a Typical Waste Audit Report

- Title Page
- Disclaimer, if any (disclaiming responsibility by the publisher for data or views expressed)
- Table of Contents
- Executive Summary (summarizing the complete audit exercise in the industry)
- Problem Statement, Objectives, and Priorities
- Adopted Approach and Reasoning

- Process Layout, Description, and Observations
- Sources and Quantities of Pollution Load Generated
- Existing Treatment Facilities and Additional Requirements
- Observations on Proposed Treatment Systems
- Integrated Pollution Prevention Strategy
- Generation of Options, Screening
- Option Grouping and Prioritization
- Identification of Candidate Systems
- System Evaluation Criteria
- Calculations
- Method and Results of System Evaluation
- Recommendations
- Implementation Strategy
- Suggestions to the Management
- Summary and Observations
- Appendices (Tables, Graphs, Assumptions, and Calculations)

Inform All Departments about Upcoming Audit Programs. Once the audit program is finalized, the audit team should notify all the respective departments about the upcoming audit. This would enable the departments to be prepared for the audit. The program should be advertised at suitable locations throughout the factory to remind the supervisors and workers about the audit schedule.

4.2 Phase II: Preassessment of Target Processes

The purpose of the preassessment is to collect all information on the target processes that may be required in detailed analysis:

- Collect baseline information
- Conduct a Plant Walkthrough
- Identify immediate implementable options

Collection of Baseline Information

Collection and compiling of information is one of the major tasks and purposes of waste auditing. While information on target processes would be of primary value, practical experience shows that various other information may be useful. As such, the scope of data collection can range over the entire cross-section of the factory:

- *Organizational data:* Organizational chart, factory layout, and site plan are included. Also collect information about the surrounding area indicating topography, water bodies, hydrology, agricultural areas, and human settlements.

- *Material and product data:* Data include specifications of feedstock, process water, product and byproducts with composition, instruction on usage and discharge, material safety data sheets, and usable and storage life. Information on quality assurance and quality control of feedstock, process water, products, and byproducts must also be collected.

- *Raw material and logistic consumption data:* Feedstock, energy, and water consumption records are important. Possible source of records of feedstock consumption can be available from stores and accounts department. Water usage can be obtained from water meters or bills, and energy usage from energy bills.

- *Process data:* Collect process flow diagrams, block diagrams, material balance diagrams, control and operational logic diagrams and instructions, manufacturer's data on each machine and process, and maintenance plans and records.

- *Environmental data:* Collect data on air emission, solid waste generation and effluent from different machineries, including waste treatment systems, environmental directives, and licenses.

- *Management data:* It is important to document the number of staff, their position and responsibilities, performance records, administrative instruction, occupational and safety procedures, quality assurance, and quality control procedures.

- *Financial data:* Product, utility and raw material cost, cost of waste treatment, operating and maintenance cost are available as part of the company's financial statements.

- *Industry data:* If possible, gather all of the data for industries of similar nature. Such information, though difficult to obtain, would however allow comparison with other plants that may help in defining realistic targets and goals.

All the audit members should be well familiar with the information that would allow them to objectively carry out the audit. To an experienced auditor, even scanning through such information can also give some clue about the opportunities, which areas need to be audited in detail, and room to improve the audit exercise. All information should be properly preserved and sources of such information should be noted to check the reliability at a later date.

All information should be checked for underlying data quality in terms of correctness and repeatability. If possible, historical data should be gathered over a period of time (suggested two to three years' records) and the process of data collection should be continued during the entire audit phase on a monthly or quarterly basis. It is suggested that statistical analysis be performed on the collected data to ensure quality of data before these are used for analysis purpose.

Plant Walkthrough Survey

A thorough walkthrough the plant is an essential part of the preassessment phase. It is recommended that all team members be involved in such audit exercise, which should generally be carried out over few days, with one or two sections covered in each day. Moreover, the audit team should be accompanied by the responsible section manager, engineer, or supervisor who is completely aware of each activity in the respective section. All types of management and technical information should be available during such surveys for the audit team to check. There are several benefits of such exercise:

- It is quite likely (especially in the case of large industries) that the members of the audit team may not be very familiar with the different activities carried out in each section other than their own. Therefore, this exercise would give them a chance to get conversant with many more activities and processes.
- It can reveal some obvious waste-minimization or cleaner-production options that may not need a detailed assessment to work out the recommendation. A common sense approach, or simple calculations with a lesser degree of accuracy or thumb rule estimates, may be sufficient for arriving at the conclusion. For instance, in situations where the steam pipes are not insulated or valves are leaking, there is no need to carry out a detailed assessment of energy or steam losses to arrive at the recommendation to insulate the steam pipes or repair the valves.
- By eliminating such obvious options, the auditors can narrow down the scope to those areas that require detailed assessment. This frees up some resources, and the auditors can then concentrate on more intriguing issues.
- With a balanced team of auditors, personnel from other sections or departments can more critically observe activities of another department. This can give rise to lateral thinking, and more avenues for improvement can be explored.

The best strategy for the walkthrough is to follow the material flow path through the industry—from the storage of raw material, through various production processes, until it is converted to the final product and stored, in absence of any other plan. During the walkthrough exercise, each member of the audit team should take detailed note of all the activities, facts, and figures, and any other information that may be useful at a later stage. It is suggested that even trivial observations are noted, as these can form some clue at a later stage. It is preferred that the auditors should prepare their own sketches, schematic arrangements, material flow diagram, block diagrams, and site plans during this walkthrough exercise. Even if some process diagrams are already present, often minor changes are carried out in the plant during operations. Notes of such changes should be taken. At the end, team members should prepare their own report of the plant walkthrough.

Identification of Readily Implementable Options

After carrying out the plant walkthrough and making detailed notes, the audit team must discuss the various observations made and should identify a number of simple and obvious measures to reduce waste generation. Such options should be simple, quickly implementable, and inexpensive. Significant waste reductions can often be achieved by such options, which are based on improved operation, better handling, and tightening up of housekeeping practices. For example, simple measures such as attending to leaking hoses or installing automatic level controllers may lead to significant water savings.

Segregation of waste is arguably one of the numerous measures that can effectively lead to waste reduction. It is the most central of such options, and is a universal issue that needs to be addressed. Segregation of waste can offer enhanced opportunities for recycling and reuse with resultant savings in raw material costs, at the same time reducing treatment costs. Concentrated simple wastes are more likely to be of value than dilute or complex wastes. The waste collection and storage facilities should be reviewed to determine if waste segregation is possible.

Such options are implemented as soon as possible without waiting for the final recommendations. Typically, implementation of such options should be completed within two to three weeks. If implemented, the performance and impacts of the changes should be closely monitored and included in the audit report. All the modifications made should be noted, and these will have to be considered in the later stage while developing the detailed process flow diagram.

It is expected that at the end of the preassessment phase, the audit team is

- Organized and aware of detailed scope
- Aware of all target process layouts for further audit
- Aware of all unit operations in each of the target processes
- Aware of sources of waste and their causes

At the end of the preassessment, the plant personnel should be well informed of audit purposes. Resources should be secured, and readily implemented waste-reduction measures should be identified and, if possible, implemented.

4.3   Phase III: Assessment

This phase can be broadly divided into two steps:

1. List the unit processes and constructing flow diagrams.
2. Prepare a mass balance diagram.

Listing of Unit Operations and Constructing Process Flow Diagrams

This step is essential for an audit program, as it gives a detailed insight into the production operations/process vis-à-vis sources of waste generation and hence enables identification of avenues for better operating practices and waste reduction.

To develop a good representative block process diagram, the audit team should undertake a detailed walkthrough in the production units and utility areas, in order to gain understanding of all the processing operations and their interrelationships. The production or plant staff should be interviewed to know about the actual operating controls, parameters, and issues. Only after conducting a detailed walkthrough and interviews with the production staff, should the audit team compile the required baseline data.

By connecting the individual unit operations in the form of a block diagram and highlighting the flow of materials, a process flow diagram can then be prepared. All the information related for example to raw materials, products/byproducts, energy, water inputs, waste discharged, material and energy flows, motion, and time should be compiled during this preassessment stage and should be presented on the block diagram.

The input and output information for each unit operation should be summarized in standard units by reference to the process flow diagram. Standardized color coding may be used to represent, say, raw material input by a black line, products by blue line, wastes by red lines and recycled stream by green lines. Intermittent operations such as cleaning, make-up, or tank dumping may be distinguished by using broken lines to link the boxes. Similar notation may be used to distinguish batch and continuous discharges.

Material Balance: Process Inputs and Outputs

In the material balance exercise, a detailed account of the process inputs and outputs is made to identify the problem areas and thus the need for improvement. Material balance is important for any waste-minimization project to identify and quantify previously unknown losses or emissions. Material balance is also useful for estimating the costs of additional installations and/or modifications.

By definition, the material balance includes materials entering and leaving a process. Inputs to a process or a unit operation may include raw materials, chemicals, water, air, and energy. Outputs include primary product, byproducts, rejects, wastewater, gaseous wastes, liquid, and solid wastes that need to be stored sent off-site for disposal and reusable or recyclable wastes (Figure 3). In its simplest form, a material balance is drawn up according to the mass conservation principle:

$$Mass\ in = Mass\ out + Generation - Consumption - Accumulation$$

If no chemical reactions occur and the process progresses in a steady state, the material balance gets simplified to

$$Mass\ in = Mass\ out$$

$$Water\ in = Water\ out + losses\ (evaporation,\ spills,\ etc.)$$

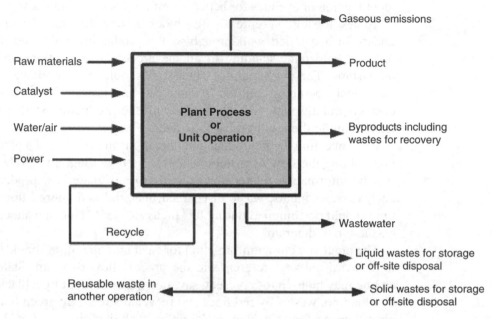

Figure 3   Schematic representation of a material balance sheet.

Sources of Information for Material Balance.   There are many sources of information in establishing material balances for the various unit operations within the plant. Data may be obtained from sample analysis and measurements of raw input materials, raw material purchase records, material and emission inventories, equipment cleaning and validation procedures, batch composition records, product specifications, operating logs, standard operating procedures, and manuals.

Material balances are easier, more meaningful, and more accurate when they are done for individual production units, operations, or production processes. For this reason, it is important to define the material balance envelope or boundary limit accurately, in addition to the tie compound. Ideally, a more accurate balance should be established for the unit operation that is more critical from the waste-generation and reduction point of view, and a less accurate balance could be established for other processes.

Although it is not possible to lay precise and complete guidelines for establishing the material balance, the following guidelines might be useful:

- In the case of an extensive and complex production system, it is better to first draw up the material balance for the whole system (or even the entire production facility as such), and then concentrate on individual operations.

- When splitting up or desegregating the total system, choose the most simple, individual subsystems that are critical from the waste reduction point of view.
- Choose the material balance envelope in such a way that the number of streams entering and leaving the process is the smallest possible.
- Always choose recycle streams within the envelope to start with.

For complex waste-minimization audit, it might be desirable first to make a preliminary or draft material balance and in the second step to evaluate and refine it. However, in case of simple audit for small plants, the steps can be merged into one.

Selection of Priority Unit Operation.   Although the material balance should be set up for all the unit operations, the unit operation most important from the point of view of waste generation must be identified and efforts are concentrated for that particular unit operation. This can be done by professional judgment and technical know-how of the audit team and specifically the production personnel.

Selection of Tie Compounds.   A *tie compound* is the parameter (or substance) for which the material balance is established around a unit operation or a process. It is important to select an appropriate tie compound. Criteria for selecting the tie compound could be:

- Expensive raw material/intermediate
- Material common in most processing stages
- Substance of hazardous nature
- Substance/compound easy to measure/estimate

A simple example of a tie compound could be water to account for most wet operations. Establishing water balance for the processes using substantial amount of water can often provide useful clues for cleaner production. In practical situations, more specific tie compounds (e.g., nickel or zinc in electroplating shops or dyestuff in textile processing) would be ideal. Another good example could be that of chromium in leather tanning.

Chemical oxygen demand (COD) is another useful tie parameter that sharpens the material balance exercise, especially to link the production areas with the effluent treatment plant. The audit team can estimate the contribution of each process department in terms of total COD load in kg/day, knowing the volume of wastewater and the COD discharged by each department and cross checking it with the COD load observed at the treatment facilities.

One need not be very particular over the accuracy (of the order of 99%) of the material balance. In practice, such high accuracies are rarely achievable. Material balance within the tolerance range of 10 percent should generally be acceptable. However, if the tie compound for material balance is hazardous, a higher order of accuracy should be targeted.

Steps for Preparing a Material Balance. There is a logical series of steps for preparing a constructive material balance:

1. *Determine inputs.* The inputs to the process and to each unit operation need to be quantified. As a first step toward quantifying raw material usage, purchasing records should be examined; this readily gives an idea of the quantities involved. The raw materials purchases and storage and handling should be recorded in a table format in order to derive the net input to the process.

   Water is frequently used in the production process, for cooling, gas scrubbing, washouts, rinsing, and steam cleaning. The water usage needs to be accurately quantified as an input. Also, some unit operations may receive recycled material from other unit operations. These also represent an input. Hence, water and recycled materials need special attention, and therefore steps 1A and 1B describe how to evaluate these two factors.

   1A. *Record water usage.* The use of water, other than for a process reaction, should be covered in all cleaner production programs. The use of water for washing, rinsing, and cooling in process and in utility operations is often overlooked, although it represents an area where effective waste reduction can frequently be achieved simply and cheaply.

   1B. *Measure current levels of material recycling.* Some materials may be transferred from one unit to another (e.g., reuse of the final rinse in a soft-drink bottle washing plant as the initial rinse); either directly or after some modifications/treatment. If recycled materials are not properly documented, double counting may occur in the material balance, particularly at the process or complete plant level; that is, a material will be quantified as an output from one process and as an input to another. Proper attention must be paid to this issue, and care must be taken to avoid any discrepancies.

2. *Quantifying outputs.* To calculate the second half of the material balance, the outputs from unit operations and the process as a whole need to be quantified. Outputs include primary product, byproducts, wastewater, gaseous wastes (emissions to atmosphere), and liquid and solid wastes that need to be stored and/or sent off-site for disposal and reusable or recyclable wastes. It is important to identify appropriate units of measurement.

   If the product is sent off-site for sale, then the amount produced is likely to be documented in company records. However, if the product is an intermediate to be input to another process or unit operation, then the output may not be so easy to quantify. Production rates will have to be measured over a period of time. Similarly, the quantification of any byproducts may require field measurement.

2A. *Account for wastewater.* On many sites, significant quantities of both clean and contaminated water are discharged to sewers or to a watercourse. In many cases, this wastewater has environmental implications and incurs treatment costs. In addition, wastewater may wash out valuable unused raw materials from the process areas. Therefore, it is extremely important to know how much wastewater is going down the drain and what the wastewater contains. The wastewater flow, from each unit operation as well as from the entire process, must be quantified, sampled, and analyzed.

2B. *Measure gaseous emissions.* To arrive at an accurate material balance some quantification of gaseous emissions associated with the process is necessary. For example, a tea drier exhaust may carry fine particles of tea dust. Measurement in such cases calls for instruments such as thimble probe dry gas meter–vacuum pump assembly. In many instances, gaseous emissions carry some amount of hazardous materials also (like VOCs). Expert assistance may be needed to determine the material/product loss through gaseous emissions.

3. *Prepare a preliminary material balance.* A material balance is designed to provide better understanding of the inputs and outputs, especially waste, of a unit operation such that areas where information is inaccurate or lacking can be identified. The initial balance should be considered as a rough assessment that must be further refined and improved.

   The units of measurement should be standardized (liter, ton or kilogram) on a per day, per year, or per batch basis. The measured values in standard units should be summarized by reference to the process flow diagram. It may be necessary to modify the process flow diagram following the in-depth study of the plant. It is highly desirable to carry out a water balance for all water inputs and outputs to and from unit operations, because water imbalances may indicate underlying problems such as leaks or spills. Similarly, a detailed material balance should be carried out for important tie compound, as agreed upon by the audit team during the planning phase.

4. *Evaluate and refine material balance.* The individual and sum totals making up the material balance should be reviewed to determine information inaccuracies. Ideally, the input should equal the outputs, but in practice, this will rarely be the case. Some judgment will be required to determine what level of accuracy is acceptable. If there is a significant material imbalance, then further investigation is needed.

When constructing material balances, watch for factors that could overstate or understate waste streams. Sometimes, all or at least a few steps of material balance may need to be repeated a few times in order to refine the material balance. These may include quantification of a few input or output streams or even hunting for some material flows that might have been totally missed in the initial stage. Additional field sampling and analysis may also be required to be carried out in certain cases, and thus the data collected should again be organized and represented so as to establish an accurate material balance.

4.4   Phase IV: Synthesis and Preliminary Analysis

Phases I to II have covered planning and undertaking waste audit, resulting in the preparation of a material balance for each unit operation. Phase IV represents the interpretation of the material balance to identify process areas or components of concern. Figure 4 represent a material balance algorithm for the textile industry in establishing waste reduction options.

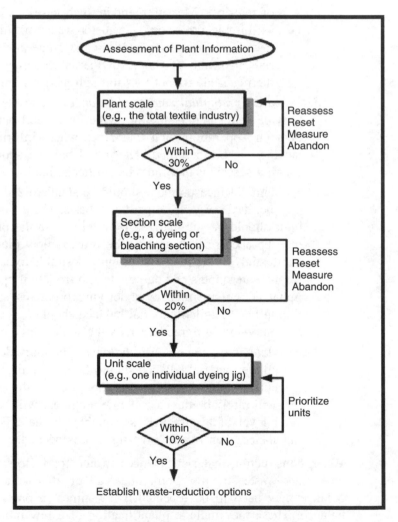

Figure 4   Material balance algorithm for textile industry.

To interpret a material balance, it is necessary to have an understanding of normal operating performance. Thus, a member of the audit team must have a good working knowledge of the process. To a trained eye, the material balance will indicate areas for concern and help to prioritize problem wastes. By using the material balance, major sources of waste may be identified, deviations from the norm in terms of waste production may be found, areas of unexplained losses may be determined, and operations that contribute to flows that exceed national or site discharge regulations may be pinpointed.

In this phase, several possible waste-reduction measures are identified that can be proceeded to the analysis phase. Different waste-minimization programs may require varying degrees of effort, time, and financial resources:

- Obvious waste-reduction measures, including improvements in management techniques and housekeeping procedures that can be implemented cheaply and quickly
- Long-term measures involving process modifications or process substitutions to eliminate problem wastes

Screening and Selecting Options for Further Study (Weighted Sum Method)

For options that require numerical evaluation, the most commonly used tool is the *weighted sum method*. Screening and selection is recommended when a large number of options have to be considered. Only those options that have sufficient merits should be carried forward. The weighted sum method provides a means of quantifying the important criteria that affect waste management in a particular industry:

1. *Determine what criteria would be considered in evaluation of the options.* The higher the degree of improvements achieve in the given criteria, the better is the result. For example, in a typical waste minimization program, such criteria can be:
   - Amount in reduction in waste quantity
   - Amount of reduction in raw material consumption
   - Amount of reduction in hazardous or toxic waste
   - Improvement in health and safety condition
   - Ease of implementation
   - Cost of implementation
   - Resource and time required

   These criteria should be determined in terms of meeting the overall waste-minimization objectives and goals. This should also take into consideration various types of constraints that may be present. Judgment would be required to select the criteria.

2. *Once the criteria are determined, each criterion should be given a weight.* The more important a criteria is, the higher is its weight. Again, a fair

amount of judgment is required to determine the weight for each criterion. Generally a relative scale of 0 to 10 is used to allot weight. For example, if reduction in waste treatment and disposal costs is more important, while time of implementation is relatively less important, then the reduction in waste treatment costs is given weightage of 8 while the time of implementation is given a weight of 2 or 3.

3. *Each option is then rated for all the criteria.* A scale of 0 to 10 could be used for rating. Marks are given according to the degree the criteria are satisfied (i.e., higher marks are given when the option fulfils or satisfies the criteria and low marks when the option doesn't suit the criteria).

4. *Each rating (for the option) is multiplied by the corresponding weight of the criterion.* An option's overall rating is the sum of all the products of ratings times the weight of the criteria.

The options that carry higher marks would be carried for further analysis. Table 2 presents an option evaluation by weighted sum method.

Preliminary Technical and Economical Evaluation
Once the options are screened and selected for further analysis, preliminary technical and economical evaluation may be required to set the priority of options and develop a preliminary action plan.

Table 2   Workout Example of Weighted-sum Method for Waste Audit

Criteria	Weight	#1 Option R	#1 Option R*W	#2 Option R	#2 Option R*W	#3 Option R	#3 Option R*W	#4 Option R	#4 Option R*W	#5 Option R	#5 Option R*W
Reduction in treatment/disposal costs	8	7	56	7	56	5	40	2	16	2	16
Reduction of input material costs	4	8	32	6	24	8	32	4	16	4	16
Extent of current use in industry	5	8	40	8	40	7	35	7	35	7	35
Extent on product quality (no effect = 10)	10	9	90	9	90	2	20	8	80	8	80
Low capital cost	5	2	10	5	25	4	20	7	35	8	40
Low operation and maintenance cost	5	5	25	6	30	5	25	8	40	8	40
Short implementation period	8	3	24	5	40	3	24	7	56	8	64
Ease of implementation	7	3	21	6	42	5	35	7	49	8	56
Reduction in energy bills	9	5	45	9	81	5	45	10	90	10	90
Improvement in OHS	7	10	70	3	21	10	70	2	14	2	14
Final Evaluation Sum of weighted ratings $\Sigma(W*R)$			413		449		346		431		451
Option ranking			4		2		5		3		1
Feasibility Analysis Scheduled for (Date)											

Technical Evaluation.   The technical evaluation determines whether a proposed waste-minimization option will be technically feasible and achievable within the framework of existing constraints. A technical evaluation often begins with examining the impacts of the proposed option on production processes, production schedule, product quality, extra resource requirement, real estate requirement, operational feasibility, and safety. Several constraints such as disruption to normal production schedule, shutdowns, match of specifications (between the existing equipment and the new one), lack of technical knowledge, and trained manpower may be presented. Moreover, psychological resistance to changes may also be an issue. It is therefore recommended that all serious changes are first tested at a laboratory-scale and pilot scale. And, trial runs with the prototypes and test products are undertaken before the change is implemented and integrated to the actual production process. It is also suggested that engineers, supervisors, and operators are suitably trained for the change.

Waste-minimization options can also have some environmental impacts. During technical evaluation, the environmental effect of implementing the option needs to be checked. For example, if an option calls for recycling of rinse water, the effect of disposal of such water needs to be evaluated (as the concentration of solid in the recycled water would increase many times due to recycling), which may impact the receiving body.

Economic Evaluation.   Economic feasibility is one of the most important criteria in determining the selection of an option. Unless forced by legislative requirement, there would be no cases where the economic merit of an option would be measured. Normally, each organization has its own economic criteria for selection of projects. However, three main criteria must be evaluated, irrespective of other criteria:

1. The capital, operation, and maintenance cost of the option
2. The benefit that it would return over and above the existing system
3. The resulting pay-back period

The relationship between capital cost, pay-back period, and likely acceptance is given in Table 3. Pay-back period can be considered as *long* if it exceeds two years; *medium* if it is between one and two years; and *short* if less than one year for common waste minimization options.

Preliminary Prioritizing Waste Reduction Options

Once waste-minimization options are evaluated, they would need to be prioritized for further analysis, design, and implementation. A preliminary prioritization at the audit phase will expedite the process of analysis in the next phase. Prioritization can be done by weighted sum method, in absence of any special criteria (e.g., where some waste-minimization measures become mandatory due to legislative requirements, these can then be classed as high priority options, without

Table 3   Relationship between Capital Cost, Pay-back Period, and Likely Acceptance

Capital Cost	Pay-back Period	Likely Acceptance
High	Long	Low
High	Medium to Short	Medium
Medium	Long	Medium
Medium	Medium	Medium
Medium	Short	Medium to high
Low	Long	Medium to high
Low	Medium	High
Low	Short	High

further analysis). Some of the criteria that can be used to prioritize the options follow:

- Technical ease in implementation
- Resource and time requirement
- Impacts on production schedule
- Short-term capital requirement
- Pay-back period

Developing Preliminary Waste Reduction Action Plan

Upon suggesting the priority, the audit report should also delineate a preliminary action plan for implementation. This can be represented as a regular bar chart or in any other form that is acceptable. In normal cases, the implementation sequence should follow the order of priority. It is suggested that implementation of waste minimization options is taken up in stages so that the cumulative impact on production processes, resources, and finance can be kept low. It is probably best to initially implement low-cost options with relatively simple technical requirements followed by progressive implementation of more complex changes that may require higher capital costs. It may also be a good idea to implement an option and test the success of it before the next one is undertaken to reduce the total risk involved in undertaking changes.

5 CONCLUSION

Waste auditing is as important as any other step in a waste-minimization or cleaner-production project. A proper waste audit should essentially provide a platform on which the rest of the project would be built. It should be a repository of all data and information that would be required to carry out the rest of the phases. As information is the strength so is a successful waste audit.

The principal intent of a waste audit is to critically assess various inputs, processes, and outputs to find methods and practices for minimizing waste and reducing the resource consumption in a sustainable and environmentally benign way without compromising the commercial interest of the company. The waste audit phase would typically be a data collection and information synthesis phase that explores the current situation. Traditionally, an industrial waste audit also develops a list of waste minimization options and undertakes some preliminary technical and economic feasibility studies on the identified options to recommend about these options.

In general, the first step of a systematic waste audit is to prepare for the audit by forming a team, defining the scope, and programming the audit timings and budget. Scope should be defined in line with the overall objective of the waste-minimization program. The next step is to collect all baseline information, conduct plant walkthrough surveys and identify readily implementable options. This would be followed by a detailed assessment of target sectors and unit processes and performing the mass balance analysis. At the end, the data should be analyzed to build up an array of information and recommendation that would set the direction of the next phase. All data collected, information derived, and recommendation made should be presented in a form of a comprehensible and well-laid-out audit report.

It should be appreciated that a waste audit forms the vital activity of data collection and investigation of the current condition of the factory and its activities. Therefore, it is the very base of the next series of activities. The waste audit should be undertaken with utmost care and should be as thorough as possible. The more effort is spend at this phase, the better will be the chance of success of the project.

REFERENCES

N. A. Aldokhin, A. I. Goncharov, M. A. Grachev, and A. N. Suturn, *Recycling of Wastewater and Solid Waste at the Selenginsk Pulp and Paper Plant*. Industry and Environment, **13**(3–4); 21–23, 1990. Bishop, *Pollution Prevention: Fundamentals and Practice*, McGraw Hill, New York, 2000.

G. C. Cushnie, *Pollution Prevention and Control Technology for Plating Operations*, National Center for Manufacturing Sciences, Ann Arbor, 1994.

P. Modak, *Waste Minimization: A Practical Guide to Cleaner Production and Enhanced Profitability*, Center for Environmental Education, Ahmedabad, India, 1995.

P. M. Modak, C. Visvanathan, and M. Parasnis, "Cleaner Production Audit," *ENSIC Review*, **32** (1995).

N. L. Nemerow, *Zero Pollution for Industry: Waste Minimization through Industrial Complexes*, John Wiley, New York, 1995.

S. Sorrell and J. Skea, *Pollution for Sale: Emission Trading and Joint Implementation*, MPG Books Ltd., Cornwall, UK, 1999.

CHAPTER 6

REVERSE PRODUCTION SYSTEMS: PRODUCTION FROM WASTE MATERIALS

I-Hsuan Hong
Department of Industrial Engineering and Management, National Chiao Tung University Taiwan

Jane C. Ammons
School of Industrial and Systems Engineering, Georgia Institute of Technology, Atlanta, Georgia

Matthew J. Realff
School of Chemical and Biomolecular Engineering, Georgia Institute of Technology, Atlanta, Georgia

1 INTRODUCTION 155

2 BACKGROUND 157
  2.1 Interdependence, Viability, and Growth of Collection and Processing Networks 157
  2.2 Uncertainties and Global Material Flows 159

3 STRATEGIC DESIGN MODELS 161

4 EXPERIMENTAL COMPARISONS 167

5 SUMMARY AND CONCLUSIONS 172

6 ACKNOWLEDGMENTS 172

"—We will recycle most effectively if our recycling programs are based on sound policy and designed for the right reasons."[1]

1 INTRODUCTION

Waste is a resource in the wrong place. It is estimated that 133,000 electronic devices are discarded daily in the United States, amounting to 3 million tons of e-scrap per year.[2] As a corresponding material and energy savings resource, the U.S. Geological Survey estimates that 1 metric ton of computer scrap contains more gold than 17 tons of ore.[3] Reverse production systems that support the recovery, processing, and resale of materials and subcomponents at the end of

155

their useful life provide significant environmental and legal compliance benefits while forming an important part of the U.S. and world economy. There are estimated to be 73,000 firms employing 480,000 people, generating revenues of US$53 billion/year in remanufacturing alone.[4] Recycling and remanufacturing can achieve significant gains in overall energy and material efficiency of a product life cycle. Worldwide remanufacturing activities annually save an estimated 120 trillion Btus.[5]

Reverse production systems (RPSs) to handle return flows of supply chain and production wastes, packaging, and end-of-life products are evolving rapidly due to increasing pressures from economic opportunities, legislation requirements, and global environmental concerns. The first of these, economic impact, is significant. A spokesperson for one industry, pertaining to the recycling and reuse of used electronics (e-scrap, e-waste, e-cycling), states it well:

> 'We see e-waste as a resource conservation issue. We think that keeping (consumer electronics) products out of landfills is important because most often they are too valuable to be there.'[6]

To understand the magnitude of the financial opportunity, considers the flow of materials among the many entities representing the RPS depicted in Figure 1. The scale and *economic development opportunity* of RPSs are enormous. In the United States, reverse logistics costs amounted to approximately $35 billion in 1997. Because Rogers and Tibben-Lembke[7] estimated that in the United States approximately 4 percent of logistics costs are spent on returns, and Bowersox and Calantone[8] give a conservative estimate of worldwide logistics costs for 1996 to be $3.43 trillion, then we estimate worldwide reverse logistics cost were in the neighborhood of $137.2 billion in 1996.

The wide-scale reuse and recycling of products to avoid the disposal of concentrated materials is a strategy that both intensifies material use and reduces disposal. However, for this strategy to reduce the net consumption of resources, the recycling system must consume less resources, material energy, and money than producing the material from virgin sources. Thus, enhanced recycling requires two types of development. First, advances in process technology to lower costs and energy input of individual material transformations will enhance overall product and material lifecycle. Second, and perhaps more importantly, the overall system configuration, the connection of discarded products to different points in new product manufacturing, will ultimately determine life-cycle efficiency and overall economic viability. This chapter is motivated by the goal of increasing the material and energy efficiency of product life cycles while maximizing economic opportunity through the design of effective, financially viable, and long-lasting collection and processing systems.

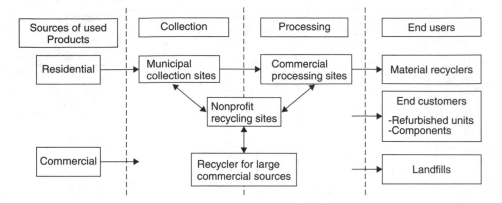

Figure 1 Material flows in a reverse production system. From Ref. 29

2 BACKGROUND

In this section we outline key factors and trends that are shaping collection systems that feed networks of processors for remanufacturing and recycling. Then we review the literature and supporting methodology.

2.1 Interdependence, Viability, and Growth of Collection and Processing Networks

Figure 2 illustrates the material flows and associated financial impacts for a generic RPS. For most industries and product flows, geographic distribution of used product supply and technological processing requirements dictate numerous collection agents (or locations) to provide sufficient material to support the economy of scale required by a much fewer number of processing agents. For example, in used carpet recycling we might need hundreds of collection locations to support one separation and depolymerization facility. There is a *chicken-and-egg* dependency between the collection and processing components: rapid growth of collection volume capacity is required to reach viable production capacity for each processor, and vice versa. This means that to achieve a stable and long-term surviving system, there is a need to concentrate the flows from several collection agents in order to feed the volume requirements for each processor.

However, conflicting objectives make it difficult to achieve a balanced system. Objectives may differ for each collector, each process, and different nongovernmental organizations (NGOs) (such as industry consortia), and these may all differ from governmental objectives, which include economic development and environmental protection. The objectives of the different entities relative to capacity and profitability may differ in the short term versus the long term. For example, companies from China are currently buying a significant amount of

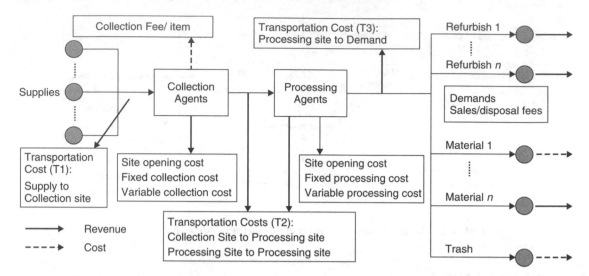

Figure 2 Material flows and financial impacts. (From Ref. 56.)

used carpet from U.S. sources and capitalizing on cheap backhaul transportation costs to ship it to China for depolymerization to raw material. Chinese plants are simultaneously building new, higher-quality polymerization capacity (for processing input streams from virgin crude oil). When these new-technology Chinese factories come online, the U.S. used-carpet collection system will have lost a major market. The loss threatens the long term viability of the U.S. system.

It is clear that for each entity in an RPS, there are trade-offs among profitability, market/supply reliability, and operational resilience and perseverance. To survive and thrive, over time each agent must find a way to grow capacity and find/expand markets while being robust to uncertainties or changes in transportation costs, technology, markets, and product obsolescence cycles. The collection system must manage network growth "smartly," by investing wisely in the recruitment of collection sites, and then sustain their retention. Recruitment decisions play out over time, regions, and among market segments (e.g., multifamily dwelling for used nylon 6 carpet collection due to higher facility turnover). Recruitment strategies range from adding individual sites (e.g., a carpet sales location) to large-scale additions (e.g., adding a retailer system, like all Walmart or BestBuy locations). In our initial work to model decisions associated with growing a collection network, we have discovered a pattern whereby the collection system can fix two dimensions from among (1) recruitment budget, (2) region size (e.g., number of retailers within), and (3) target collection volume capacity, and then determine the third.[9]

Strategic and operational decisions affect the retention of collection network entities in the face of their defection opportunities to other regional processors and other materials and global markets. Retention may depend on many factors, such as system service levels (e.g., allocation of trucks, logistics, and inventory storage capabilities), profitability, and economies of scales (e.g., volume that justifies a baler). Recycling network infrastructure and logistics are critical to financial viability and long-term survival, because in many systems the transportation costs can typically compose about half of the overall system cost. As fuel prices and corresponding transportation costs rise in the future, efficient logistics may be even more significant to RPS viability. Processing capabilities are driven by technological developments, product life cycles, and end-use markets. A key feature of many processing systems is the ability to sort and extract high-value items while removing toxic/damaging items (e.g., used PVC tile left in a used latex-backed carpet stream can increase corrosion in extrusion equipment). Investment costs may be significant for processing equipment as technological upgrades become available.

2.2   Uncertainties and Global Material Flows

Each of the different entities, as well as segments of the collection and processing network, faces critical uncertainties or ambiguities. These may occur at multiple scales. A key near-term uncertainty is the price of oil, and its resulting impact on transportation costs, raw materials cost for the forward production system, and end product sales prices for the reverse system. Another critical ambiguity is the collection network agent behavior associated with competition versus cooperation at the local and regional levels. Similar uncertainties can be faced by processor agent networks at the regional and global levels, due to their larger capacity and economic scale. All of these factors are influenced by ambiguity in technological change and changes in world markets and material substitutability.

Key questions exist in how to structure the relationships among the agents in the collection and processing network. There are trade-offs between private (e.g., each company builds its own collection network) versus public (industry collaboration or government collection network) endeavors. Current and pending governmental legislation may affect the feasibility, scope, and transaction capability of these relationships. One key question faced by a growing collection network is related to the enhanced economy of scale induced by the growth—how will the gains be shared among the entities in the network as more are recruited?

Models for addressing many of these questions do not currently exist. In the following sections, we outline the current state of research for these problems as a foundation for the research tasks that follow.

Reverse Logistics—Network Design Models
Broad overviews of models for RPS design are given by Dowlatshahi, Flapper, and Fleischmann et al..[10-15] Carter and Ellram, Dowlatshahi, and Ferguson

and Browne provide a review of the literature on reverse logistics.[16–18] Models to support decisions around the location and allocation of tasks to recycling sites and flows between and within sites for general problems are developed in Ammons et al. and Guide et al.[19,20] Spengler et al. propose a location-allocation model in determining infrastructure of reclamation facilities and demanufacturing plants.[21] Examples of logistics network models include Barros et al., Chouinard et al., Kusumastuti et al., Marin and Pelegrin, Pati et al., and Thierry.[22–27] Like many realistic problem, the real-world design of RPS infrastructure is large-scale as shown in Realff et al. and Spengler et al.[21,28] With challenges in problem scale, the decompositions and relaxation methodologies is applied to the RPS strategic design in Pas et al., using concepts from Dantzig and Wolfe, Geoffrion, Lasdon, Mulvey and Crowder, and Olaf et al.[29–34] The concept of clustering in determining collection network is also applied in Pas et al.[29] In evaluating the RPS models, Hong et al. construct case studies of electronics.[35] Key results on recycling and resource recovery for materials such as paper, plastics, steel, iron, and sand include Huttunen, Pohlen and Farris, Wang et al., Spengler et al., Russell and Vaughan, and Barros et al., respectively.[21,22,36–39]

Decision Making for Reverse Production Systems—Decentralized Framework

Most RPS design efforts are from the perspective of a centralized system such as Ammons et al., Assavapokee et al., Barros et al., and Shih.[19,22,40,41,42] The drawbacks of the centralized system are discussed in Wang et al.[43] Models and designs for decentralized forward supply chains can be found in sources such as Bernstein and Federgruen, Fan et al., Lee and Whang, Anupindi et al., Granot and Sosic, Wang et al., Nagurney et al., and Walsh and Wellman.[43–50] A growing number of papers that address the modeling of independent decision-making processes in reverse supply chains, especially the interaction between pricing decisions and material flow volume, such as the work of Guide et al., Ferguson and Toktay, Majumder and Groenevelt, Savaskan et al., Corbett and Karmarkar, and Hong et al.[54–56]

Nagurney and Toyasaki and Hong et al. present multitiered network models for decentralized reverse logistics systems.[57,58] Nagurney and Toyasaki construct an equilibrium model that yields the material flow solution and endogenous prices of recycled materials.[57] Hong et al. explicitly proposed a decentralized decision making and protocol design of recycled material flows in a multitiered RPS constituted by several privately owned entities in different tiers within the network.[58] Furthermore, Hong et al. contrast potential outcome differences between centralized and decentralized system models and examine government-subsidized effects on price and flow decisions in a decentralized RPS.[59,60]

Supply Chain Coordination—Growing Network and Collection Recruitment

A supply chain is defined as a system of suppliers, manufacturers, distributors, retailers, and customers where material, financial, and information flows connect participants in both (forward and reverse) directions.[61] In creating coordination, Giannoccaro and Pontrandolfo discuss ideas to make the chain behave as if it operated in a centralized fashion.[62] A framework of control and coordination mechanisms is described in Kumar and Wainer.[63] Integrated analyses of production, distribution, and inventory planning can be found in Bhatnagar et al. and Thomas and Griffin.[64,65] The importance of the coordination and integration mechanisms is discussed by Romano.[66] Wongthatsanet Korn addresses the complex set of decisions that surround the growth of reverse supply chain networks over time.[9] This work uses simulation-based optimization and dynamic programming algorithms to solve multistage recruitment problems in reverse production networks. The decision levels are strategic, tactical, and operational. The operational problem deals with the recruiting of agents in the local scale. The strategic problem deals with the growth in the long term. Simulation-based optimization concepts can be found in Abdelfatah and Mahmassani, Benyoucef et al., Fu, Gosavi and Subramaniam, and Homen-de-Mello et al.[67–71] Dynamic programming concepts can be found in Bertsekas and Tsisiklis, Ross, Rummery and Niranjan, Sutton, and Watkins.[72,76]

3 STRATEGIC DESIGN MODELS

An RPS is a network of transportation logistics and processing functions that collect, recycle, refurbish, and demanufacture end-of-life products. In this chapter, we model the RPS as a multitiered network, depicted in Figure 3, which consists of an upstream boundary tier, several intermediate tiers, and a downstream boundary tier. We consider N_1 entities in the upstream boundary tier as represented by the top tier of nodes in Figure 3, $N_2, \ldots, N_{M-1}$ entities in intermediate tiers $2, \ldots, M - 1$ respectively, and N_M downstream boundary tier entities associated with the bottom tier in the network. In addition, we let sources of recycled products and demand markets be the two end exogenous tiers of the network, which may be represented as several independent and possibly geographically distinct sources of end-of-life products and demand markets for secondary used products or raw materials.

Typical upstream boundary tier entities represent municipal collection sites, nonprofit collection organizations, and private collectors, for example. The entities in the upstream boundary tier collect recycled materials from the source supply, which can include, for example, residential households, businesses, schools, or the government. The amount collected depends on the collection fee between the upstream boundary tier and the source. We note that the sites in the upstream boundary tier—specifically, collectors in the e-scrap recycling industry—may pay or charge for collecting or processing recycled items. The intermediate tiers may contain several levels of entities. These include the tier of consolidation sites and material brokers. They also include processing sites that bid for collected items from their preceding tier and conduct some value-added processes

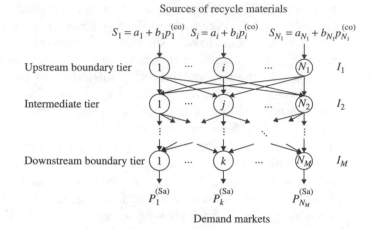

Sources of recycle materials

Figure 3 A general multitiered RPS network structure. (From Ref. 58.)

such as sorting or disassembling operations, or simply act as an intermediary broker between tiers. Downstream boundary tier entities associated with nodes in the bottom tier in the network can be seen as the final stage of the entire RPS, where they purchase recycled items from their preceding tier and conduct further dismantling/mechanical fragmentation of items or refurbish end-of-life products for consumption purposes. Hence, downstream boundary tier entities may convert the recycled items into raw materials or refurbished products and sell them to the specific demand markets. In general, recycled items flow from the upstream tier to the downstream tier of entities, but financial incentives are driven from the downstream tier back to the upstream tier of entities. For simplicity, we assume that materials must move through each tier sequentially, and may not be transported directly across two or more tiers within the network.

We let $I_m = \{1, \cdots, j, \cdots, N_m\}$ denote the set of sites in tier m. The entities in the upstream boundary tier collect recycled items from the source, and the source supplies the upstream boundary tier site on the basis of a fee paid by the upstream boundary tier site. We let S_i denote the collection amount in upstream boundary tier site $i \in I_1$ and $p_i^{(Co)}$ be the collection fee per unit of the recycled item paid by site $i \in I_1$. We characterize the collection amount for the upstream boundary tier site $i \in I_1$ by a linear function $S_i = a_i + b_i p_i^{(Co)}$, where a_i and b_i are parameters and $a_i, b_i > 0$. The collection fee, $p_i^{(Co)}$, of site $i \in I_1$ is without sign restriction. The use of a linear function allows the analysis of the problem to be simplified. It captures a qualitative market behavior of increased flow with either an increased payment or a decreased collection fee (charged by the collector). In other words, the upstream boundary tier site may *pay* or *charge* for collecting recycled items if $p_i^{(Co)}$ is positive or negative, respectively.

At the other end of the system, we assume that the amount of raw materials resulting from the decomposition of end-of-life products and used products is relatively small compared to the quantity in the virgin raw material and brand-new product markets. This observation leads to the assumption that the selling prices of raw materials or used products in final demand markets are fixed amounts, not affected by the sales quantities. We let $P_k^{(Sa)}$ denote the selling price obtained in downstream boundary tier site $k \in I_M$. In other words, the two main exogenous information streams to the system are the source supply functions of the collected recycled item amount in the upstream boundary tier, represented by $S_i = a_i + b_i p_i^{(Co)}$ for site $i \in I_1$, and the selling price obtained in the downstream boundary tier, denoted by $P_k^{(Sa)}$ for site $k \in I_M$. Here, a_i, b_i, and $P_k^{(Sa)}$ are known parameters, but S_i and $p_i^{(Co)}$ are unknown variables of the system.

Decentralized Reverse Production Systems

The principle of the *decentralized* set-up is that the network system is composed of several independent entities individually operated by self-interested parties. Each independent entity has its own profit function subject to its own processing or transportation constraints, and is not willing to reveal its own information to other entities or the public. Often the decision variables for each entity in a decentralized system are also influenced by other entities' decisions. The foundations of the decentralized RPS models are derived from our recent work in a multitiered RPS network.[56,58,59] Using this decentralized RPS network framework, we obtain the equilibrium collection fee paid by the upstream boundary tier site and the resulting material flow allocation, as well as the internal transaction prices within the network.

We assume the upstream tier designs the price-flow contract, shown in the equation (1), below which is a mechanism describing the correspondence between the acquisition prices offered by downstream sites and the flow amount supplied by the upstream sites to its subsequent downstream sites. The material flow from site $i \in I_m$ to site $j \in I_{m+1}$, denoted by $x_{ij}^{(Tr)}$, is a function of the acquisition prices, represented by p_j, $j \in I_{m+1}$, to be offered by the sites in tier $m + 1$. We let $V_{ij}^{(Tr)}$ denote the unit transportation cost from site $i \in I_m$ to site $j \in I_{m+1}$. The format of the price-flow contract implies that any particular arc of material flows is not only a function of the price offered by its destination downstream site, but also the relative price offers of other downstream sites.

$$x_{ij}^{(Tr)} = \sum_{j' \in I_{m+1}} \alpha_{ijj'}(p_{j'} - V_{ij'}^{(Tr)}) \quad \forall i \in I_m, j \in I_{m+1}, m = 1, \cdots, M - 1 \quad (1)$$

The downstream tier determines the equilibrium prices to acquire the recycled items from its preceding upstream tier. We refer to the price-flow contract as the *flow function*. We assume that the transportation cost for the shipment of the recycled item between any two tiers is paid by the downstream tier entity. The price the downstream tier entity pays for transportation is taken into account by the upstream tier entity in the flow function. In this chapter, we specifically focus on the flow of valuable items transacted within the network; as a result, we assume the prices obtained in entities from their subsequent tier are positive. Decisions of the entities in intermediate tiers are the flow functions for their subsequent downstream tier and the acquisition prices for their preceding upstream tier, but the decisions in the upstream and downstream boundary tiers are slightly different from intermediate tiers. The entity in the upstream boundary tier determines the optimal collection fee to obtain end-of-life products and communicates the flow function to the subsequent tier. We also assume that the amount of raw materials resulting from the decomposition of end-of-life products is small relative to the quantity of available virgin raw materials and brand-new products in the market, and hence does not affect market price. As a result, the entity in the downstream boundary tier decides the acquisition price for its preceding upstream tier.

The decision timeline for an M-tiered problem is shown in Figure 4, where the upper arrows indicate the entity tasks and the lower arrows show the information disclosure timeline. The flow functions are independently designed by the upstream tier sites and communicated to the subsequent downstream tier sites. The algorithm starts at the sites in tier 1 to determine the flow function between itself and the second tier sites, given the source supply functions, which describe the variation of the collected amount with the collection fee between the sites in tier 1 and the sources. The sites in tier 1 communicate flow functions to the sites in intermediate tier 2. Each intermediate tier site independently determines the associated flow functions and communicates them to its next tier sites. This proceeds sequentially until the last tier is reached. The sites in the downstream

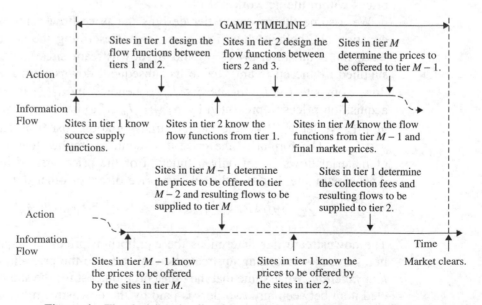

Figure 4 The decision timeline for an M-tiered problem. (From Ref. 58.)

boundary tier determine the equilibrium acquisition price on the basis of the flow functions given by the sites in the preceding tier and the final market price. This completes the upper part of Figure 4. Then, the resulting flow into the downstream boundary tier site can be obtained by substituting the equilibrium price into the flow function.

Acquisition prices are set by the downstream tier and passed back to the upstream tier sequentially from the downstream to upstream boundary tier, as shown in the lower part of Figure 4. In this chapter, we focus on a single horizon of the transaction problem where entities do not collect items in excess of the amount they need to supply to the next tier. Due to this flow conservation rule, the resulting flows can be determined as the acquisition prices are realized. Finally the sites in the first tier decide the collection fees to acquire recycled items from sources.

In the decentralized decision-making framework, each entity within the RPS concentrates on optimizing its own profit subject to its own transportation and processing capacity constraints.[56,58] The upstream entities in one tier provide the price-flow contract that connects the downstream price information to the flow they will provide. We refer to this price-flow contract as the flow function. Each upstream entity acts individually to determine the flow function used to contract with each member of the next tier. The flow function is determined using a robust optimization formulation that captures the idea that the upstream entity does not have exact price information from the downstream entities, and wants to minimize the worst outcome it can have.

The downstream tier sites are assumed to reveal their bids for the items from the preceding tier until they have no incentive to change them. This allows a Nash equilibrium to be reached within the tier.[58] The algorithm for finding this equilibrium respects the structure of the system by only having the previous bids of each entity available for inspection when the next bid is being determined by each independent entity. Within this framework, entities in the system reach the equilibrium of the acquisition prices, as well as the resulting material flow allocation in the network. The decentralized model contains this set of *internal* equilibrium acquisition prices, which are not present in the centralized problem setting. In summary, the decentralized model solves for the equilibrium collection fee and resulting material flow, while each entity determines its own associated decision variables of acquisition prices and price-flow contract mechanism.

Centralized Reverse Production Systems

As an alternative to a decentralized system, consider a setup in which management is *centralized*. A single decision maker (e.g., the state or local government) has the requisite information about all the participating entities and seeks the optimal solution for the entire system. The underlying assumption of the centralized problem setting is that the decision maker has the authority to manage associated operations or processes of all entities within the network. In a centralized set-up, the decision maker determines the optimal level of the collection amount from the source, and the most efficient way of material flow allocation through the network, so that the system net profit is maximized. In addition, there are some internal transaction variables among entities in the network such as internal transaction prices; however, these are not relevant in the centralized setting. With these assumptions, the centralized model finds the optimal collection fees and the material flow allocation so that the system profit function is maximized subject to the individual entity and system constraints.

We let $x_{ij}^{(\text{Tr})}$ denote the material flow from site $i \in I_m$ to site $j \in I_{m+1}$ and $x_k^{(\text{Tr})}$ denote the aggregate flow shipped from the sites in tier $M - 1$ to downstream boundary tier site $k \in I_M$. The decision variables for the system are the optimal material flow allocation, $x_{ij}^{(\text{Tr})}$ and $x_k^{(\text{Tr})}$, within the system and the collection fee, $p_i^{(\text{Co})}$, in the upstream boundary tier. We also define the following system parameters:

$V_{ij}^{(\text{Tr})}$ Transportation cost per standard unit from site i to site j

$C_{ij}^{(\text{Tr})}$ Maximum amount of materials that can be shipped from site i to site j

$C_i^{(\text{Pr})}$ Maximum amount of materials can be processed in site i

These parameters are known to the decision maker. The centralized RPS optimization model for the entire system can be stated as follows:

Maximize

$$\sum_{k \in I_M} x_k^{(\text{Tr})} P_k^{(\text{Sa})} - \sum_{i \in I_1} p_i^{(\text{Co})}(a_i + b_i p_i^{(\text{Co})}) - \sum_{m=1}^{M-1} \sum_{i \in I_m} \sum_{j \in I_{m+1}} V_{ij}^{(\text{Tr})} x_{ij}^{(\text{Tr})} \qquad (2)$$

Subject to

$$\sum_{j \in I_2} x_{ij}^{(\text{Tr})} = a_i + b_i p_i^{(\text{Co})} \qquad \forall i \in I_1 \qquad (3)$$

$$\sum_{i \in I_{m-1}} x_{ij}^{(\text{Tr})} = \sum_{k \in I_{m+1}} x_{jk}^{(\text{Tr})} \qquad \forall j \in I_m, \forall m = 2 \cdots M - 1 \qquad (4)$$

$$\sum_{j \in I_{M-1}} x_{jk}^{(\text{Tr})} = x_k^{(\text{Tr})} \qquad \forall k \in I_M \qquad (5)$$

$$x_{ij}^{(\text{Tr})} \le C_{ij}^{(\text{Tr})} \qquad \forall i \in I_m, j \in I_{m+1}, \forall m = 1 \cdots M - 1 \qquad (6)$$

$$\sum_{j \in I_{m+1}} x_{ij}^{(\text{Tr})} \le C_i^{(\text{Pr})} \qquad \forall i \in I_m, \forall m = 1 \cdots M - 1 \qquad (7)$$

$$\sum_{j \in I_{M-1}} x_{jk}^{(\text{Tr})} \le C_k^{(\text{Pr})} \qquad \forall k \in I_M \qquad (8)$$

$$x_{ij}^{(\text{Tr})} \ge 0 \qquad \forall i \in I_m, j \in I_{m+1}, \forall m = 1 \cdots M - 1 \qquad (9)$$

$$x_k^{(\text{Tr})} \ge 0 \qquad \forall k \in I_M. \qquad (10)$$

The objective function (2) maximizes the system net profit, which is the sum of the sales profit from the destination demand markets, collection fees incurred between the upstream boundary tier and sources, and transportation costs of all shipments through the system. Constraints (3), (4), and (5) are the flow conservation among sites within the network. Constraints (6), (7), and (8) are the transportation and processing capacity limitations, respectively. We also intuitively require all material flow variables, x, to be nonnegative in the centralized model in constraints (9) and (10).

The centralized model has a concave quadratic objective function and a convex constraint set since we require b_i to be positive and the model itself is subject to a linear constraint set. Several algorithms can be used to solve quadratic programming problems.[77] Constraint (3) specifies the recycled item amount from sources to the system. The volume between the source and the upstream boundary tier site is increasing as the upstream boundary tier site increases the collection fee. Obviously, because the total amount collected is a linear function of the unit collection fee, the corresponding unit fee must be large when a large amount is collected. Consequently, recycled items flowing into the system are limited to either the system capacity itself or the optimal acquisition amount determined by the concave quadratic net profit objective function. In the latter case, the system limits its input because the marginal cost of acquiring more flow exceeds the marginal value derived from it. In the following sections, we investigate numerical results of the comparison between the centralized and decentralized approaches.

4 EXPERIMENTAL COMPARISONS

An example, depicted in Figure 5, demonstrates the mathematical behavior of the centralized and decentralized models and provides several insights to compare these two models. We follow the numerical example presented in Hong et al. to compare the system and individual behaviors of the RPS network in the centralized and decentralized problem settings.[58,59] We consider a three-tier RPS with collection, consolidation, and processing sites. There are five collection sites, $i = 1, \ldots, 5$, in tier 1, three consolidation sites, $j = 1, 2, 3$, in tier 2, and four processing sites, $k = 1, \ldots, 4$, in tier 3. The transportation costs per unit flow between any two associated sites are given in Table 1.

The final market prices for processing sites, $k = 1, \ldots, 4$, are \$155, \$145, \$147, and \$150, respectively. The collection amount functions in collection sites, $i = 1, \ldots, 5$, are given by $S_1 = 400 + 5p_1^{(Co)}$, $S_2 = 420 + 6p_2^{(Co)}$, $S_3 = 440 + 6p_3^{(Co)}$, $S_4 = 430 + 6p_4^{(Co)}$, and $S_5 = 410 + 5p_5^{(Co)}$. We consider two cases of capacitated and uncapacitated settings for the arc transportation and site processing capacities. In the capacitated case, we limit the arc transportation capacity to 200 units, the collection site capacity to 600 units, the consolidation site capacity to 800 units, and the processing site capacity to 800 units.

The centralized model solution is derived from solving the quadratic programming problem presented in Section 3 and the decentralized model solution is obtained using solution methodology described in our previous work.[58] We

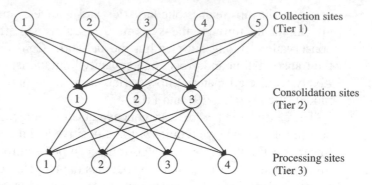

Figure 5   The reverse production system for the example. (From Ref. 59.)

Table 1   The Unit Transportation Costs between Sites

Unit Transportation Cost		$j \in I_2$			Unit Transportation Cost		$k \in I_3$			
		1	2	3			1	2	3	4
	1	10.0	15.0	18.0		1	8.0	8.0	10.0	12.0
	2	10.0	13.0	16.0	$j \in I_2$	2	10.0	8.0	7.0	11.0
$i \in I_1$	3	13.0	10.0	14.0		3	12.0	10.0	8.0	7.0
	4	15.0	13.0	11.0						
	5	17.0	14.0	9.0						

Note: From Ref. 59.

examine the decision variables of the optimal collection fees paid by collection sites, $i = 1, \ldots, 5$, in tier 1 and the material flow allocations within the network as well as the net profit values for the centralized and decentralized problems. Figure 6, Figure 7, Table 2, and Table 3 summarize the numerical solutions of the net profits, collection fees, and material flow allocations for centralized and decentralized problems in capacitated and uncapacitated cases.

There are several insights that can be drawn from the numerical results of this example. As expected, both capacitated and uncapacitated cases show that the net profit of the centralized model outperforms the net profit of the decentralized model. The net profit of the centralized model serves an upper bound on the net profit of the decentralized model for both the capacitated and uncapacitated cases. For the centralized model, the net profit of the capacitated case is bounded by arc capacities of transportation, but the net profit of the uncapacitated case is constrained by the first-order condition of its quadratic concave objective function.

Figure 7 indicates that the collection sites in the centralized problem pay a positive collection fee to sources to acquire recycled items, but sources pay the collection fee to collection sites for discarding end-of-life products in the decentralized problem. This implies that the centralized approach acquires more

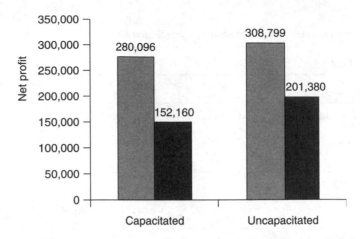

Figure 6   Net profits of centralized (■) and decentralized (■) models. (From Ref. 59.)

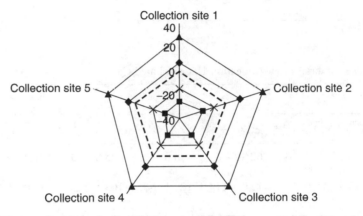

Figure 7   The collection fees of the decentralized-capacitated (■), decentralized-uncapacitated (×), centralized-capacitated (♦), centralized-uncapacitated (▲) cases, and the zero reference line (----). The negative collection fee indicates the collection site charges the sources a positive fee for collecting items. (From Ref. 59.)

recycled items compared to the decentralized problem. Moreover, the net profit ratios of the decentralized to centralized problem settings are 54.4 percent and 65.2 percent in the capacitated and uncapacitated cases, respectively. In other words, especially in the capacitated case, one may overestimate the system profit and/or the volume of recycled items processed by the system if it is assumed that the decisions are made centrally in a system of independent entities.

Our results also capture the notion of *double marginalization* of the vertical supply chain where two independent firms, upstream and downstream, may end up with lower profits in the decentralized setting.[78] The decentralized model also considers *price ambiguity* since the price information is not revealed between

Table 2   The Material Flow Allocation of Capacitated Case

$x_{ij}^{(Tr)} : i \in I_1, j \in I_2$

Flows	$x_{11}^{(Tr)*}$	$x_{12}^{(Tr)*}$	$x_{13}^{(Tr)*}$	$x_{21}^{(Tr)*}$	$x_{22}^{(Tr)*}$	$x_{23}^{(Tr)*}$	$x_{31}^{(Tr)*}$	$x_{32}^{(Tr)*}$	$x_{33}^{(Tr)*}$	$x_{41}^{(Tr)*}$	$x_{42}^{(Tr)*}$	$x_{43}^{(Tr)*}$	$x_{51}^{(Tr)*}$	$x_{52}^{(Tr)*}$	$x_{53}^{(Tr)*}$
Centralized	200.0	120.7	116.0	200.0	153.1	147.0	179.2	200.0	136.9	179.0	126.1	200.0	41.8	200.0	200.0
Decentralized	79.7	99.7	90.7	106.4	107.3	90.7	101.7	110.3	94.3	90.4	107.3	107.6	90.3	106.0	80.6

$x_{jk}^{(Tr)} : j \in I_2, k \in I_3$

Flows	$x_{11}^{(Tr)*}$	$x_{12}^{(Tr)*}$	$x_{13}^{(Tr)*}$	$x_{14}^{(Tr)*}$	$x_{21}^{(Tr)*}$	$x_{22}^{(Tr)*}$	$x_{23}^{(Tr)*}$	$x_{24}^{(Tr)*}$	$x_{31}^{(Tr)*}$	$x_{32}^{(Tr)*}$	$x_{33}^{(Tr)*}$	$x_{34}^{(Tr)*}$
Centralized	200.0	200.0	200.0	200.0	200.0	200.0	200.0	200.0	200.0	200.0	200.0	200.0
Decentralized	106.6	126.4	131.6	103.9	148.0	118.3	123.8	140.7	109.8	112.1	125.4	116.6

Note: From Ref. 59.

Table 3   The Material Flow Allocation Uncapacitated Case

$x_{ij}^{(Tr)} : i \in I_1, j \in I_2$

Flows	$x_{11}^{(Tr)*}$	$x_{12}^{(Tr)*}$	$x_{13}^{(Tr)*}$	$x_{21}^{(Tr)*}$	$x_{22}^{(Tr)*}$	$x_{23}^{(Tr)*}$	$x_{31}^{(Tr)*}$	$x_{32}^{(Tr)*}$	$x_{33}^{(Tr)*}$	$x_{41}^{(Tr)*}$	$x_{42}^{(Tr)*}$	$x_{43}^{(Tr)*}$	$x_{51}^{(Tr)*}$	$x_{52}^{(Tr)*}$	$x_{53}^{(Tr)*}$
Centralized	542.5	0	0	621.0	0	0.0	0	625.0	0	191.2	185.3	234.5	0	0	540.0
Decentralized	110.5	99.0	115.1	121.7	114.4	129.1	140.1	103.1	123.6	139.0	93.3	134.1	119.9	101.1	109.8

$x_{jk}^{(Tr)} : j \in I_2, k \in I_3$

Flows	$x_{11}^{(Tr)*}$	$x_{12}^{(Tr)*}$	$x_{13}^{(Tr)*}$	$x_{14}^{(Tr)*}$	$x_{21}^{(Tr)*}$	$x_{22}^{(Tr)*}$	$x_{23}^{(Tr)*}$	$x_{24}^{(Tr)*}$	$x_{31}^{(Tr)*}$	$x_{32}^{(Tr)*}$	$x_{33}^{(Tr)*}$	$x_{34}^{(Tr)*}$
Centralized	1354.7	0	0	0	810.3	0	0	0	386.5	0	0	388.1
Decentralized	212.5	152.0	120.7	145.9	148.6	122.3	120.7	119.4	159.2	125.7	140.0	186.9

Note: From Ref. 59.

two independent entities or to the public.[58] Price ambiguity in decentralized problems is another factor, which leads to the difference of net profit values in centralized and decentralized settings. Price ambiguity essentially plays a more critical factor in the capacitated case since the price-flow dependence is sensitive to the flow capacity, which is limited in the capacitated case. As a result, we make the observation that the net profit value difference between the centralized and decentralized problems in the capacitated case is larger than the net profit value difference in the uncapacitated case.

The collection fees paid by the upstream boundary tier sites are not determined in the centralized problem. As a next step, we examine the most efficient material flow allocation under the centralized model given the equilibrium collection fees from the decentralized model. Here we are interested in the optimal material flow allocation within the network, from a centralized perspective, given the same source amount as that in the decentralized problem. This investigation provides us with, given the same amount of source supply, the comparison between the best case the system can achieve in a centralized setting, and all independent entities can obtain in the decentralized problem setting. The decision variables of the collection fees in the centralized model are substituted by the equilibrium collection fees derived by the decentralized model. Under this setting, the total amounts of recycled items are identical in the centralized and decentralized problems, and the centralized model is essentially a linear programming model. The material flow allocations under this framework are listed in Table 4.

The net profits of the centralized model given the equilibrium collection fees under the capacitated and uncapacitated cases are 224,850 and 258,543, respectively. The optimal net profit difference between the decentralized model and the centralized model given the equilibrium collection fees can be interpreted as the system gain due to the *efficiency* of material flow allocation in the centralized problem setting. This demonstrates that the loss of surplus in the decentralized model is due both to a failure to accept the economically optimal total amount, and to inefficiently allocate it among the network participants.

Table 4 The Material Flow Allocations in the Centralized Model Given Equilibrium Collection Fees

$x_{ij}^{(Tr)} : i \in I_1, j \in I_2$

Flows	$x_{11}^{(Tr)*}$	$x_{12}^{(Tr)*}$	$x_{13}^{(Tr)*}$	$x_{21}^{(Tr)*}$	$x_{22}^{(Tr)*}$	$x_{23}^{(Tr)*}$	$x_{31}^{(Tr)*}$	$x_{32}^{(Tr)*}$	$x_{33}^{(Tr)*}$	$x_{41}^{(Tr)*}$	$x_{42}^{(Tr)*}$	$x_{43}^{(Tr)*}$	$x_{51}^{(Tr)*}$	$x_{52}^{(Tr)*}$	$x_{53}^{(Tr)*}$
Cap	200.0	70.1	0	200.0	104.4	0	0	200.0	106.3	0	105.3	200.0	0	77.0	200.0
Uncap	324.6	0	0	365.3	0	0	0	366.8	0	0	0	366.4	0	0	330.7

$x_{jk}^{(Tr)} : j \in I_2, k \in I_3$

Flows	$x_{11}^{(Tr)*}$	$x_{12}^{(Tr)*}$	$x_{13}^{(Tr)*}$	$x_{14}^{(Tr)*}$	$x_{21}^{(Tr)*}$	$x_{22}^{(Tr)*}$	$x_{23}^{(Tr)*}$	$x_{24}^{(Tr)*}$	$x_{31}^{(Tr)*}$	$x_{32}^{(Tr)*}$	$x_{33}^{(Tr)*}$	$x_{34}^{(Tr)*}$
Cap	200.0	0	0	200.0	200.0	0	200.0	156.8	200.0	0	106.3	200.0
Uncap	689.9	0	0	0	366.8	0	0	0	697.2	0	0	0

Cap: capacitated case Uncap: uncapacitated case *Note*: From Ref. 59.

5   SUMMARY AND CONCLUSIONS

There are considerable differences in the results of net profits and material flow allocations derived from the centralized and decentralized RPS models. This chapter demonstrates the comparison of the individual and system behavior between the centralized and decentralized decision making for a RPS network. We develop a centralized framework for the recycling network system where a single decision maker is acquainted with all system information, including transportation capacities, processing capabilities, and associated sales prices of recycled materials. In a centralized system, the planner also has authority to determine system decision variables of the material flow allocation throughout the entire network, and the collection fees paid by the upstream boundary tier sites to acquire recycled items from sources. The centralized RPS model presented in this chapter can be used to generate results to compare the equilibrium solution obtained from the decentralized multitiered RPS model analyzing and predicting the individual behavior of independent participants. As expected, the centralized solution is superior to the decentralized solution in terms of the net profit, especially in the capacitated case. However, most entities in recycling networks are self-interested parties instead of centrally controlled agents. Our analysis demonstrates that one may overestimate the system profit if the decision maker utilizes the centralized approach to model a decentralized RPS network. The difference of the results between the centralized and decentralized solutions is mainly attributed to price ambiguity and double marginalization, which are two common characteristics in real-world decentralized systems.

A key extension of this work is to incorporate additional types of related recycled items or conduct a more complicated network such that the materials may or may not move through all tiers sequentially. Another extension of the research in this chapter is to examine the individual or the system behavior of the *semi-centralized* or *semi-decentralized* network, which may contain several independent recycling organizations or firms and several municipal collection sites or recyclers.

6   ACKNOWLEDGMENTS

This research has been partially supported by the National Science Foundation under grants DMI#0200162 and SBE-0123532. The authors are grateful for the generous interaction and guidance provided from many industry experts, including Julian Powell of Zentech, Carolyn Phillips and the staff of Reboot, Nader Nejad of Molam, Ken Clark of MARC5R, and Bob Donaghue and Chuck Boelkins of P2AD.

REFERENCES

1. C. Miller, *Waste Age*, www.wasteage.com. Accessed on October, 14, 2005.

2. "'E-cycling' Puts New Life in Electronic Junk," http://www.msnbc.msn.com/id/10642954. Accessed in 2006.

3. U.S. Geological Survey, http://pubs.usgs.gov/fs/fs060-01/ fs060-01.pdf. Accessed 2001.

4. W. Hauser and R. Lund, "Remanufacturing: An American Resource," http://www.bu.edu/reman/RemanSlides.pdf, Boston University, 2003.

5. NC3R: National Center for Remanufacturing and Resource Recovery, "Remanufacturing Industry Timeline," www.reman.rit.edu/timeline.aspx, Rochester Institute of Technology, 2006.

6. K. Taylor, *Greenwire*, www.greenwire.com. Accessed November 2005.

7. D. S. Rogers and R. S. Tibben-Lembke, *Going Backwards: Reverse Logistics Trends and Practices*, Reverse Logistics Executive Council, Pennsylvania, 1999.

8. D. J. Bowersox and R. Calantone (1998), "Executive Insights: Global Logistics," *Journal of International Marketing*, **6**(4), 89–93 (1998).

9. W. Wongthatsanekorn, "Strategic Network Growth with Recruitment Model," Ph.D. Dissertation, Georgia Institute of Technology, 2006.

10. S. Dowlatshahi, "Developing a Theory of Reverse Logistics," *Interfaces* (May–June 2000).

11. S. Dowlatshahi, "A Strategic Framework for the Design and Implementation of Remanufacturing Operations in Reverse Logistics," *International Journal of Production Research*, **43**(16), 3455–3480 (2005).

12. S. D. P. Flapper, "On the Operational Aspects of Reuse," Proceedings of the Second International Symposium on Logistics, Nottingham, U.K., 109–118, July 11–12, 1995.

13. S. D. P. Flapper, "Logistic Aspects of Reuse: An Overview," *Proceedings of the First International Symposium on Reuse*, Eindhoven, The Netherlands, 109–118, November 11–13, 1996.

14. M. Fleischmann, J. M. Bloemhof-Ruwaard, R. Dekker, E. van der Laan, J.A.E.E. van Nunen, and L.N. van Wassenhove, "Quantitative Models for Reverse Logistics: A Review," *European Journal of Operation Research*, **103**, 1–17 (1997).

15. M. Fleischmann, H. R. Krikke, R. Dekker, and S.D.P. Flapper, "A Characterization of Logistics Networks for Product Recovery," *Omega*, **28**, 653–666 (2000).

16. C. R. Carter and L. M. Ellram, "Reverse Logistics: A Review of the Literature and Framework for Future Investigation," *Journal of Business Logistics*, **19**(1), 85–102 (1998).

17. S. Dowlatshahi, "A Framework for Strategic in Reverse Logistics," Proceedings Annual Meeting of the Decision Sciences Institute, 425–430, 2002.

18. N. Ferguson and J. Browne, "Issues in End-of-Life Product Recovery and Reverse Logistics," *Production Planning and Control*, **12**(5), 534–547 (2001).

19. J. C. Ammons, M. J. Realff, and D. E. Newton, "Decision Models for Reverse Production System Design," *Handbook of Environmentally Conscious Manufacturing*, Kluwer Academic Publishers, 2001, pp. 341–362.

20. V. Guide Jr. R. Daniel V. Jayaraman, R. Srivastava, and W. C. Benton, "Supply-Chain Management for Recoverable Manufacturing Systems," *Interfaces* (May–June 2000).

21. T. Spengler, H. Puchert, T. Penkuhn, and P. Rentz, "Environmental Integrated Production and Recycling Management," *The Journal of Polymer-Plastics Technology and Engineering*, **97**(2), 308–326 (1997).

22. A. I. Barros, R. Dekker, and V. Scholten, "A Two-Level Network for Recycling Sand: A Case Study," *European Journal of Operational Research*, **110**, 199–214 (1998).

23. M. Chouinard, S. D'Amours, and A.-K. Daoud, "Integration of Reverse Logistics Activities within a Supply Chain Information System," *Computers in Industry*, **56**(1), 105–124 (2005).

24. R. D. Kusumastuti, R. Piplani, G. H. Lim, "An Approach to Design Reverse Logistics Networks for Product Recovery," *Proceedings of the 2004 IEEE International Engineering Management Conference*, 3, Vol. 3, 1239–1243, 2004.

25. A. Marin and B. Pelegrin, "The Return Plant Location Problem: Modelling and Resolution," *European Journal of Operational Research*, **104**, 375–392 (1998).

26. R. K. Pati, P. Vart, and P. Kumar, "Cost Optimisation Model in Recycled Waste Reverse Logistics System," *International Journal of Business Performance Management*, **3–4**(6), 245–261 (2004).

27. M. C. Thierry, "An Analysis of the Impact of Product Recovery Management on Manufacturing Companies," Ph.D. Dissertation, Erasmus University, Rotterdam, The Netherlands, 1997.

28. M. J. Realff, J. C. Ammons, D. E. Newton, "Robust Reverse Production System Design for Carpet Recycling," *IIE Transactions*, **36**, 767–776 (2004).

29. J. W. Pas, J. C. Ammons, M. J. Realff, "Effective Mathematical Programming Formulations for Reverse Production System Design," working paper, Georgia Institute of Technology, 2006.

30. G. B. Dantzig and P. Wolfe, "Decomposition Principle for Linear Programs," *Operations Research*, **8**(1), 101–111 (1960).

31. A. M. Geoffrion, "Large-Scale Linear and Nonlinear Programming," in *Optimization Methods for Large-scale Systems... with Applications*, D.A. Wismer (ed.), McGraw-Hill, New York, 1971, 47–74.

32. L. S. Lasdon, *Optimization Theory for Large Systems*, Dover Publications, New York, 1970.

33. J. M. Mulvey and H. P. Crowder, "Cluster Analysis: An Application of Lagrangian Relaxation," *Management Science*, **25**(4), 329–340 (1979).

34. E. F. Olaf, H. G. Alexander, and R. Kan, "Decomposition in General Mathematical Programming," *Mathematical Programming*, **60**, 361 (1993).

35. I.-H. Hong, T. Assavapokee, J. C. Ammons, C. Boelkins, K. Gilliam, D. Oudit, M. J. Realff, J. M. Vannicola, and W. Wongthatsanekorn, "Planning the E-Scrap Reverse Production System under Uncertainty in the State of Georgia: A Case Study," accepted, *IEEE Transactions on Electronics Packaging Manufacturing* (2006a).

36. A. Huttunen, "The Finnish Solution for Controlling the Recovered Paper Flows," Proceedings of the First International Seminar on Reuse, Einhoven, The Netherlands, 177–187, November 11–13, 1996.

37. T. L. Pohlen, M. Farris II, "Reverse Logistics in Plastic Recycling," *International Journal of Physical Distribution & Logistics Management*, **22**(7), 35–47 (1992).

38. C.-H. Wang, J. C. Even Jr., and S. K. Adams, "A Mixed-Integer Linear Model for Optimal Processing and Transport of Secondary Materials," *Resources, Conservation, and Recycling*, **15**, 65–78 (1995).

39. C. Russell and W. J. Vaughan, "A Linear Programming Model of Residuals Management for Integrated Iron and Steel Production," *Journal of Environmental Economics and Management*, **1**, 17–42 (1974).

40. T. Assavapokee, J. C. Ammons, and M. J. Realff, "A New Min-Max Regret Robust Optimization Approach for Interval Data Uncertainty," submitted to *Journal of Global Optimization* (2005a).

41. T. Assavapokee, J. C. Ammons, M. J. Realff, "A New Relative Robust Optimization Approach for Full Factorial Scenario Design of Data Uncertainty and Ambiguity," submitted to IMA *Journal of Management Mathematics* (2005b).

42. L.-H. Shih, "Reverse Logistics System Planning for Recycling Electrical Appliances and Computers in Taiwan," *Resources, Conservation, and Recycling*, **32**, 55–72 (2001).

43. H. Wang, M. Guo, and J. Efstathiou, "A Game-Theoretical Cooperative Mechanism Design for a Two-Echelon Decentralized Supply Chain," *European Journal Operational Research*, **157**(2), 372–388.

44. F. Bernstein and A. Federgruen, "Decentralized Supply Chains with Competing Retailers under Demand Uncertainty," *Management Science* (2001).

45. M. Fan, J. Stallaert, and A. B. Whinston, "Decentralized Mechanism Design for Supply Chain Organizations Using an Auction Market," *Information Systems Research*, **14**(1), 1–22 (2003).

46. H. Lee, and S. Whang "Decentralized Multi-Echelon Supply Chains: Incentives and Information," *Management Science*, **45**(5), 633–640 (1999).

47. R. Anupindi, Y. Bassok, and E. Zemel, "A General Framework for the Study of Decentralized Distribution Systems," *Manufacturing and Service Operations Management*, **3**(4), 349–368 (2001).

48. B. Granot, and G. Sosic, "A Three-Stage Model for a Decentralized Distribution System of Retailers," *Operations Research*, **51**(5), 771–784 (2003).

49. A. Nagurney, J. Dong, and D. Zhang, "A Supply Chain Network Equilibrium Model," *Transportation Research* Part E, **38**, 281–303 (2002).

50. W. E. Walsh and M. P. Wellman, "Decentralized Supply Chain Formation: A Market Protocol and Competitive Equilibrium Analysis," *Journal of Artificial Intelligence Research*, **19**, 513–567 (2003).

51. V. D. R. Guide, R. H. Teunter, and L. N. Van Wassenhove, "Matching Demand and Supply to Maximize Profits from Remanufacturing," *Manufacturing & Service Operations Management*, **5**(4), 303–316 (2003).

52. M. E. Ferguson and L. B. Toktay, "The Effect of Competition on Recovery Strategies," working paper, Georgia Institute of Technology, 2005.

53. P. Majumder and H. Groenevelt, "Competition in Remanufacturing," *Production and Operations Management*, **10**(2), 125–141 (2001).

54. R. C. Savaskan, S. Bhattacharya, and L. N. Van Wassenhove, "Closed-Loop Supply Chain Models with Product Remanufacturing," *Management Science*, **50**(2), 239–252 (2004).

55. C. J. Corbett, U. Karmarkar, "Competition and Structure in Serial Supply Chains with Deterministic Demand," *Management Science*, **47**(7), 966–978 (2001).

56. I-H. Hong, J. C. Ammons, and M. J. Realff, "Decentralized Decision-making and Protocol Design for Recycle Material Flows," submitted to *Manufacturing & Service Operations Management* (2006b).

57. A. Nagurney and F. Toyasaki, "Reverse Supply Chain Management and Electronic Waste Recycling: A Multitiered Network Equilibrium Framework for E-Cycling," *Transportation Research Part E*, **41**, 1–28 (2005).

58. I-H. Hong, J. C. Ammons, and M. J. Realff, "Decentralized Decision-making and Protocol Design for Recycle Material Flows for a Multitiered Network," submitted to *IIE Transactions* (2006c).

59. I-H. Hong, J. C. Ammons, and M. J. Realff, "Centralized vs. Decentralized Decision-making for Recycle Material Flows," submitted to *ES&T* (2006d).

60. I-H. Hong, J. C. Ammons, and M. J. Realff, "Modeling Subsidy and Price Fluctuation Impacts on Recycled Material Flows," submitted to *Journal of Industrial Ecology* (2006e).

61. P. Fiala, "Information Sharing in Supply Chains," *Omega*, **33** (5), 419–423.

62. I. Giannoccaro, and P. Pontrandolfo, "Supply Chain Coordination by Revenue Sharing Contracts," *International Journal of Production Economics*, **89**, 131–139 (2004).

63. A. Kumar, and J. Wainer, "Meta Workflows as a Control and Coordination Mechanism for Exception Handling in Workflow Systems," *Decision Support Systems*, **40**(1), 89–105 (2005).

64. R. Bhatnagar, P. Chandra, and S. K. Goyal, "Models for Multi-Plant Coordination," *European Journal of Operational Research*, **67**, 141–160 (1993).

65. D. J. Thomas and P. M. Griffin, "Coordinated Supply Chain Management," *European Journal of Operational Research*, **94**, 1–15 (1996).

66. P. Romano, "Co-Ordination and Integration Mechanisms to Manage Logistics Processes across Supply Networks," *Journal of Purchasing and Supply Management*, **9**(3), 119–134 (2003).

67. A. S. Abdelfatah and H. S. Mahmassani, "A Simulation-Based Signal Optimization within a Dynamic Traffic Assignment Framework," IEEE Intelligent Transportation Systems Conference Proceedings, 2000.

68. L. Benyoucef, H. Ding, C. Hans, and X. Xie, "On a New Tool for Supply Chain Network Optimization and Simulation," Proceedings of the 2004 Winter Simulation Conference, 2004.

69. M. C. Fu, "Optimization for Simulation: Theory vs. Practice," *Journal on Computing*, **14**(3), 192–215 (2002).

70. A. Gosavi and G. Subramaniam, "Simulation-Based Optimization for Material Dispatching in a Retailer Network," Proceeding of the 2004 Winter Simulation *Conference*, 2004.

71. T. Homem-de-Mello, A. Shapiro, and M. L. Spearman, "Finding Optimal Material Release Times using Simulation-based Optimization," *Management Science*, **45**(1), 86–102 (1999).

72. D. P. Bertsekas and J. Tsisiklis, "Dynamic Programming: Deterministic and Stochastic Models," Prentice-Hall, Englewood Cliffs, NJ, 1987.

73. S. Ross, *Introduction to Stochastic Dynamic Programming*, Academic Press, New York, 1983.

74. G. A. Rummery and M. Niranjan, "On-line Q-learning Using Connectionist Systems," *Technical Report CUED/F-INFENG/TR 166*, Cambridge University, England, 1994.

75. R. Sutton, "Reinforcement Learning," *Machine Learning (Special Issue),* **8**(3) (1992).

76. C. J. Watkins, "Learning from Delayed Rewards," Ph.D. Dissertation, Kings College, England, 1989.

77. M. S. Bazaraa, M. D. Sherali, and C. M., Shetty, *Nonlinear Programming, Theory and Algorithms*, 2nd edition, John Wiley, New York, 1993.

78. Y. Durham, "An Experimental Examination of Double Marginalization and Vertical Relationships," *Journal of Economic Behavior & Organization*, **42**, 207–229 (2000).

CHAPTER 7

ENVIRONMENTAL LIFE-CYCLE ANALYSIS OF ALTERNATIVE BUILDING MATERIALS

A Key to Better Environmental Decisions

Bruce Lippke
Rural Technology Initiative, University of Washington, Seattle, Washington

Jim L. Bowyer
Department of Bio-based Products, University of Minnesota St. Paul, Minnesota

1	**INTRODUCTION**	**180**
2	**LIFE-CYCLE ASSESSMENT—BASICS**	**180**
	2.1 The Life-cycle Inventory	181
	2.2 The Impact Assessment (LCIA)	182
	2.3 Practical Applications of LCA	182
	2.4 Findings of Recent LCA Studies	184
3	**LIFE-CYCLE INVENTORY—FRAMEWORK DESIGN AND DATA COLLECTION**	**187**
	3.1 Data Collection	187
	3.2 Industry Description	188
	3.3 Unit Processes	188
	3.4 Material Flow and Allocation Methodology	189
	3.5 Scope, Objectives, and Boundaries	189
4	**LIFE-CYCLE INPUTS AND OUTPUTS**	**190**
	4.1 Extractive versus Renewable versus Purchased Inputs	190
	4.2 Purchased versus Recycled Materials	191
	4.3 Output Mass Balance	191
	4.4 Output Categorization	192
	4.5 Total versus Site-generated Emissions	195
5	**LIFE-CYCLE ASSESSMENT**	**196**
	5.1 Developing Impact Categories	196
	5.2 Input- versus Output-Oriented Categories	197
	5.3 Worst-offending Toxin Example	198
	5.4 Weighting of Categories	200
6	**IMPROVEMENT ANALYSIS: COMPARING DIFFERENT HOUSE DESIGNS**	**201**
7	**STRENGTHS AND WEAKNESSES OF LCI/LCA**	**204**
	7.1 Capturing System Impacts	204
	7.2 Cross Sectional versus Dynamic	204
	7.3 Time Value	205

1   INTRODUCTION

An environmental manager is faced with the task of identifying areas in which her company's environmental performance can be improved, but she does not have trustworthy data with which to make an evaluation. A homebuilder committed to environmentally responsible building construction needs a way to identify construction materials and building designs that minimize environmental impacts, but finds available information to be limited, conflicting, confusing, and often based on a single attribute. A government organization wishes to mount a preferred purchasing program for all of its paper products with the intent of minimizing environmental impacts and providing environmental leadership for society at large, but is faced with pressure to focus only on recycled content. As society becomes more and more interested in environmental attributes of products, those involved in all aspects of product manufacture, selection, use, maintenance, and end-of-life disposal need definitive, scientifically based tools for evaluating environmental impacts and potential mitigation strategies.

Environmental life-cycle assessment, or LCA, has become the tool of choice for leading organizations in both the public and private sectors. Sometimes referred to as *cradle-to-grave* analysis, LCA provides a mechanism for systematically evaluating the environmental impacts linked to a product or process and in guiding process or product improvement efforts. LCA-based information also provides insights into the environmental impacts of raw material and product choices, and maintenance and end-of-product-life strategies. Because of the systematic nature of LCA and its power as an evaluative tool, the use of LCA is increasing as environmental performance becomes more and more important in society. It is likely that LCA will soon become widely used within American industry and by those involved in crafting national and regional environmental policy.

2   LIFE-CYCLE ASSESSMENT—BASICS

An LCA typically begins with a careful accounting of all the measurable raw material inputs (including energy), product and co-product outputs, and emissions to air, water, and land. This part of an LCA is called a life-cycle inventory (LCI). Examination of energy use is particularly revealing, since a number of serious environmental problems are related to consumption of energy, including acid deposition, oil spills, air pollution (SO_2, NO_x), and increasing concentrations of atmospheric carbon dioxide. An LCI may deal with product manufacture only, or the study boundaries may be defined more broadly to include product use, maintenance, and disposal. In a subsequent stage of the LCA, factors are considered that are currently not precisely measurable, such as impacts of an industrial activity on the landscape, flora, fauna, air, or water.

2.1   The Life-cycle Inventory

As depicted in Figure 1, a life-cycle inventory may involve all stages in production, use, and disposal, including raw material extraction, transportation, primary processing, conversion to finished products, incorporation into finished products, maintenance and repair, and disposal. The system boundary (indicated by the dashed line) defines those operations to be included in the inventory of environmental impacts.

In the life-cycle inventory of wood-framed construction illustrated in Figure 1, the inventory would begin with the harvesting of trees and would include an accounting of the use of gasoline, oil, lubricants, saw blades, tires, and so on consumed in that process. All of the impacts associated with producing and transporting items consumed would also be considered. Included as well would be regeneration of the forest harvest site. Since the construction of wooden houses typically involves concrete foundations, the use of steel nails and other fasteners, glass, and other nonwood materials, all environmental impacts associated with the mining and processing of limestone, sand and gravel, iron ore, and other raw materials must be determined. Next, the processes involved in converting wood to lumber, panels, or other wood products are considered, as are industrial processes for converting limestone to cement, iron ore to steel nails, silica to glass, and so on. Energy directly consumed in the industrial processes is accounted for, as is the energy needed to provide heating of manufacturing plants.

Since we are focused on the environmental assessment of buildings, all activities involved in the building construction process are also considered, and

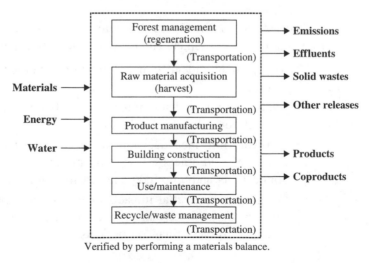

Verified by performing a materials balance.

Figure 1   Schematic of a life-cycle inventory. (From Ref. 1.)

again, all emissions, effluents, solid wastes, and other releases associated with consumption of energy and all other materials are accounted for. Finally, all materials and processes involved in the use and maintenance of the building are accounted for, as are processes involved and/or materials recovered at the time of building demolition at the end of the useful life of the structure.

A life-cycle inventory can be conducted for a period of time that is less than the full product life. For instance, a number of recent analyses have examined all of the steps involved up to completion of the shell of a residential structure. In this instance the system boundaries (dashed line in Figure 1) would encompass only the top four boxes; the use and maintenance and recycle/waste management stages would not be considered. A number of international protocols as published by the International Organization for Standards (ISO)[2,3] and the Society for Environmental Toxicology and Chemistry (SETAC) guide LCI practitioners,[4,5] allowing analyses to be conducted using a uniform set of guidelines. These protocols help to eliminate bias on the part of analysts and ensure that results of like assessments from various regions can be compared.

2.2   The Impact Assessment (LCIA)

Figure 2 illustrates how the life-cycle inventory fits within a life-cycle assessment. The LCI, shown as the large box at the top center of Figure 2, provides essential data regarding resource use and emissions to air, water, and ground. The impact assessment examines aspects of product production and use that are not considered in the LCI: impacts on ecosystem and human health, implications for long-term resource availability, and considerations relative to social equity and well being.

The bottom box of Figure 2 provides examples of how information from the LCI and impact assessment can be used. Such information is key to the following:

- Systematically identifying environmental burdens associated with a product or process
- Evaluating the probable impacts of a change in product or process design, product durability, or product life
- Allowing informed decision making on the part of designers, architects, engineers, and others who specify materials used in construction and other applications and who have interest in minimizing environmental impacts
- Gauging the potential impacts of government policies, such as those that favor or disfavor certain products or materials in government purchasing or in government-financed projects

2.3   Practical Applications of LCA

The Athena Sustainable Materials Institute[6] and the National Renewable Energy Laboratory provide several examples of current applications of LCA in North

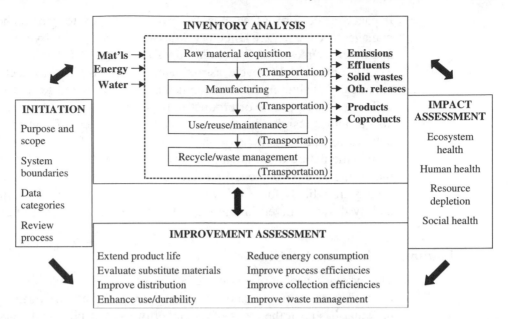

Figure 2   Life-cycle analysis: Steps in the process and applications of findings. (Adapted from Ref. 6.)

America. Other examples can be found throughout the U.S. and Canadian industrial sector, where a number of corporations are actively involved in the use and development of LCA. The Athena Sustainable Materials Institute is a Canadian-based organization that is recognized for its extensive contributions toward building a Canadian and U.S. LCI/LCA database for a broad array of wood and nonwood building materials. Among many programs of Athena is one that allows direct use of Athena software to carry out assessments of various design options for a building, thereby allowing designers to minimize environmental burdens.

The National Renewable Energy Laboratory is working with the U.S. Environmental Protection Agency and with Athena on an initiative known as the U.S. LCI Database Project.[7] The objective is to create a publicly available, national LCI database for commonly used materials, products, and processes. The purpose is to (1) support public- and private-sector efforts to develop environmentally oriented decision support systems and tools; (2) provide regional benchmark data for use in assessing environmental performance of companies, manufacturing plants, and production processes, and in evaluating the environmental attributes of new technologies or products; and (3) provide a firm foundation to subsequent life-cycle assessment tasks such as impact assessment. Ultimately, the database could also provide the foundation for a national product-labeling program in which building materials and other products would bear a label—very similar to the label found today on food packages—that would summarize environmental impacts in the form of as many as five to ten easy-to-understand indices. EPA manages a Global LCI Directory to assist users in identifying completed LCIs and other sources on LCI data but does not store LCI data directly.[8]

Environmental life-cycle assessment (LCA) provides a mechanism for systematically evaluating the environmental impacts linked to a product or process and in guiding process or product improvement efforts. LCA-based information also provides insights into the environmental impacts of raw material and product choices, and maintenance and end-of-product-life strategies. Because of the systematic nature of LCA and its power as an evaluative tool, the use of LCA is increasing as environmental performance becomes more and more important in society. It is likely that LCA will soon become widely used within U.S. industry and by those involved in crafting national and regional environmental policy.

2.4   Findings of Recent LCA Studies

The primary use of a life-cycle assessment is to guide product and process improvement for purposes of improving environmental performance. One "product" that has been the focus of a number of LCA studies is the residential house. There is a remarkable similarity of findings of research groups from all over the world that have studied the relative environmental impact of various construction materials. In every case, wood has been shown to have a substantial advantage in relation to other materials in terms of energy consumption per unit of finished product and vastly lower emissions in the process of raw material to product conversion. For example, a 1992 Canadian assessment of alternative materials for use in constructing a 110,000-square-foot building showed all-wood construction on a concrete foundation to require only 35 percent as much energy as steel construction on a concrete foundation. Furthermore, the liberation of carbon dioxide associated with building the steel structure was over 3.1 times that when building with wood. In the same year, a New Zealand study found office and industrial buildings constructed of timber to require only 55 percent as much energy as steel construction and approximately 66 to 72 percent as much energy as concrete construction. When residential buildings were considered, wood-frame construction with wood-framed windows and wood fiberboard cladding was found to require only 42 percent as much energy as a brick-clad, steel-framed dwelling built on a concrete slab and fitted with aluminum-framed windows. Accordingly, large differences in carbon dioxide emission were noted. A 1993 Canadian comparison of wood and steel-frame construction for light frame commercial structures, which examined a wide range of factors in addition to energy, again showed low environmental impacts of wood construction relative to steel (Table 1).

The values shown in Table 1 are dramatic, and they show that although wood construction clearly has environmental impacts, these impacts are small compared to those of steel. When use of recycled steel is considered, the differences between wood and steel narrow, but wood retains a significant advantage. As part of the

Table 1   Comparative Emissions in Manufacturing Wood vs. Steel—Framed Interior Wall (From Ref. 9)

Emission/Effluent	Wood Wall	Steel Wall
Energy Consumption (GJ)	3.6	11.4
Air Emissions		
Carbon dioxide (kg)	305	965
CO (g)	2,450	11,800
SO_x (g)	400	3,700
NO_x (g)	1,150	1,800
Particulates (g)	100	335
VOCs (g)	390	1,800
CH_4 (g)	4	45
Water and Effluents		
Water Use (L)	2,200	51,000
Suspended solids (g)	12,180	495,640
Nonferrous metals (mg)	62	2,532
Cyanide (mg)	99	4,051
Phenols (mg)	17,715	725,994
Ammonia (mg)	1,310	53,665
Halogenated organics (mg)	507	20,758
Oil and grease (mg)	1,421	58,222
Sulfides (mg)	13	507
Solid Wastes (kg)	125	95

wood versus steel wall comparison, load-bearing wood and steel-framed walls were examined in which the steel contained 50 percent recycled steel content. (Currently the maximum recycled content that technology allows in steel studs is about 40 percent.) In this case the steel-framed wall was found to be some four times as energy intensive, and correspondingly at least that much more environmentally damaging, despite its recycled steel content.

A 2004 study by the Consortium for Research on Renewable Industrial Materials (CORRIM) compared wood and steel houses built to Minneapolis code standards and wood and concrete houses built to Atlanta code standards.[10] Table 2 provides comparative data for above-grade walls (the focus is above-grade since the foundations were made of the same materials in each location regardless of the method of framing). The results mirror those of earlier studies, showing substantial environmental advantages of wood construction. Another difference that is not shown in Table 2 is that in the case of wood framing, most of the energy used in product manufacture is bio-energy from wood and other plant-derived material, whereas no bio-energy is used in producing steel or concrete elements. As a result, a typical steel-framed house in Minneapolis uses 281 percent more non-bio-energy (fossil energy) in the steel and insulation that displaces wood studs and insulation in a comparable wood-framed house. Similarly, a typical concrete house constructed in Atlanta uses 250 percent more fossil energy in the

Table 2 Environmental Performance Indices for Above-grade Wall Designs (From Ref. 10)

Minneapolis House	Wood Frame	Steel Frame	Difference	Steel vs. Wood (% Change)
Embodied energy (GJ)	250	296	46	18
Global warming potential (CO_2 kg)	13,009	17,262	4,253	33
Air emission index (index scale)	3,820	4,222	402	11
Water emission index (index scale)	3	29	26	867
Solid waste (total kg)	3,496	3,181	−315	−9

Atlanta House	Wood Frame	Concrete	Difference	Concrete vs. Wood (% Change)
Embodied energy (GJ)	168	231	63	38
Global warming potential (CO_2 kg)	8,345	14,982	6,637	80
Air emission index (index scale)	2,313	3,373	1,060	46
Water emission index (index scale)	2	2	0	0
Solid waste (total kg)	2,325	6,152	3,827	164

block and mortar, rebar, and wood that displaces the wood framing materials in a comparable wood-frame house.

These results do not indicate that wood should be used to the exclusion of all other materials, but rather, that production and use of all materials have environmental impacts that must be considered when formulating environmental policies. In the future, it can be expected that development of building design and construction technology will seek to take maximum advantage of the properties of each raw material, thereby designing buildings so as to minimize the total environmental impact. What comparative studies do show is that as environmental performance increases in importance, wood will clearly play a key role in buildings of the future.

Numerous studies of paper production, use, and disposal have also been done. One of the more interesting is an extensive examination of paper recycling conducted by a team in England.[11] In this study, that included consideration of LCI data as well as an economic assessment, the following options were compared:

- Recycling to make a similar grade of paper
- Recycling to make a lower grade of paper

- Incineration of recovered paper to generate energy
- Composting of recovered paper
- Land filling of recovered paper with recovery of methane to produce electricity

The findings were surprising to many. The research team concluded that if environmental externalities are given little value (i.e., if the environmental costs assigned to release of pollutants such as carbon dioxide, sulfur dioxide, nitrogen dioxide, or other pollutants are low), then it makes sense to recycle as much paper as possible. By contrast, if environmental costs are valued more highly, then the best course is to incinerate as much paper as possible for purposes of generating energy. The reason for this result lies in the fact that production of energy from paper reduces the need for energy production from petrochemicals, while also generating much lower quantities of pollutants. As environmental costs of pollutants rise, the value of pollution avoidance increasingly favors the generation of energy from waste paper incineration. This result indicates that blind pursuit of increased paper recycling is not necessarily the best environmental strategy.

3 LIFE-CYCLE INVENTORY—FRAMEWORK DESIGN AND DATA COLLECTION

3.1 Data Collection

The ability to assess environmental performance differences between different products, designs, or processes begins with the assembly of quality data. The objective is to measure all the inputs and outputs for every stage of processing leading up to the production unit that is the focus of the analysis—a house, an item of furniture, or a car. This is a daunting task and will generally require collection of primary data (data gathered through observations and measurements at manufacturing process centers). Primary data cover all of the inputs and outputs associated with a product or process, including emissions, from each stage of manufacturing leading up to the final product or assembly. Secondary data are sometimes also used. Secondary data, such as nationally collected or other published data as obtained from the literature may be suitable for LCI but the data collection procedures, conversion of units, and methods of compiling and aggregating data most often preclude the precision and specificity desired for an LCI analysis.

The process of collecting data for an LCI analysis when production processes are not uniform across regions, or when process efficiencies or emissions differ substantially between different production methods, must provide for accounting practices that maintain these categorical differences. Using wood as an example, since production processes differ by wood species, with each region producing different species, region and species differences become defining attributes of the product being produced. Thus, kiln-dried southern pine lumber from the southeastern United States (SE) is different both physically and in terms of manufacturing inputs and outputs, from kiln-dried hemlock lumber produced in the Pacific Northwest (NW) region. Kiln-dried hemlock is, in turn, different from green (undried) fir lumber produced in the NW region, as the undried product requires much less energy for processing. Similarly, even if the process for producing a specific steel product such as steel beams were fairly homogeneous across the country, regional energy sourcing differences such as the mix of natural gas, coal, nuclear, or hydro vary enough that it is better to maintain regional differences associated with a product in order to reflect differences resulting from energy sources.

As a consequence, in collecting quality input data, the list of distinctly different inputs can become quite large. Energy is not just energy, but purchased energy from a specific regional electrical grid, specific fuel source, and/or internally processed energy. By the same token, products are not just products but, as already noted, may differ according to a number of measures, including different processing methods.

3.2 Industry Description

In conducting an LCI that is intended to represent an entire industry or family of products, it is important to begin with a description of the industry, including an estimation of volumes produced or shares of each distinct product category. With this information and with knowledge of the number of mills representing each category, a LCI developer can determine how many producing mills need to be surveyed. Standards for surveys have generally required not fewer than 10 percent of the manufacturing facilities if there are a large number of mills, and up to a 100 percent sample if there are just a few. The industry description should also provide insight as to the age or level of technology adapted. Frequently new mills incorporate higher technology than the industry average, and this new technology will likely be representative of average practice some years in the future. While data for an individual mill will generally be regarded as proprietary, and therefore not reported on a mill-by-mill basis, statistics across mills provide insights on best practices and improvement processes already developing. The data also provide a statistical source measure for error, although not a very precise one, since from a practical perspective the surveys of mill data will rarely be random samples and will likely be biased to larger mills.

3.3 Unit Processes

The importance of defining mill categories adequately leads to the next hurdle in data collection. At what stage of processing should data be collected? There are clear benefits from collecting data at the level of each unit process rather than for an entire manufacturing facility that is made up of many unit processes. One benefit of collecting data at the unit process level is that it becomes possible to examine the probable impacts of changes to practices or adoption of new technology related to any particular process center. In this case, development of a new product LCI requires only that additional information for that process be collected.

Another and more compelling logic to collect data at the unit process level is based on the reality that every manufacturing process produces certain wastes that are recycled internally and/or co-products that are sold. Since material inputs as well as emission outputs are assigned to both products and coproducts on the basis of mass or some other surrogate, and because co-products may not depend on every unit process in the mill, the most accurate way to allocate inputs to units of output is by allocating input flows to each unit process. For example, a mill that produces green (undried) and dried lumber should not average the energy input across the two product outputs. All the drying energy uses should be allocated to the dried lumber. Collecting data at the unit process level correctly allocates the drying energy to the products that come out of the drier.

3.4   Material Flow and Allocation Methodology

Although collection of data at the unit process level solves much of the input allocation problem there is still room for variation. Most input inventories are allocated on a mass basis because whatever the input units, the inputs can be accurately converted to mass units, and mass balances between mill inputs and outputs can be easily checked for accuracy. The mass balance between input and output can be used as a sound validity check on the data; a difference of a few percent is generally considered acceptable.

3.5   Scope, Objectives, and Boundaries

Stages of processing that may need to be considered in addressing a specific objective include:

- Extraction
- Renewable regeneration, growth management, and harvesting
- Energy purchase (or internal production)
- Material purchase (or internal production)
- Transport with each stage and delivery to the next
- Processing primary products
- Processing secondary products or assemblies (e.g., roof trusses from lumber or steel)
- Construction or final assembly of the service unit
- Use
- Maintenance
- Disposal or recycling
- Landfill and decomposition

Boundaries may be drawn to exclude consideration of any part of this sequence. For example, many early LCI analyses have excluded consideration of use, maintenance, and disposal, leaving these issues to subsequent analyses.

Collecting data at the unit process level does not simplify the need to define and keep track of boundaries. It should be clear that developing an LCI requires tracking many material flows and environmental burdens across many process boundaries, some that are internal to a system and others that define an overall system. The relevancy of where boundaries are placed will depend on the objectives of the study. Objectives will generally be targeted at improving some dimension of environmental performance such as extension of product life, enhancement of use and durability, evaluation of preferred substitute products, evaluation of unit process alternatives, improvement of process efficiency, reduction of energy consumption, reduction of toxic releases, or improvement of waste management.

For any of these or other objectives, the project scope still needs to be defined. How is the final unit of service defined? If, for example, the objective is to extend product life, and if the end unit of interest is a building, it will be necessary to determine what goes into the building (structure only or also cabinets, appliances, millwork, and so on), what determines the life of the structure and its components, and what makes or limits its usefulness. The objective might be defined to include reduction of a comprehensive list of environmental burdens for a complex assembly such as a building or defined more narrowly to focus the impacts of substituting one component for another. For the building structure analogy the boundaries need to encompass all stages of processing for all products and services used to construct the building, as well as all use phases of the building, including maintenance, heating and cooling, demolition, and recycling or landfilling of wastes.

It is common practice to draw boundaries for an LCI such that the burdens generated by the production of the manufacturing facility itself are ignored. Since the burdens resulting from the construction of a facility and its internal equipment are spread over a large number of products and many years of useful life of the facility, they will generally have no significant impact on product burdens. Even here there may be exceptions. For example, production of a disposable fuel cell with a very short life could potentially produce as large a burden as the emissions created in producing energy from the fuel cell. The process of suppressing any inputs or outputs should be defensible and fully transparent in documentation.

4   LIFE-CYCLE INPUTS AND OUTPUTS

4.1   Extractive versus Renewable versus Purchased Inputs

In the spirit of collecting all of the inputs for any stage of processing that may result in some environmental burden, inputs measured include all raw materials, energy sources, and air, water, and land used, whether purchased, extracted, or grown. If the material comes from the ground, air, or water, the main focus of LCA is on what is done to the resource that results in emissions, effluents, products and co-products, as though these resources are infinite. If a scarce source of water is consumed rather than modified and not replaced, such a measure of consumption may be an important local environmental burden. Extraction from the ground is just another manufacturing process (i.e., a stage of processing). Harvesting of renewable resources such as wood is also just another manufacturing process, starting with the regeneration of trees and ending with harvesting them. A separate analysis should be performed to the degree that extracted sources are finite (i.e., their use results in drawdown of known reserves), as this is a useful environmental measure in itself. Each subsequent stage of processing leading up to a final product carries forward the outputs of these beginning stages as a basis for accumulating all of the inventories needed for the final product or assembly.

Assessment of the nature of materials extracted or used may be an important aspect of impact assessment (see Figure 2), but LCI methods are of limited use since there are no accepted methods for directly comparing the extraction of a nonrenewable resource with the harvesting of renewable resources involving specific land use plans. In contrast, environmental burdens associated with emissions and effluents are directly comparable using LCI data and analysis methods.

Table 3 shows inputs to produce 1,000 square feet of plywood (3/8-inch thickness basis).[12] This table is typical of those that result from an LCI. Note the substantial number of material and energy inputs.

4.2 Purchased versus Recycled Materials

Accounting for emission and effluent outputs must include not only the direct outputs from processing in a manufacturing facility but also the outputs associated with delivery of raw materials and energy input sources. Thus, an LCI usually includes a detailed accounting of the extraction activities needed to provide raw materials and to support the utility that produces and delivers energy to the manufacturing site.

Purchases or use of waste that occur at the end of product life are generally considered free of burdens at the point of pickup, since the entire burden for that waste was allocated to prewaste product use up until the point of pickup for secondary use. The burdens arising from recycling of manufacturing wastes within a plant get allocated through the stages of processing just like other resources to the output products and coproducts.

4.3 Output Mass Balance

Table 4 provides an example of the mass balance of wood inputs and outputs for producing 1,000 square feet of plywood; balancing inputs and outputs provides a check on the accountability of materials flows. Wood outputs are either products

Table 3 Inputs to Produce 1.0 MSF (3/8-inch) Plywood in the Pacific Northwest. (From Ref. 12)

PNW Plywood—INPUTS

Materials[2]	Units	MSF 3/8-in. Basis	SI Units	MSM 9-mm Basis
Roundwood	ft^3	6.56E+01	m^3	1.89E+01
	lb	1.79E+03	kg	8.23E+03
Phenol-formaldehyde adhesive	lb	1.59E+01	kg	7.30E+01
Extender and fillers	lb	8.90E+00	kg	4.09E+01
Catalyst[3]	lb	1.11E+00	kg	5.11E+00
Soda ash[3]	lb	3.30E−01	kg	1.52E+00
Bark[4]	lb	9.90E+01	kg	4.55E+02
Purchased				
Dry veneer	lb	6.40E+00	Kg	2.96E+01
Green veneer	lb	1.42E+01	Kg	6.54E+01
Electrical Use				
Electricity	kWh	1.39E+02	MJ	5.08E+03
Fuel Use				
Hogged fuel (produced)	lb	1.92E+02	Kg	8.801.76E+02
Hogged fuel (purchased)	lb	1.703.40E+01	Kg	7.80E+01
Wood waste	lb	5.00E−01	Kg	2.30E+00
Liquid propane gas	gal	3.59E−01	L	1.38E+01
Natural gas	ft^3	1.63E+02	M^3	4.71E+01
Diesel	gal	3.95E−01	L	1.52E+01
Water Use				
Municipal water source	gal	8.28E+01	L	3.19E+03
Wellwater source	gal	2.94E+01	L	1.13E+03
Recycled water source	gal	3.00E−01	L	1.16E+01

[1]All information comes from primary survey data collected in 2000.
[2]All materials are given as an oven-dry basis or solids weights.
[3]These materials were not included in the SimaPro LCI analysis; excluded based on the 2% rule.
[4]Oven-dry Weight. Assumed to be 50% moisture content on wet-basis in survey data.

and co-products that are sold, or wood wastes used as a fuel for an on-site boiler that provides energy for heating or running process equipment. In rare cases wastes may go to a landfill.

4.4 Output Categorization

Although the direct inputs to an extractive or manufacturing process are generally fairly small in number, the chemical processes that take place often produce a large number of emission and effluent outputs. The number of products, co-products, and waste products is again usually small. Table 5 provides a list of inputs to and outputs from the production of 1,000 square feet of plywood.

The system LCI output requires that the energy purchased at a manufacturing facility be treated like any other input product, including the initial energy sources

Table 4 Wood Mass Balance for Plywood Production in the Pacific Northwest Region (From Ref. 12)

Inputs	lb./MSF 3/8-in. Basis	kg/MSM 9-mm Basis
Round wood (logs)	1,788[1]	8,226
Purchased dry veneer	6	30
Purchased green veneer	14	65
Total	1,809	8,321

Outputs	lb./MSF 3/8-in. Basis	kg/MSM 9-mm Basis
Plywood (wood only)	916[2]	4,214
Wood chips	425	1,956
Peeler core	95	438
Green clippings	31	143
Veneer downfall	3.4	16
Panel trim	107	491
Sawdust	9.6	44
Wood waste to boiler	0.25	1.2
Sold wood waste	21	97
Sold dry veneer	63	290
Unaccounted for wood	137[3]	630
Total	1,809	8,321

[1]Based on Douglas fir, spruce, western hemlock and western larch weighted average wood density of 27.26 lb/ft3 for 65.6 cu. ft. of wood in logs to produce MSF 3/8-inch basis.

[2] Plywood (wood only) based on estimated weight of plywood, 991 lb. minus 80% of resin, filler, soda ash, and catalyst total use.

[3]7.6% unaccounted for wood; may have been included in the hogged fuel.

Note: All weights are on an oven-dry basis.

serving the utility and distribution losses through the grid, as well as all of the emissions and effluents produced in the power generation and distribution process.

Tables 3 and 4 detail raw material and energy inputs to the plywood producing process, as well as product and co-product outputs from the process; shown in Table 5 are all emissions to air, water, and ground. Results were generated in SimaPro 5.0.9 version 5 LCA software with the Franklin Database employed for LCI on fuel use, and electricity production burdens assuming production serving the Pacific Northwest region. Data shown in Table 5 are allocated total emissions, which include emissions for the production and delivery of electricity, fuel, and adhesives. Emissions for production of the phenolic adhesive were obtained from Athena™. All other inputs and outputs were based on manufacturing plant surveys.

Table 5   Emissions to Air, Water, and Ground Resulting from Production of 1.0 MSF 3/8-inch Basis Plywood in the Pacific Northwest Region (From Ref. 12)

Emissions to Air		
	lb./MSF	kg/MSM
Substance	3/8 in.	9 mm
Acetaldehyde	1.19E−02	5.49E−02
Acetone	5.11E−03	2.35E−02
Acrolein	8.75E−07	4.03E−06
Aldehydes	8.56E−04	3.94E−03
Alpha-pinene	7.69E−02	3.54E−01
Ammonia	4.85E−04	2.23E−03
As	1.26E−05	5.81E−05
Ba	5.82E−04	2.68E−03
Be	1.02E−07	4.69E−07
Benzene	4.86E−04	2.23E−03
Beta-pinene	2.99E−02	1.37E−01
Cd	5.69E−07	2.62E−06
Cl_2	1.03E−03	4.74E−03
CO	2.08E+00	9.55E+00
CO_2 (fossil)	7.78E+01	3.58E+02
CO_2 (non-fossil)	2.85E+02	1.31E+03
Cobalt	7.44E−07	3.42E−06
Cr	7.44E−06	3.42E−05
Cumene	7.44E−05	3.42E−04
Dichloromethane	1.37E−06	6.30E−06
Dioxin (TEQ)	1.83E−12	8.42E−12
Fe	5.82E−04	2.68E−03
Formaldehyde	3.74E−02	1.72E−01
HCl	1.73E−03	7.96E−03
HF	2.40E−04	1.10E−03
Hg	7.13E−07	3.28E−06
K	1.03E−01	4.74E−01
Kerosene	1.09E−05	5.03E−05
Limonene	8.63E−03	3.97E−02
Metals	1.19E−05	5.46E−05
Methane	2.13E−01	9.80E−01
Methanol	1.36E−01	6.24E−01
Methyl ethyl ketone	6.81E−04	3.13E−03
Methyl i-butyl ketone	5.58E−04	2.57E−03
Mn	1.19E−03	5.49E−03
N-nitrodimethylamine	7.31E−08	3.36E−07
N_2O	1.96E−04	9.03E−04
Na	2.38E−03	1.10E−02
Naphthalene	3.18E−04	1.46E−03
Ni	8.19E−05	3.77E−04
Non methane VOC	3.29E−01	1.51E+00
Nox	6.50E−01	2.99E+00
Organic substances	2.28E−02	1.05E−01
Particulates	3.81E−01	1.75E+00
Particulates (PM10)	2.27E−01	1.04E+00
Particulates (unspecified)	2.52E−02	1.16E−01
Pb	1.60E−04	7.36E−04
Phenol	3.02E−02	1.39E−01
Sb	2.98E−07	1.37E−06
Se	2.71E−06	1.25E−05
SO_2	8.25E−04	3.80E−03
SOx	1.06E+00	4.86E+00
Tetrachloroethene	3.30E−07	1.52E−06
Tetrachloromethane	5.85E−07	2.69E−06
THC as carbon	1.65E−01	7.59E−01
Trichloroethene	3.27E−07	1.50E−06
VOC	6.69E−01	3.08E+00
Zn	5.82E−04	2.68E−03

Emissions to Water		
	lb./MSF	kg/MSM
Substance	3/8 in.	9 mm
Acid as H+	1.23E−08	5.66E−08
B	9.19E−04	4.23E−03
BOD	1.44E−03	6.61E−03
Ca	1.03E−07	4.72E−07
Calcium ions	9.31E−06	4.28E−05
Cd	6.23E−05	2.87E−04
Chromate	4.43E−07	2.04E−06
Cl-	6.24E−02	2.87E−01
COD	1.67E−02	7.68E−02
Cr	6.23E−05	2.87E−04
Cyanide	9.31E−08	4.28E−07
Dissolved solids	1.38E+00	6.33E+00
Fe	1.35E−03	6.21E−03
Fluoride ions	4.36E−05	2.01E−04
H_2SO_4	2.30E−04	1.06E−03
Hg	4.89E−09	2.25E−08
Metallic ions	2.61E−04	1.20E−03
Mn	7.56E−04	3.48E−03
Na	1.73E−05	7.96E−05
NH_3	5.45E−05	2.51E−04
Nitrate	4.11E−06	1.89E−05
Oil	2.45E−02	1.13E−01
Other organics	4.08E−03	1.88E−02
Pb	2.24E−08	1.03E−07
Phenol	8.50E−07	3.91E−06
Phosphate	1.15E−04	5.29E−04

Table 5   (*continued*)

Substance	lb./MSF 3/8 in.	kg/MSM 9 mm	Nonmaterial Emissions	Ci/MSF 3/8 in.	Bq/MSM 9 mm
Sulphate	5.43E−02	2.50E−01			
Suspended solids	3.27E−02	1.50E−01	Substance	Ci/MSF 3/8 in.	Bq/MSM 9 mm
Zn	2.16E−05	9.95E−05	Radioactive substances to air	1.21E−05	4.57E+06
Solid Waste Emissions					
Substance	lb./MSF 3/8 in.	kg/MSM 9 mm			
Solid waste	1.88E+01	8.63E+01			

The length of the list should make it obvious that it poses a difficult problem to decide how to evaluate so many outputs. Hence, a useful step is to collect outputs in categories that can be evaluated for their human health or ecosystem risk. Output data are typically grouped to focus on (1) those outputs that contribute to greenhouse gas emissions and therefore to global warming potential (GWP), (2) air pollution, (3) water pollution (4) solid waste generation, and perhaps other toxic categories that provide additional detail to 2, 3, or 4. Energy or nonrenewable energy is often treated as a separate category recognizing the dominant role that energy plays in contributing to pollution, as well as the fact that reducing the consumption of energy is one of the most effective strategies to reduce environmental burdens.

Table 5 illustrates the grouping of effluents and emissions to water, air, and solid waste.[12] It is important to note that the mass in these categories cannot be simply added to produce a meaningful environmental index because some are associated with much more significant environmental impacts than others. The process of converting all of these outputs to useful burden index measures is a part of the inventory assessment process.

4.5   Total versus Site-generated Emissions

The relative importance of site-generated emissions versus total emissions integrated back to all fuel sources is of interest when looking for sources of improvement. Continuing the plywood production example, Table 6 shows total emissions as well as only those generated on site that exclude all emissions associated with purchased inputs. Notice that the total fossil CO_2 emissions are more than six times larger than the site-generated fossil CO_2 emissions. Even total phenols and formaldehyde are substantially larger than those generated on site, with much of this related to purchased resins. The take-home message is that looking only at site-generated emissions is not sufficient in an LCI and can produce misleading results. Knowing the true source of each emission allows one to focus improvement so as to bring about the greatest impact. The LCI methodology incorporates all emissions associated with all inputs.

Table 6 Summary of Life-cycle Inventory Results for 1.0 MSF 3/8-inch Basis Plywood Production in the Pacific Northwest Region—a Comparison of Total to Site-generated Emissions (From Ref. 12)

Substance	LCI Total lb./MSF 3/8-in. Basis	LCI Site-Generated lb./MSF 3/8-in. Basis
Acetaldehyde	1.19E−02	1.19E−02
Acrolein	8.75E−07	5.28E−07
CO	2.08E+00	1.94E+00
CO_2 (fossil)	7.78E+01	1.20E+01
CO_2 (non-fossil)	2.85E+02	2.85E+02
Formaldehyde	3.74E−02	2.06E−02
Methane	2.13E−01	7.13E−05
Methanol	1.36E−01	1.36E−01
Nonmethane VOC	3.29E−01	2.32E−02
NO_x	6.50E−01	3.79E−01
Particulates	3.81E−01	3.75E−01
Particulates (PM10)	2.27E−01	2.22E−01
Particulates (unspecified)	2.52E−02	0.00E 00
Phenol	3.02E−02	8.44E−03
SO_2	8.25E−04	8.25E−04
SO_x	1.06E+00	1.80E−02
VOC	6.69E−01	6.69E−01

Note: Summary data for site-generated emissions for plywood production only; do not include production or transportation emissions for fuel, electricity, and resin.

5 LIFE-CYCLE ASSESSMENT

5.1 Developing Impact Categories

With typically 30 or more different emissions to each of air, to water, and in the form of solid waste, it is extremely difficult to draw meaningful conclusions based on examination of lists of outputs that are developed as part of the LCI. To get beyond numbers overload, each LCI inventory value is assigned to a small number of different impact categories so that a composite of a large number of LCI values can be reduced to just a few key impact measures. The characterization of how much impact each compound has on a given category is based on models that estimate relative importance to human or ecological health or landscape impacts.

Impact category indicators may be defined to indicate the ultimate impact on human health. An example is expected deaths per year resulting from a cancer-inducing toxin. Thus far, such estimates are highly controversial. Expected deaths provide a direct risk measure for human health, and measurements of this kind are characterized as endpoint indicators. Global warming potential fits the criteria of an impact category, but the science is not far enough advanced to compute an equivalent measure of health risk. Global warming potential is believed to have health impacts, but when measured in CO_2 equivalent contributions to greenhouse gas emissions it does not provide an endpoint measure and is considered an intermediate variable. Only when the impact categories can be measured in equivalent units can a relatively simple model be used to develop a single integrated risk across several categories. Weighting *across* assessment categories raises a much higher degree of uncertainty than *aggregation* of values into a single category. Many assessments therefore do not attempt to produce a single weighted index defined by a less-than-transparent weighting system. Instead, each category is quantified separately and the more important indicators are displayed side by side like the labels provided on cereal boxes, allowing the user to incorporate his or her own weighting system in making a selection using several environmental categories.

5.2 Input- versus Output-Oriented Categories

Another way to group impact categories is into subsets such as those related to emission output and those related to the efficiency of input use:

Output Categories:

- Global warming
- Ozone depletion
- Acidification
- Nitrification
- Human toxicity to air, water, land
- Ecotoxicity
- Solid waste
- Vegetation diversity

Input Categories:

- Resource depletion (fossil fuels, minerals)
- Resource use efficiency (renewables, nonrenewables, or energy)
- Land use (may be considered an input (efficiency) or output (ecological) measure)

Inputs might also be grouped so as to focus on depletion of nonrenewable resources or the use rates of renewable resources.

Developing categories under which to group environmental burdens is essential and is considered by ISO 14040, the international standard for LCA, a mandatory aspect of LCI/LCA. LCI values are assigned to one or more categories, and the indicator for each category must be defined in terms of some intermediate variable if not an endpoint measure. Although an analysis can be conducted with intermediate variables, there needs to be at least a theoretical link to categorical endpoints such as human life expectancy, ocean or land vegetation, or forest health or diversity.

For example, EPA has developed safe standards for the toxicity of many emissions. Relative safety can then be defined in terms of safe emission levels of each. A method of accounting known as the *critical volume method* is used to estimate the volume of ambient air or water that would be required to dilute contaminants to acceptable safety standards. Athena™ EIE calculates and reports indices based on the worst offender—that is, the substance requiring the largest volume of air and water to achieve dilution to acceptable levels.[13,14] The hypothesis is that the same volume of air or water can contain a number of pollutants.

The equation for the air pollution equivalence index (arbitrary scaling) is:

$$\text{Air pollution equivalence index} = \text{MAX of}[SO_2/.03(mg/m^3)\ldots$$

$$SPM/.06(\text{suspended particulate mater, } mg/m^3)\ldots CO/6(mg/m^3)\ldots$$

$$NO_2/.06(mg/m^3)\ldots VOC/6(mg/m^3)\ldots Phenol/2(mg/m^3)]$$

The equation for the water pollution equivalence (arbitrary scaling) is:

$$\text{Water pollution equivalence index} = \text{MAX of } [TDS/5(mg/L)\ldots$$

$$PAH/.0000001(mg/L)\ldots Non\text{-}Ferr/.003(mg/L)\ldots CN/.00005(mg/L)\ldots$$

$$Phenol/.00001(mg/L)\ldots Nitr\text{-}Ammn/.02(mg/L)\ldots$$

$$Halg.Org/.0002(mg/L)\ldots Chlor/3.5(mg/L)(mg/L)\ldots Alum/.001(mg/L)\ldots$$

$$Oil.Gr./.01(mg/L)\ldots Sulphate/5(mg/L)\ldots Sulphide/.0005(mg/L)\ldots$$

$$Iron/.003]$$

If the dilution principle is used to develop an air or water pollutant index, it is only important to measure the worst-offending substance, since the other substances will be diluted more in the process of diluting the worst offender to a safe level. However dilution of air pollutants is recognized as a separate process, meaning that the air pollution index is in addition to the water pollution index, even though each are derived as the single worst-offending compound in their respective category.

No system is perfect for every situation, so the LCI/LCA analyst has to select those input and output categories that are most important to the objectives sought, select or otherwise determine models that produce a numerical index for each category and that are important to the endpoints of interest, and finally to display the indicators for judgmental weighting or develop a weighting system reducing the measurements to a single numerical indicator.

5.3   Worst-offending Toxin Example

By diluting each toxic compound to the EPA safe standard, the diluted mass provides a measure of its relative toxicity. For example, cyanide (CN) emissions

Table 7 Emissions to Water by Process Stage for Minneapolis and Atlanta Alternative House Designs (From Ref. 14)

Manufacturing	Minneapolis Steel	Wood	Construction Transportation	Minneapolis Steel	Wood
Total dissolved solids (g)	156,861	162,184	Total dissolved solids (g)	40	43
Nonferrous metals (g)	127	99	**Total**		
Cyanide (CN) (g)	3,528	477	Total dissolved solids (g)	156968	162299
Phenols (g)	138	126	Nonferrous metals (g)	135	107
Nonhalogenated Org. (g)	636	582	Cyanide (g)	3528	477
Chlorides (g)	168,380	144,521	Phenols (g)	138	126
Aluminum & alumina (g)	18	18	Nonhalogenated Org. (g)	640	586
Oil & grease (g)	102,691	17,069	Chlorides (g)	168386	144528
Sulphates (g)	32,192	32,524	Aluminum & alumina (g)	18	18
Sulphides (g)	10,401	1,429	Oil & grease (g)	102694	17072
Iron (g)	2,125	973	Sulphates (g)	32196	35528
Manufacturing Transportation			Sulphides (g)	10401	1429
Total Dissolved Solids (g)	16	19	Iron (g)	2151	1000
Construction					
Total dissolved solids (g)	49	53			
Nonferrous metals (g)	7	7			
Nonhalogenated Organics (g)	3	3			
Chlorides (g)	4	4			
Sulphates (g)	2	2			
Iron (g)	25	27			

Note: Emissions less than 1 gram were omitted.

to water associated with construction of a typical steel-framed house constructed in Minneapolis are 3,528 grams (Table 7), a much smaller mass than many other emissions. However because cyanide is so much more toxic than other emissions, thus requiring considerably more dilution to reach the EPA safe standard, it becomes the worst-offending water toxin linked to construction of a steel-framed structure (Figure 3). A similar comparison of offending toxins from an equivalent wood-framed structure shows much less cyanide; here, the worst-offending emission is phenol. A comparison of the two worst-offending emissions shows that cyanide is some 400 percent worse than the worst-offending wood-frame emission (phenol). This provides a direct comparison of the water-pollution toxicity between the two structures as the worst-offending toxins are both diluted to the same safety standard. Table 7, which details the effects associated with construction of structures made of various materials, also shows that the vast majority of water emissions associated with construction are generated in the product-manufacturing process, with relatively insignificant contributions from transportation and the construction process itself. However, the same cannot be said for all assessment categories.

Note that the worst-offending safety standard is still an intermediate measure of health risk because it is not measured directly in terms of human deaths or impairment. But to the degree that EPA's estimates of safe standards are consistently

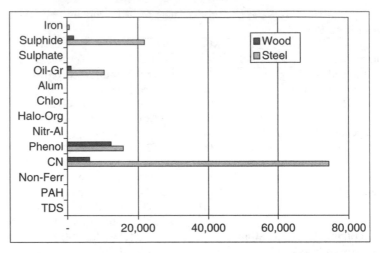

Figure 3   Emissions to water on equal health risk index—Minneapolis house. (From Ref. 14.)

developed across categories, they justify equal weighting. The development of safe air standards and safe water standards are quite different processes, however, so the uncertainty associated with an indicator for a single category will likely be much smaller than that of a composite of the indicators across several categories.

The reliance on worst offenders instead of more complicated models is a somewhat arbitrary simplification of a complex problem and may strictly apply only at the component level. When several components are used in an assembly, the technique of using dilution of the emissions to produce an indicator based on safe standards should be applied at the component level, not at the level of emissions from the assembly of components, since different components are often manufactured in different mills. The worst offender for one component may well be different from that of another component.

5.4   Weighting of Categories

The process of developing LCI measures generally relies on primary data and can be characterized with some statistical measure on the degree of accuracy. The subsequent step of identifying categories for integrating many of these LCI inventory values into an indicator for an environmental risk category, while still quite rigorous, does involve a greater degree of uncertainty, especially if accepted models for quantifying index levels for the various categories are lacking. Establishing safe health risk standards for toxicity is itself an evolving science, such that toxicity-related models can be expected to change with new research findings.

There may be the desire to weight each different environmental category in order to come up with a single performance indicator across all categories. An integrated weighting scheme across all categories should be viewed with caution. If the objective is to identify opportunities to improve environmental performance, a single integrated indicator may suppress the visibility of many important improvements that might be made. It was noted in the introduction that just by displaying indexes for several categories it becomes quite obvious which of two designs produces the least burden. Therefore, weighting across categories may not be that useful for most applications. When there are trade-offs between categories, the better emphasis may be to look for process product or design changes that can improve both.

Different methods for deriving weights that can be applied across categories have been developed. They may rely on expert opinion[15] or estimates based on how different publics value different categories. Even without using weights to develop a single index across categories, a comparative analysis looking across categories quite readily reveals substantially different burdens and where changes can be made to improve environmental performance.

6 IMPROVEMENT ANALYSIS: COMPARING DIFFERENT HOUSE DESIGNS

Comparing alternative assemblies that use different products, designs, or processes is a straightforward way to make a selection of a preferred alternative. Table 8 provides an example of an environmental assessment across five environmental categories for wood- and steel-framed housing designs (structure, envelope, and interior partitions) typical of a two-story Minneapolis house. The two designs share a number of the same structural assemblies (e.g., foundation footings and basement block wall, support beams/jack-posts, and roof) and therefore, the difference in their environmental impact can be traced to their respective wall and floor assemblies. The wall assembly includes the below-grade concrete wall and above-grade exterior wood sheathing and vinyl siding, which is common to both designs. Within this assembly, only the above-grade wall materials—wood versus steel studs—contribute to the differences. The floor and roof assembly includes the concrete basement floor, and first and second floor wooden subflooring, as well as the roof assemblies as common elements. Only the wood I-joists versus steel joists in the floor contribute to the differences.

Overall and relative to the wood design, the steel design embodies 17 percent more energy (113 GJ), produces 26 percent more global warming potential (10 metric tons), emits 14 percent more air pollution and 311 percent more water pollution, but produces 1 percent less solid wastes (0.1 metric tons). Focusing on only the assembly groups affected by the change in material (i.e., walls and floors and roofs), the differences in the two designs are more pronounced. For example, when using the wood results as the baseline, the steel design's floors and roof embody 68 percent more energy (74 GJ), produce 156 percent more global warming potential (6 metric tons), emit 85 percent more air pollution and 322 percent more water pollution but produce 31% less solid wastes (0.4 metric

Table 8   Environmental Results Summary for the Minneapolis Designs (From Ref. 14)

Design by Assembly Group	Embodied Energy MJ	Global Warming Potential Equiv. CO_2 kg	Air Pollution Index Critical Vol. Measure	Water Pollution Index Critical Vol. Measure	Solid Wastes kg
Wood Design					
Foundations	68,487	6,041	1,001	1	2,548
Walls	457,130	26,468	6,472	4	9,701
Floors & roof	108,552	3,763	981	9	1,383
Extra basic mat.	16,424	775	112	3	134
Total	**650,593**	**37,047**	**8,566**	**17**	**13,766**
Steel Design					
Foundations	68,487	6,041	1,001	1	2,548
Walls	496,437	30,360	6,803	28	9,145
Floors & roof	182,207	9,650	1,813	38	1,814
Extra basic mat.	16,424	775	112	3	134
Total	**763,555**	**46,826**	**9,729**	**70**	**13,641**

Note: The Extra Basic Materials grouping includes the supporting steel jack-posts and beams for the structure.

tons). In effect, the steel design is relatively less efficient in the floor, where stiffness is required, than in the walls, where narrower steel wall studs with an extra 1-inch expanded polystyrene (EPS) insulation layer on the outside are contrasted with 2 × 6 wood wall studs with fiberglass batts.

Although this analysis of alternative house designs involves a rather complex assembly of products involving many processes, the results provide a clear indicator of how environmental impacts might be reduced by careful materials selection. The analysis was extended to include use, maintenance, and disposal and only minimal differences were found in these stages of the life of the structure.

Table 9 shows the fossil fuel energy per square foot consumed by three different exterior wall designs for a Minneapolis house. The estimates for fossil energy were developed by using Athena EIE software to design 2,000 square feet of exterior wall area, average for a Minneapolis house, under three different design scenarios. Each of the exterior wall designs are built to Minneapolis local code (R19 insulation).

Columns two through four in Table 9 represent the wall designs. The final row, "Total," shows the total fossil fuels consumed to produce each design expressed on a per square foot basis. The three preceding rows, "Structural" (frame and sheathing), "Insulation" (fiberglass and polystyrene), and "Covering" (cladding and interior wall covering) represent assemblies of building materials that, when added up for each design, yield total fossil fuel consumed.

The first exterior wall design in Table 9, "Lumber Wall," represents a typical wood-based exterior wall frame design. The lumber wall design uses 2 × 6 kiln-dried lumber studs, plywood sheathing, vinyl siding, fiberglass insulation, and

Table 9 Fossil Fuels Consumed by Three Exterior Wall Designs in a Cold-climate Home (From Ref. 16)

	Exterior Wall Designs for Cold-climate Home		
	Lumber Wall (MJ/ft^2)	Steel Wall (MJ/ft^2)	Lumber Wall with Additional Wood Product Substitutes (MJ/ft^2)
Structural[1]	9.54	15.22	4.82
Insulation[2]	12.63	21.02	5.45
Covering[3]	22.42	22.42	2.91
TOTAL[4]	**44.59**	**58.66**	**13.18**

[1]Includes studs and plywood sheathing. For the third design, lumber wall with substitutes, lumber and plywood are produced with higher than average levels of biofuels.
[2]Includes fiberglass (lumber wall and steel wall designs), extruded polystyrene (steel wall design only), and insulation created with recycled paper products (lumber wall with substitutes). All three designs include a six mil polyethylene vapor barrier.
[3]Includes interior and exterior wall coverings. Exterior wall coverings for lumber wall and steel wall are vinyl cladding and interior wall coverings are gypsum. Half-inch thick plywood (produced with higher than average levels of biofuels) serves as the exterior wall covering for Lumber wall with substitutes. Quarter-inch thick plywood (produced with higher than average levels of biofuels) replaces gypsum as the interior wall covering for the lumber wall with additional substitutes. [4]Includes subtotals from structural, insulation, and covering categories.

gypsum board covering of the interior wall. "Steel Wall" represents a typical steel-based exterior wall design. The steel wall design uses 2×4 steel studs, plywood sheathing, vinyl siding, fiberglass and extruded polystyrene insulation, and gypsum on the interior wall. The Lumber wall uses a wider stud to house the fiberglass insulation, but the narrower steel design requires a layer of extruded polystyrene to achieve the same thermal rating. The designs' different studs and insulation creates a significant difference in the quantity of fossil fuels required. Selecting the lumber wall instead of the steel wall achieves a 24 percent reduction in fossil fuel consumption per square foot.

The design presented in the fourth column of Table 9, "Lumber Wall with Additional Wood Product Substitutes," represents a hypothetical wall design with drastically reduced fossil fuel consumption achieved through product and process energy substitution. This design includes 2×6 kiln-dried wood studs produced using above-average levels of bio-fuels (generally available from scraps and low-valued products such as beauty bark) and substitutes half-inch plywood for vinyl siding and quarter-inch plywood panels for the gypsum covering on the interior wall. Recycled paper-based insulation replaces fiberglass. The plywood used in this design, like the kiln-dried lumber, is produced using above-average levels of bio-fuels. The substitutions incorporated into this hypothetical wall design add up to a substantial reduction in fossil fuel consumption (70 to 80 percent) if selected in place of either of the two more traditional designs. Comparison of the three different exterior wall designs in the Minneapolis house suggests that substantial reductions in fossil fuel use and related environmental burdens are possible through product and process substitution. For brevity, only the fossil fuel category is shown, but similar reductions also occur in global warming potential, and air and water pollution.

These examples suggest that identifying attractive environmental performance improvement opportunities can be approached rather systematically by first looking at comparison alternatives across all stages of processing from cradle to grave for a total systems view, and then by successively leaving out life stages that are not major contributors (use, maintenance, disposal) and by looking more closely at those components that are contributing the most to performance improvement. While this example focused on structural walls, one cannot prejudge that the findings would be the same for floors or roofs. Each material has different properties that will have different impacts in different applications. For example engineered wood I-Joists are preferred in floor joists for a number of reasons, but their increased stiffness over dimension lumber is an advantage in floor designs. In contrast, steel is at a disadvantage where stiffness is important. These differences produce substantial environmental performance differences.

7   STRENGTHS AND WEAKNESSES OF LCI/LCA

7.1   Capturing System Impacts

LCI/LCA provides a systemwide measure of environmental impacts. The findings rest on the quality of the LCI database, which attempts to capture the impact of every input and output in the production chain for analyzing everything from a single product to a complex assembly of products with a specific use, such as a house, car, or appliance. The methodology provides a degree of quality control so that independent LCI developers can produce data that can be used in comparative studies. The USLCI database project is a repository for properly reviewed LCI data.

Looking at any single impact such as recycling content can miss the more important contributors to environmental burdens such as energy efficiency. The LCA/LCI allows flexibility to make product, design, or process changes and to determine their impact. Although the LCI database is key to the strength of LCI/LCA it is also a weakness in that it takes a large effort to develop the data, which subsequently will need to be updated to accommodate new products, processes, or designs.

7.2   Cross Sectional versus Dynamic

LCI information is collected as a cross-section of all processes as though they are in steady state. The useful life of many products will be measured in the decades. The actual recycling of disposed products may look very different in the future than currently. It would be rational to predict higher recycling of many products in the future, but this opens the analysis to speculation in contrast to a systematic evaluation of current performance. Similarly, for renewable resources there may be many decades between the planting of trees and the manufacturing of products from those trees making it impossible to replicate historic data for regeneration or to account for the many management changes that are likely to occur in the intervening years. The gains of increased productivity that have been so pronounced with agricultural commodities are, in effect, left out of an LCI/LCA analysis unless inserted as an override in order to demonstrate sensitivity to change. Similarly, the cumulative impacts of performance differences may not receive enough attention in a steady-state analysis.

7.3   Time Value

Since the ultimate objective of environmental performance improvement is to invest more in improvement, the steady-state feature of LCI can cause problems. For gate-to-gate analysis, such as a direct comparison between two manufacturing processes where the time dimension is minimal, one can make a direct environmental benefit comparison and divide by the investment cost for each alternative to derive a benefit to cost ratio. The prudent investor would gain more environmental benefit by investing where the ratio of benefit to cost is highest.

However, for evaluations where the product life is long and maintenance or use costs are high over time, the time value of money can be important and can make analysis more complicated. Although the energy for heating a house over its life is typically much higher than the energy used to construct the house, the two categories of energy expenditure do not produce benefits over the same time period. If for a $1,000 investment you can get the same reduction in energy consumed to build the house as the amount of heating/cooling energy over the life of the house, which one is preferred? From a steady-state LCI perspective they are equal; however, if time is brought into consideration, it will take perhaps decades for the investment in energy saving to pay off compared to the investment in reducing energy immediately. The usual method to equate these differences is to assume that the value of environmental improvements should be discounted to the present with the time value of money, which for low-risk investments might be 3 to 5% without inflation. Then the investment in reducing energy over the life of the house is of much less benefit per dollar invested.

REFERENCES

1. J. K. Meil, "Environmental Measures as Substitution Criteria for Wood and Nonwood Building Products," in *The Globalization of Wood: Supply, Processes, Products, and Markets*, Forest Products Society Proceedings 7319, pp. 53–60 1994.

2. International Organization for Standardization, *Environmental Management—Life-cycle Assessment: Principles and Framework*, ISO XXX, Geneva, Switzerland, 1997.

3. International Organization for Standardization, *Environmental Management—Life-cycle Assessment: Goal and Scope Definition and Inventory Analysis*, ISO XXX, Geneva, Switzerland, 1998.

4. J. Fava, R. Denison, B. Jones, M. Curran, B. Vigon, S. Selke, and J. Barnum, "A Technical Framework for Life-Cycle Assessment. Society for Environmental Toxicology and Chemistry," (1991).

5. J. Fava, A. Jensen, L. Lindfors, S. Pomper, B. De Smet, J. Warren, and B. Vigon, "Life-cycle Assessment Data Quality: A Conceptual Framework," Society for Environmental Toxicology and Chemistry (SEATAC) and SEATAC Foundation for Environmental Education, 1994.

6. Athena Sustainable Materials Institute, "The Summary Reports: Phase I and Phase II Research," FORINTEK Canada and Wayne B. Trusty & Associates, Ottawa (http://www.athena.smi.ca), 1997.

7. National Renewable Energy Laboratory, "U.S. LCI Database Project—Phase I Final Report," NREL/SR-550-33807, http://www.nrel.gov/lci/pdfs/33807.pdf. Accessed August 2003.

8. EPA Global LCI Directory (http://www.epa.gov/ORD/NRMRL/lcaccess/dataportal.htm)

9. J. Bowyer, J. Howe, P. Guillery, and K. Fernholz, 2005. Life Cycle Analysis: A Key to Better Environmental Decisions. Dovetail Partners, Inc.

10. B. Lippke, J. Wilson, J. Perez-Garcia, J. Bowyer, and J. Meil, "CORRIM: Lifecycle Environmental Performance of Building Materials," *Forest Products Journal*, **54**(6): 8–19 (2004), (http://www.corrim.org/reports/pdfs/FPJ_Sept2004.pdf).

11. M. Leach, A. Bauer, and N. Lucas, "A Systems Approach to Materials Flow in Sustainable Cities," *Journal of Environmental Planning and Management*, **40**(6): 705–723 (1997).

12. J. Wilson and E. Sakimoto, "Softwood Plywood Manufacturing: Module D," in *Life-cycle Environmental Performance of Renewable Materials in Context of Residential Building Construction: Phase I Research Report*, J. Bowyer, D. Briggs, B. Lippke, J. Perez-Garcia, and J. Wilson (eds.), Consortium for Research on Renewable Industrial Materials CORRIM Inc., Seattle, www.corrim.org, 2004.

13. Athena Sustainable Materials Institute (ATHENA), 1993. Raw material balances, energy profiles and environmental unit factor estimates for structural wood products building materials in the context of sustainable development Ottawa, Canada, pp. 42.

14. J. Bowyer, D. Briggs, B. Lippke, J. Perez-Garcia, J. Wilson, L. Johnson, J. Marshall, M. Puettmann, E. Sakimoto, E. Dancer, B. Kasal, P. Huelman, M. Milota, I. Hartley, C. West, E. Kline, J. Meil, P. Winistorfer, C. Zhanging, C. Manriquez, J. Comnick, and N. Stevens, *Life Cycle Environmental Performance of Renewable Building Materials in the Context of Residential Building Construction*, CORRIM Inc., Seattle, 2004.

15. M. Goedkoop and R. Spriensma, *The Eco-Indicator 99: A Damage-oriented Method for Life-cycle impact assessment—Methodology report*, Pre Consultants B.V. Plotterweg 12 Amersfoort, 2001.

16. L. Edmonds and B. Lippke, "Reducing Environmental Consequences of Residential Construction through Product Selection and Design," *CORRIM FS4*, www.corrim.org/factsheets. Accessed in 2004.

CHAPTER 8

WASTEWATER ENGINEERING

Say Kee Ong, Ph.D
Department of Civil, Construction, and Environmental Engineering, Iowa State University, Ames, Iowa

1	**INTRODUCTION**	**207**
2	**WASTEWATER TREATMENT STRATEGIES AND REQUIREMENTS**	**208**
3	**PHYSICAL TREATMENT TECHNOLOGIES**	**213**
	3.1 Air Stripping	213
	3.2 Solids Removal: Clarification and Sedimentation	214
	3.3 Filtration	216
	3.4 Flotation	216
	3.5 Oil and Grease Removal	217
	3.6 Evaporation	218
4	**CHEMICAL TREATMENT TECHNOLOGIES**	**219**

	4.1 Neutralization	219
	4.2 Precipitation	219
	4.3 Chemical Oxidation and Reduction	222
	4.4 Wet Air Oxidation and Supercritical Water Oxidation	224
	4.5 Electrochemical Process	225
	4.6 Adsorption and Ion Exchange	226
5	**BIOLOGICAL WASTE TREATMENT**	**228**
	5.1 Treatment Processes	228
	5.2 Activated Sludge Process	229
	5.3 Sequencing Batch Reactors	229
	5.4 Membrane Bioreactors	230
	5.5 Biological Aerated Filters	231
6	**SUMMARY**	**232**

1 INTRODUCTION

There are more than 16,000 sewage treatment plants in the United States treating more than 32 billion gallons of municipal and industrial wastewater per day. Industries that are major contributors of industrial wastewaters include chemicals, petrochemicals, petroleum refining, food and consumer products, metals, and pulp and paper industries. To minimize environmental impact and to protect human health, these wastewaters must be treated to remove various chemical constituents and toxic/hazardous compounds before they are released to the environment, or they must be pretreated before release into the municipal sewer system. As required by the Clean Water Act (1987), wastewaters from industries discharging directly to a body of navigable water (receiving streams, etc.) require a National Pollution Discharge Elimination System (NPDES) permit. The effluent quality to be discharged is based on several factors, including the concentrations of specific pollutants in the wastewaters and the quality, quantity, and assimilative capacity of the receiving water body. To avoid obtaining a NPDES permit, many industries discharge their wastewaters into municipal sewers, where they are treated in municipal wastewater treatment plants along with municipal wastewaters. Under these circumstances, municipal wastewater treatment plants require the industries to comply with the National Pretreatment Program (40 CFR 403).

The National Pretreatment Program controls the discharge of 126 priority pollutants from industries into sewer systems, as described in the Clean Water Act. These priority pollutants fall into two categories, metals and toxic organics.

1. *Metals:* Examples include arsenic, mercury, lead, chromium, and cadmium. Metals cannot be degraded or broken down through treatment.

2. *Organic compounds:* Toxic organics can be categorized as volatile organics, semivolatile organics, and persistent organic compounds (POPs). Examples include solvents such as trichloroethylene, aromatics such as toluene and benzene, pesticides, and polychlorinated biphenyls (PCBs). Many can be degraded to innocuous compounds or to carbon dioxide and water. However, some compounds, such as PCBs, can be recalcitrant.

Other pollutants include:

- *Acid/alkalis:* Examples include nitric acid, sulfuric acid and caustic soda. They have low or high pH, are corrosive but can be neutralized.

- *Inorganics salts:* Examples include innocuous ions such as sodium, potassium, chloride, and sulfates that are harmless, but they increase the total dissolved solids of the receiving water.

- *Suspended solids:* These may come in various forms—inorganic or organic solids from processing of raw materials such as ores or raw products.

2 WASTEWATER TREATMENT STRATEGIES AND REQUIREMENTS

Wastewater from an industrial facility may be categorized as (1) process wastewaters from the manufacturing processes, (2) utility wastewaters from blowdowns of boilers and cooling towers, or (3) wastewaters from sanitary activities. All three categories of wastewaters are significantly different in their quality, with varying wastewater characteristics. One of the treatment strategies for large industrial facilities is to separate the different wastewater streams from each process and treat the wastewater streams individually before discharging into a sanitary sewer or directly discharging into a body of water. There are instances where it is more economical to combine the process and utility wastewaters and treat them at a central wastewater treatment plant before they are discharged. In some industrial complexes, several industries may combine their wastewaters and treat them together at a central wastewater facility. Some of the disadvantages of a central wastewater treatment plant treating wastewaters from several industries include designing and operating various unit treatment processes to remove the pollutants sequentially, along with careful monitoring of wastewater pollutants from different industries that may upset the treatment plant. Sanitary wastewaters from industrial facilities are typically not mixed with the process wastewaters and are discharged separately into the municipal sewers.

If the wastewater is to be pretreated before it discharges into the sanitary sewer, the effluents are regulated by National Pretreatment Program. The National Pretreatment Program identifies specific requirements that apply to different categories of industrial users (see 40 CFR 403.12(a) and 40 CFR 403.3(h)) that discharge industrial wastewaters to municipal wastewater treatment plants (see Table 1). For each of the different categories of users, three types of discharge standards are applied and enforced:

1. *Prohibited discharge standards:* General, national standards are applicable to all industrial users to a municipal wastewater treatment plant, designed to protect against pass through and interference, protect the sewer collection system, and to promote worker safety and beneficial biosolids use (40 CFR 403.5). A listing of the effluent guidelines and standards is presented in Table 2.

2. *Categorical pretreatment standards:* These are national, technology-based standards that limit pollutant discharges for specific process wastewaters of particular industrial categories, called *categorical industrial users*. The standards applicable to industrial discharges are designated in the Effluent Guidelines & Limitations (40 CFR Parts 405–471) (Table 1 provides a list of pollutants regulated for certain selected industries).

3. *Local limits:* Developed to reflect specific needs and capabilities at individual municipal wastewater treatment plants, these are designed to protect the publicly operated treatment work (POTW) receiving waters (40 CFR 403.8(f)(4)).

Within each industry category, there are subparts in the 40 CFR that refer to specific manufacturing processes. For example, the Iron and Steel Manufacturing category has a total 13 subparts for different processes, with subpart A addressing wastewaters from the coke-making process while subpart F addresses the continuous casting process. The effluent guidelines and limits for an industrial facility are based on whether it is an existing source or a new source and the use of best available technology (BAT) or best practicable control technology (BPT). Effluent limits are either production-based with daily maximums and monthly averages or concentration-based with daily maximums and monthly averages, or mass-based (concentration-based standards multiplied by process flow) with daily maximums and monthly averages. In some cases, the guidelines will prohibit discharges of process wastewater at a flow rate or mass loading rate that is excessive over any time period during the peak period. For example, for the continuous casting process, the pretreatment effluent limits are production-based

Table 1   Effluent Guidelines and Standards Title 403–471

Category	40 CFR Part	Pollution of Concern (partial listing only)
Aluminum Forming	467	
Asbestos Manufacturing (COD)	427	Suspended solids, pH, organic loading
Battery Manufacturing	461	
Canned and Preserved Fruits and Vegetables Processing	407	Organic loading, suspended solids, pH, oil and grease
Canned and Preserved Seafood Processing	408	Organic loading, suspended solids, pH, oil and grease
Carbon Black Manufacturing	458	
Coil Coating	465	
Copper Forming	468	
Cement Manufacturing	411	Suspended solids, temperature, pH
Centralized Waste Treatment	437	
Concentrated Aquatic Animal Producing	451	
Concentrated Animal Feeding Operations (CAFO)	412	Organic loading (BOD5), pH, solids, nitrogen and phosphorus
Dairy Products Processing	405	Organic loading, suspended solids, pH
Electrical and Electronic Components	469	
Electroplating	413	Cyanide, heavy metals (copper, nickel, chromium, zinc, lead, cadmium, silver), total solids, pH, total toxic organics (e.g., acrolein, benzene chlorinated compounds), chelating agents
Explosives Manufacturing	457	
Ferroalloy Manufacturing	424	suspended solids, chromium, manganese, pH, cyanide, phenols, ammonia,
Fertilizer Manufacturing	418	Ammonia, Organic nitrogen, nitrate, Phosphorus, fluoride, suspended solids, pH, organic loading,
Glass Manufacturing	426	Organic loading (BOD5, COD), phenol, suspended solids, pH, oil (mineral, animal and vegetable), phosphorus, lead, fluoride, ammonia
Grain Mills	406	Organic loading, suspended solids, pH
Gum and Wood Chemicals Manufacturing	454	
Hospital	460	
Ink Formulating	447	
Inorganic Chemicals Manufacturing	415	Organic carbon, Suspended solids, pH. Metals (antimony, arsenic, barium, cobalt, mercury, lead, copper, nickel, chromium, iron, selenium, silver, zinc), fluoride, residual chlorine, cyanide, sulfide, ammonia, oil and grease,
Iron and Steel Manufacturing	420	Suspended solids, oil and grease, ammonia, cyanide, phenols, pH, aromatics (benzo(a)pyrene, naphthalene, 2,3,7,8-TCDF, tetrachloroethylene), lead, zinc, chromium, nickel
Landfill	445	

Table 1   (*continued*)

Category	40 CFR Part	Pollution of Concern (partial listing only)
Leather Tanning and Finishing	425	Organic loading (BOD5), suspended solids, oil and grease, chromium, pH, sulfide
Meat Poultry Products	432	
Metal Finishing	433	
Metal Molding and Casting	464	
Mineral Mining and Processing	436	
Metals Products and Machinery	438	
Nonferrous Metals Forming and Metal Powders	471	
Nonferrous Metals Manufacturing	421	Fluoride, suspended solids, pH, benzo(a)pyrene, metals (arsenic, antimony, beryllium, cobalt, nickel, aluminum, indium, molybdenum, mercury, copper, cadmium, lead, zinc, selenium, silver, tantalum, tin, titanium. tungsten) ammonia, cyanide, oil and grease, organic loading (COD), total phenolics, fluoride, hexachlorobenzene
Oil and Gas Extraction	435	
Ore Mining and Dressing	440	
Organic Chemicals, Plastics, and Synthetic Fibers	414	Organic loading, suspended solids, pH, specific organic compounds (benzene, anthracene, chlorobenzene, phenols, etc.) metals (chromium, copper, lead, nickel, zinc), cyanide
Paint Formulating	446	
Paving and Roofing Materials (Tars and Asphalt)	443	
Pesticide Chemicals	455	
Petroleum Refining	419	Organic loading (BOD5, COD), suspended solids, oil and grease, phenolic compounds, ammonia, sulfide, chromium, pH
Pharmaceutical Manufacturing	439	
Photographic	459	
Plastics Molding and Forming	463	
Porcelain Enameling	466	
Phosphate Manufacturing	422	Total phosphorus, fluoride, suspended solids, pH
Pulp, Paper, and Paperboard	430	
Rubber Manufacturing	428	Organic loading, oil and grease suspended solids, pH, lead, chromium, zinc
Soap and Detergent Manufacturing	417	Organic loading (BOD5), suspended solids, oil and grease, pH, surfactants
Steam Electric Power Generating	423	Suspended solids, oil and grease, copper, iron, free available chlorine, priority pollutants (cooling tower), chromium and zinc

(*continued overleaf*)

Table 1   (*continued*)

Category	40 CFR Part	Pollution of Concern (partial listing only)
Sugar Processing	409	Organic loading (BOD5), suspended solids, coliform, temperature, pH
Textile Mills	410	Organic loading (BOD5), suspended solids, oil and grease, sulfide, phenol, chromium, pH
Timber Products Processing	429	
Transportation Equipment Cleaning	442	
Waste Combustors	444	

Table 2   Prohibited Discharge Standards

General	• Any pollutant(s) that cause passthrough or interference
Specific	• Pollutants that create a fire or explosion hazard, waste streams with a closed cup flashpoint of less than 60°C or 140°F
	• Pollutants that will cause corrosive structural damage, in no case discharges with pH lower than 5.0, unless the works is specifically designed to accommodate such discharges
	• Solid or viscous pollutants that will cause obstruction to the flow in the POTW
	• Any pollutant, including oxygen-demanding pollutants (BOD, etc.), released that will cause interference with the POTW
	• Heat that will inhibit biological activity in the POTW, in no case heat discharges in such quantities that the temperature at the POTW treatment plant exceeds 40°C (104°F) unless approved
	• Petroleum oil, nonbiodegradable cutting oil, or products of mineral oil origin that will cause interference or passthrough
	• Pollutants that result in the presence of toxic gases, vapors, or fumes within the POTW in a quantity that may cause acute worker health and safety problems
	• Any trucked or hauled pollutants, except at discharge points designated by the POTW

with maximums for any day equal to 0.00730, 0.00313, 0.0000939, and 0.000141 pounds per 1,000 pounds of product for total suspended solids, oil and grease, lead and zinc, respectively. pH of the wastewater is permissible between 6 and 9. For the metal-finishing industries (Part 433), the pretreatment effluent limits for a new source are concentration-based and include maximum concentrations for any day for cyanide (1.2 mg/L), oil and grease (52 mg/L), total suspended solids

(60 mg/L), pH (6 to 9), and seven heavy metals such as cadmium (0.11 mg/L), copper, (3.38 mg/L), and lead (0.69 mg/L).

3 PHYSICAL TREATMENT TECHNOLOGIES

3.1 Air Stripping

Air stripping is the process of transferring volatile organic compounds (VOCs) and semi-VOCs from the wastewater (liquid phase) into vapor phase by passing a high volume of stripping medium through the wastewater. Two common stripping media used to remove VOCs are air and steam. Air stripping is cost effective for treatment of waste streams with low concentrations of the hazardous VOCs, especially for constituents that can be volatilized under room temperature. When semivolatile compounds are present, steam is often used to increase the temperature of the liquid phase to enhance volatilization of the organic compounds. Stripping can be performed by using packed towers, tray towers, spray systems, diffused aeration systems, or mechanical aeration. In a typical packed tower system, water is distributed evenly at the top of the tower through a packing material with a high specific surface area. Water in the tower is broken into small droplets and made to spread over the surface of the packing, resulting in a high surface area for the exchange of the volatile organics from the liquid phase to air or steam. Air is introduced from the bottom and leaves from the top of the tower. Factors affecting the efficiency of an air-stripping system include volatility of the compounds as measured by Henry's Law constant, air flow rate, water loading rate, mass transfer at the air–water interface, packing material and depth of packing.[1]

The height of the packing tower needed for the removal of a pollutant to a certain effluent concentration can be estimated using the following equations:

Tower packing height, $Z = HTU \times NTU$

NTU is defined as the number of transfer units and is given by:

$$NTU = \left(\frac{S}{S-1}\right) \ln\left[\frac{(S-1)}{S}\frac{C_{L,in}}{C_{L,out}} + \frac{1}{S}\right]$$

S is called the stripping factor and is given by:

$$S = K_H^1 \frac{Q_Q}{Q_w}$$

HTU is defined as the height of transfer units and is given by:

$$HTU = \frac{Q_w}{K_L a A} = \frac{L_M}{p_L K_L a}$$

K_H' is the Henry's Law constant (dimensionless), $K_L a$ is the overall volumetric mass transfer coefficient (m^3/s-m^2), a is the interfacial surface area for mass transfer (m^2/m^3), Q_G is the air flow rate (m^3/s), L_M is the liquid mass flux (kg/m^2–s), Q_w is the liquid flow rate (m^3/s), $C_{L,in}$ is the influent liquid solute concentration (mg/L), $C_{L,out}$ is the effluent liquid solute concentration (mg/L), ρ_L is the liquid density (kg/m^3).

3.2 Solids Removal: Clarification and Sedimentation

Settling of solids in wastewaters by gravity is a common and inexpensive technology for separation and removal of solids. The process is termed as *sedimentation* or *clarification*. Sedimentation is carried out in a settler or sedimentation tank called a *clarifier* or *thickener*. The clarified liquid that is low in suspended solids may be reused as process water or discharged while the concentrated suspension or sludge may be concentrated further to produce a drier product before disposal. In the design of a clarifier, two processes are occurring: clarification as demonstrated by hindered settling, and thickening as demonstrated by compression settling at the bottom of the clarifier. Overflow rates are typically used in the design of the sedimentation tanks (see Table 3 for design clarification rates).

Conventional settling tanks have surface overflow rates ranging from 1.0 to 2.5 m/h and are typically used along with an organic polymer as a coagulant. To increase the sedimentation rates, lamella tube settlers have been used giving surface overflow rates in the range of 2.5 to 7.5 m/h.[2] In recent years, sedimentation technologies with high processing rates and, consequently, a smaller equipment footprint are becoming more attractive than conventional sedimentation tanks. There are a variety of high-rate clarification technologies for the separation of solids. High-rate clarification systems may rely on the addition of a ballast (such as fine sand) or recycling of flocculated solids to enhance the formation of microflocs and settling of the flocs. Most high-rate clarification systems use tube settlers to maximize the settling surface area and increase the settling velocities. Examples of ballasted flocculation system are the Actiflo® and Microsep® process. A ballasted system consists of a mixing zone, a maturation zone for the formation of the flocs, and a settling zone using lamella plate settling. The microflocs settle rapidly to the bottom of the clarifier. Sand and flocs are removed from the clarifier and pumped into a cyclone separator for sand

Table 3 Typical Surface Overflow Rates for Clarifier Design

Type of Waste	Specific	Surface Gravity (m^3/m^2-hr)	Average Hydraulic Overflow Retention Time Rates (hr)
Precipitation treatment			
Aluminum and iron floc	1.002	1.0–3.1	2–8
Calcium carbonate precipitates	1.2	1.0–4.4	1–4
Metals precipitation (using NaOH)	1.1	0.87–2.6	1–4
Biological solids (activated sludge)	1.005	0.7–2.0	1–4

Table 4 High Rate Processes from Various Vendors

Technology Name	Surface Overflow Rates		Description	Vendors
	(gpm/ft^2)	(m/h)		
Lamella Settler	Up to 3.0	Up to 7.5	Inclined tubes or parallel plates to increase sedimentation	Various
Actiflo®	20–70	50–175	Micro-sand ballasted flocculation and lamella clarification	Kruger
Microsep®	20–40	50–100	Micro-sand ballasted flocculation/solids contact and clarification	Veolia Water
Densadeg®	10–50	25–125	Two-stage flocculation with chemically conditioned recycled sludge followed by lamella clarification	Infilco-Degremont
Trident® HS	5–25	12–65	Adsorption of flocs onto floating media followed by clarification	US Filter
CONTRAFAST™	>6	>15	Flocculation enhanced with chemically conditioned recycled sludge, followed by lamella clarification, within one tank	US Filter

removal. The clean sand is returned to the injection tanks and solids from the cyclone are sent to a solids-handling system for disposal. Surface overflow rates ranging from 50 to 150 m/hr (20–60 gpm/ft^2) have been achieved. Examples of some of the more recent clarification technologies are presented in Table 4.

3.3 Filtration

Filtration removes solid particles in fluid by passing the fluid through a filtering medium on which the solids are deposited or trapped. The concentration of the solids in the fluid that can be removed may vary from trace concentrations to very high percentages, depending on the filtration medium. Removal of solids by filtration may be enhanced by adding a filter aid such as an organic polymer. Filters are divided into three main categories: cake filters, clarifying filters, and cross-flow filters.

Cake filters separate relatively large amounts of solids by forming a cake on the surface of the filtration medium. Cake filters may be operated by applying pressure on the upstream section of the filter medium or vacuum on the downstream section of the filter medium. The operation of the filters may be continuous or discontinuous. However, most pressure filters are discontinuous since the operation of the filter under positive pressure needs to be stopped to facilitate the removal and discharge of solids. Examples of cake filters include filter press, vacuum filter, and centrifugal separator.

In a typical *clarifying filter*, solids are removed within the filter media through straining, inertial impact of the solids to the media, and adhesion of solids to the media. Removal efficiency is fairly constant throughout the filter run before the solids breakthrough, resulting in a sharp increase in the effluent solids concentrations. The pores of a clarifying filter media are typically larger than that of a cake filter. Examples of clarifying filters are sand filters for water treatment and cartridge filters.

In *cross-flow filtration*, the wastewater flows under pressure at a fairly high velocity tangentially or across the filter medium. A thin layer of solids form on the surface of the medium, but the high liquid velocity keeps the layer from building up. At the same time, the liquid permeates the membrane producing a clear filtrate. Filter media may be ceramic, metal (e.g., sintered stainless steel or porous alumina), or a polymer membrane (cellulose acetate, polyamide, and polyacrylonitrile) with pores small enough to exclude most suspended particles. Examples of cross filtration are microfiltration with pore sizes ranging from 0.1 to 5 μm and ultrafiltration with pore sizes from 1 μm down to about 0.001 μm.

3.4 Flotation

Dissolved air flotation (DAF) is a proven, robust technology for removal of oil and grease, fibers, low-density materials, and suspended solids, and is used for thickening of wastewater sludges or chemical sludges.[3] Solids in wastewaters are removed by attaching fine gas bubbles to the suspended solids to increase their buoyancy. Suspended solids then rise to the surface and are removed using scrapers or overflow weirs. Most air-flotation systems operate with a recycle where a portion of the clarified liquid is pumped to a retention tank and is pressurized (5–10 atm) with air. The recycled clarified liquid containing dissolved air

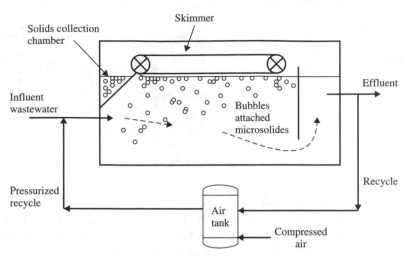

Figure 1 Schematic of a dissolved air flotation unit.

is mixed with fresh wastewater at the entrance of the flotation unit. The reduction in pressure at the entrance results in the release of the dissolved air as air bubbles. Figure 1 shows a typical air flotation unit. Recycle ratios varies from 20 percent and above. Bubbles sizes are between 30 and 120 microns. Surface overflow rates that can be achieved by DAF are between 10 and 20 m/hr and solids loading rates are in the range of 5 to 20 kg/m^2/hr. Solids removal can be as high as 98 percent and the solids concentration can be as high as 5 percent. Some of the advantages of DAF include: (1) rapid start-up (in minutes) and withstanding of periodic stoppages, (2) high solids capture, particularly of finer solids (80–90%), (3) reduced chemical usage, and (4) mechanical float removal that can produce a relatively thick sludge (2–4% sludge dry solids).

3.5 Oil and Grease Removal

Oil/water separators employ various separation methods to separate the oil from the aqueous phase. The most common oil/water separators employ the principle of flotation with or without physical coalescing of the oil droplets for the removal of oils and greases from industrial wastewater.[4] The application of a particular separation process depends on the properties of the oil in the oil/water mixture. Figure 2 shows a simplified diagram of a typical oil/water separator system. The oil layer at the surface of the water is then skimmed to an oil holding tank, where it is recycled or disposed of accordingly. The treated wastewater is passed under a baffle to the outlet chamber and is discharged for further treatment or into the sewer.

Depending on the sizes of the oil droplets in the wastewater, a simple oil/water flotation separator may not be sufficient to remove oil to meet the regulatory

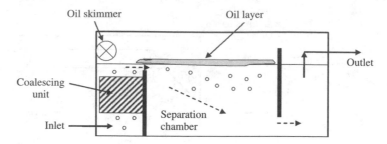

Figure 2 Schematic diagram of an oil/water separator.

discharge standards. Under such situations, oil removal may be enhanced by coalescing the oil droplets to form larger oil droplets, making them more buoyant and causing them to rise faster. For example, the time needed for a 100-micron-diameter oil droplet to rise 15 cm in water is approximately 10 minutes, while the time needed for a 20-micron diameter oil droplet to travel the same distance is approximately 2 hours. In a typical oil/water separator, the minimum water depth of the separator may be between 1.2 and 1.5 meters, meaning that oil droplets of certain diameters may pass through the oil/water separator uncollected.

Inclined plates placed within the separation chamber of the oil/water separators is one of the approaches used where the wastewater with the oil droplets is made to travel short vertical distances (approximately 0.6 cm). As the oil droplets encounter the fixed surface, they coalesce and rise along the plates to the water's surface. Another approach is to use a filter media made of fine oleophillic (oil "loving") fibers such as polypropylene. As wastewater flows through the filter, fine oil droplets are attached to the fibers. The attached droplets get larger with time and become buoyant and detach from the fibers and rise to the surface. The use of detergents and soaps for the removal oil and grease from equipment surfaces can adversely affect the operation of an oil/water separator. These detergents and soaps, known as *emulsifying agents*, are specifically formulated to increase the dispersal of oil into tiny drops in water. The parameters affecting the effectiveness of an oil/water separator are the residence times of the wastewater in the oil/water separator and the surface area of the chamber for the accumulation of oil on the surface. If too much oil is accumulated and not removed in a timely manner, oil may flow out of the oil/water separator with the treated wastewater.

3.6 Evaporation

Evaporation is used to concentrate a particular wastewater by evaporating the solvent or the aqueous phase. The targeted/valuable product is the concentrated solution of the solute while the vapor is condensed and reused or discarded. If the targeted/valuable product is the vapor or the condensed solvent, then the process is known as *distillation*. An example of distillation is the production of pure water (solute-free solvent) from seawater. Single- and multiple-effect evaporators are used for evaporation or distillation. Steam is typically used to heat the liquid waste to the required boiling point. Factors to be considered for evaporators include: (1) *concentration* — as concentration increases, the density and viscosity of the solution may become saturated or viscous, resulting in crystallization/precipitation/scaling that may clog the heat transfer tubes, or heat transfer may not be adequate because the properties of the viscous solution are different from the starting material; and (2) *foaming*, which may result in entrainment and carryover of the valuable product in the evaporation process or contamination of the valuable product, as in distillation.

4 CHEMICAL TREATMENT TECHNOLOGIES

4.1 Neutralization

Neutralization is a process where acid regents are added to an alkaline wastewater or alkaline regents are added to an acidic wastewater to adjust the pH of the wastewater to a more acceptable pH for subsequent treatment or disposal into municipal sewer. The typical acceptable pH range before it can be discharged into the sewer is between 6.0 and 9.0. Typical acid reagents used in neutralization are sulfuric acid and nitric acid, and in some cases waste acid streams from the manufacturing process. Alkaline reagents used are sodium hydroxide, potassium hydroxide, or alkaline waste streams from the manufacturing process. The process for neutralization is fairly simple and is accomplished in a mixing tank with a pH sensor. The pH sensor monitors the pH of the treated wastewater and adds the needed amount of acid or base according to neutralize the wastewater. The amount of reagent needed is usually determined by conducting an acid- or base-titration curve.

4.2 Precipitation

Precipitation is a common treatment method for metal-finishing waste streams. The physical state of dissolved metals is altered by adding precipitating chemicals such that the solubility of a metal compound is exceeded, resulting in an insoluble phase termed as the *precipitate*. The precipitation reaction can be generalized as follows:

$$A_zB_y \text{ (s)} \Longleftrightarrow zA^{Y+} + yB^{Z-}$$

where A is the metal or cation, B is the anion, and z and y are number of molecular units in the compound. The product of the activities of the species involved in the precipitation is represented by the solubility product, K_{sp}, which provides an indication of the extent of the solubility of the compound:

$$K_{sp} = \{A^{y+z}\{B^{z-}\}^y$$

The process can be reversed, with the precipitate dissolving in the aqueous phase when the activities of the precipitate species in the aqueous phase are less than the solubility of the precipitate or when the environmental conditions such as pH and redox are changed. For example, typical K_{sp} values of various lead compounds (e.g., $Pb(OH)_2 = 10^{-14.3}$, $PbSO_4 = 10^{-7.8}$, $PbS = 10^{-27.0}$ and $PbCO_3 = 10^{-13.1}$) indicate that the least soluble of the four compounds is PbS, while the most soluble is $PbSO_4$. Although K_{sp} provides an indication of the solubility of the compound, metal ions in wastewater interact with other ions or molecules to form complex ions or coordination compounds, and this may also affect their solubilities. Examples of complexes formed include hydro-, cyano-, and ammonium complexes when cyanide and ammonium ions are present.

A common approach in precipitating metals from wastewater is to precipitate the metals as metal hydro-complexes. Lime $(Ca(OH)_2)$ and sodium hydroxide (NaOH) are typically used. Lime is widely used and is the less expensive of the two. The formation of various metal hydro-complexes is dependent on the solution pH. Figure 3 shows the solubilities of several metal hydroxides at different pH in equilibrium with the metal precipitate. Hydro-complexes of metals are amphoteric resulting in a minimum solubility between pH 9 and 11 but are fairly soluble at elevated pHs (>11) and at low pH or acidic conditions (<6). The

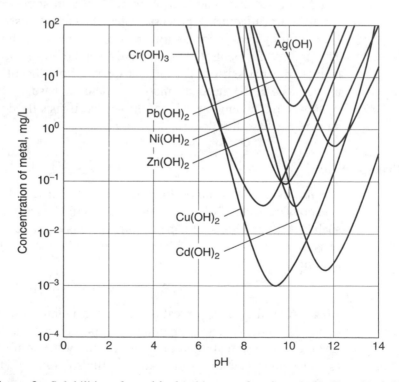

Figure 3 Solubilities of metal hydroxides as a function of pH. (From Ref. 32.)

pH range of the minimum solubility depends on the metal and other constituents present in the wastewater.

As indicated earlier (using lead as an example), metals removal from wastewater by precipitation can be maximized by precipitating the metals in the form of metal sulfide. Figure 4 shows the solubilities of several metal sulfide compounds at different pH in the equilibrium with the metal precipitate. Chemicals used for sulfide precipitation are sodium sulfide or bisulfide (Na_2S or $NaHS$). Extremely low concentrations of the metals can be obtained using sulfide precipitation. Since it is possible that hydrogen sulfide (H_2S) gas may be generated during treatment, sulfide precipitation is generally operated under alkaline conditions.

A typical metal precipitation system is shown in Figure 5. Some systems have equalization basins to equalize flow into the treatment system, along with a prereaction tank for pH adjustment. The central part of the treatment plant is the rapid mixing tank with a detention time of between 1 and 5 minutes, where coagulating chemicals are added and a slow mix tank or flocculation tank with a detention time of 20 to 30 minutes for the agglomeration of metal precipitates or flocs. Organic polymers may be added to aid in the flocculation. The precipitates are settled out in a settling tank or clarification tank with typical overflow rates

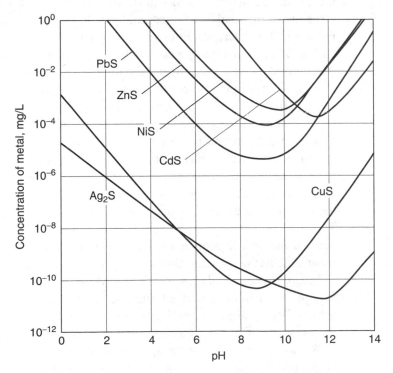

Figure 4 Solubilities of metal sulfides as a function of pH. (From Ref. 32.)

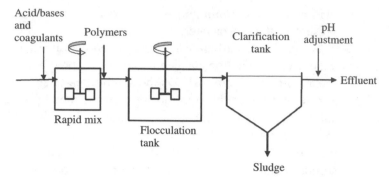

Figure 5   Schematic diagram for chemical precipitation system.

ranging from between 1 and 5 m/hr. The precipitates are either reprocessed to recover the metals or dewatered and stabilized before disposal.

4.3   Chemical Oxidation and Reduction

Chemical oxidation is a well-established technology for the destruction of a wide range of compounds to harmless or less toxic pollutants by an oxidizing agent. Compounds that can be degraded include chlorinated hydrocarbons, aromatic compounds and inorganics such as cyanide and ammonia. Chemical reduction, however, has limited applications but is effective in the destruction of certain halogenated compounds such as tetrachloroethene, trichloroethene, carbon tetrachloride, and is used to alter the ionic charge of metal ions to facilitate precipitation. An example is the reduction of dichromate anions (Cr(VI)) to chromium ion (Cr(III)), which can be precipitated as an hydroxide.

Common oxidizing agents used in the oxidation of hazardous waste include ozone (O_3), hydrogen peroxide (H_2O_2), and chlorine. In recent years, the application of chemical oxidation has advanced by using a combination of oxidants to enhance oxidation and the use of various techniques such as ultraviolet light (UV), ultrasound (US) and inorganic catalysts to increase the oxidizing power of the oxidants by generating oxidizing radicals such as hydroxyl radicals (OH·).[5] The oxidizing power of radicals is several orders of magnitude more powerful than ozone and hydrogen peroxide, but the hydroxyl radicals have extremely short half-lives. Systems that are engineered to produce hydroxyl radicals are called *advanced oxidation processes (AOPs)*. The more common AOP applications are Fenton's reagent (use of hydrogen peroxide and Fe(II) catalyst), ozone/UV, hydrogen peroxide/ozone/UV, and ozone or hydrogen peroxide in the presence of a metal catalyst such as titanium dioxide (TiO_2).

A classical example of the application of chlorine in industrial waste treatment is the alkaline destruction of cyanide wastes. The process is maintained at pH greater than 10 and cyanide is destroyed in a two step process using separate tanks. The reactions where cyanide is oxidized to cyanate (CNO^-), followed complete destruction to carbon dioxide and nitrogen is written as follows:

$$NaCN + Cl_2 + 2NaOH \Rightarrow NaCNO + 2NaCl + H_2O$$

Cyanate is oxidized to carbon dioxide and nitrogen:

$$2NaCNO + 3\,Cl_2 + 4\,NaOH \Rightarrow 2CO_2 + N_2 + 6NaCl + 2H_2O$$

However, in a typical wastewater, cyanides are complexed with copper, nickel, and precious metals, which slow the destruction of cyanide as compared to free cyanide. Excess chlorine is needed to oxidize these cyanide complexes. Reaction times for the complete destruction of cyanide ranged from 60 to 120 minutes.

AOPs using a combination of O_3 and/or H_2O_2 with UV irradiation have been successfully applied for the oxidation of organic compounds. Commercially available systems include Calgon perox-pure UV/H_2O_2, US Filter UV/O_2/H_2O_2, Calgon Rayox® UV/H_2O_2, Magnum CAV-OX® UV/H_2O_2, Wedeco UV/H_2O_2, Wedeco UV/O_3, and Matrix UV/TiO_2. The flow diagram of the US Filter UV/O_2/H_2O_2 is presented in Figure 6. The hydraulic retention time of the system is about 5 minutes. More than 99 percent destruction of chlorinated compounds have been reported as summarized in Table 5 for a pilot study treating water contaminated with chlorinated compounds. Reactions for a UV/O_2/H_2O_2 system are assumed to proceed by H_2O_2 initiating the decomposition of O_3 to form hydroperoxide ion (HO_2^-) which, in turn, react with ozone to produce ozonide ions (O_3^-) and hydroxyl radicals (HO·).

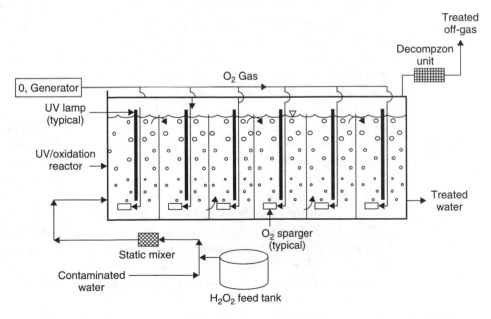

Figure 6 Flow diagram of U.S. filter ozone/H_2O_2/UV system. (From Ref. 30.)

Table 5   Pilot Study Results for Calgon UV/H_2O_2 and US Filter O_3/H_2O_2/UV Systems

	Compounds	Initial Concentration (μg/L)	Percent Removal (%)
Calgon UV/H_2O_2	Benzene	52	>96
	Chlorobenzene	3,100	>99.9
	Chloroform	41–240	93.6 to >97
	1,1-Dichloroethane	120–400	>95.8 to >99.5
	1,2-DCA	22	>92
	Tetarchloroethylene	63–2,500	>98.7 to > 99.9
US Filter O_3/H_2O_2/UV	1, 1-Dichloroethane	9.5 to 13	65
	1,1,1- trichloroethane	2 to 4.5	87
	Trichloroethylene	50 to 520	>99

Note: From Ref. 32.

4.4   Wet Air Oxidation and Supercritical Water Oxidation

For wastewater containing high organic carbon content in the range of 10,000 to 200,000 mg/L and with refractory content, the use of chemical oxidation or biological treatment may not be cost effective. Under such circumstances, the organic wastes may be oxidized in the liquid phase using wet air oxidation. For wet air oxidation systems, the wastewater is oxidized in the presence of air at elevated pressures and temperatures but below the critical point of water (374°C and 218 atm).[6] Temperatures and pressures of wet air oxidation systems are in the range of 150 to 325°C and 100 to 200 atm, respectively. When the system is operated at temperatures and pressures above the critical point, the system is called *supercritical water oxidation*.[11] Beyond the critical point, the liquid and gas phases exist as a single phase fluid where the solubility of organics is enhanced while the solubility of inorganics in the fluid is decreased by three to four orders of magnitude. The gaslike properties of the fluid enhance contact between the target organics and the oxidizing radicals, maximizing degradation of the target organics. A typical wet air oxidation system is shown in Figure 7.

In wet air oxidation, COD removal between 75 and 90 percent can be achieved. The end products consist of simpler forms of biodegradable compounds such as acetic acid and inorganic salts, along with the formation of carbon dioxide and water. Depending on the pollutants in the wastewater, further treatment of the waste stream may be needed. Residence time of reactor is typically 15 to 120 minutes. The system is adaptable to a wide variety of oxidizable materials and water acts as a heat sink assisting in the control of the temperature within the reactor. Special alloy materials are needed for the reactor due to the high corrosivity of the reactions resulting in high maintenance costs of the system.

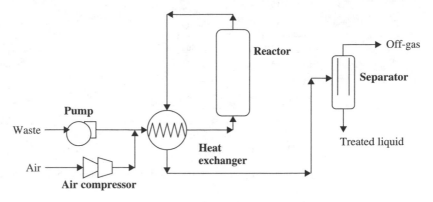

Figure 7 Schematic diagram of a wet air oxidation system.

Residence time for supercritical water oxidation systems may be as short as several minutes at temperatures of 600 to 650°C. More than 99.9 percent conversion of EPA priority pollutants such as chlorinated solvents has been achieved in a pilot-scale plant with retention time less than 5 minutes.[7] The system is limited to treatment of liquid wastes or solids less than 200 microns in diameter. Char formation during reaction may impact the oxidation time of the organics, while separation of inorganic salts during the process may be a problem. Typical materials for the reactor are Hastelloy C-276 and Iconel 625 (high nickel alloys), which can withstand high temperatures and pressures and the corrosive conditions.

4.5 Electrochemical Process

In an electrochemical process, pollutants in the liquid wastes are chemically oxidized and reduced by applying electricity across appropriate electrodes to create the oxidation and reduction potential instead of using an external oxidizing agent. Reactions are conducted in an electrochemical cell and may be enhanced by adding oxidizing chemicals. Electrodes used are of special materials allowing for selectivity in pollutants removed and may at the same time prevent the production of unwanted byproducts.[8] In some processes, a separation membrane may be used in the electrochemical cell to improve removal of specific pollutants. Chemical reactions in an electrochemical cell can be controlled by controlling the electrode potential and the environment at the surface of the electrodes. Electrochemical processes can be viewed as reactions due to direct electrolysis (reactions at the cathode or the anode) or reactions due to indirect electrolysis. Figure 8 illustrates the two processes.

Examples of direct electrolysis reactions include removal of specific metals by cathodic deposition where a metal ion is reduced by accepting electrons at

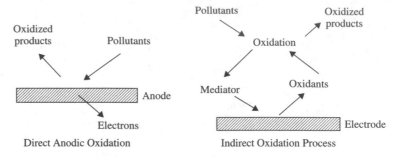

Figure 8   Direct and indirect electrochemical reactions.

the cathode. To minimize side reactions and to provide a certain level of selectivity, ion selective membranes of thin polymeric materials may be used within the electrochemical cell. Indirect electrolysis electrochemically generates redox reagents as a chemical reactant to convert pollutants to less harmful products. The redox reagent acts as an intermediary for shuttling electrons between the pollutant substrate and the electrode. An example is the generation of chlorine (Cl_2) from chloride (Cl^-) in the waste solution at the anode, which in turn is used as an oxidizing agent to oxidize the pollutants.

A commercial application of electrochemical cells is the CerOx process, a mediated electrochemical oxidation (MEO), or catalyzed electrochemical oxidation (CEO), which has been found to successfully destroy PCBs at a concentration of 2 mg/L in alcohol in a patented electrochemical cell (called the T-Cell). The process uses, a cerium metal ion, Ce^{4+}, placed in contact with an organic compound, which reduces the cerium ion to Ce^{3+}. The process operates at low temperature (90 to 95°C) and near atmospheric pressure. The CerOx process may also treat wastes containing refractory pesticide compounds such as DDT, silvex and chlordane, and pharmaceuticals wastewaters.[9] Advantages of electrochemical processes include versatility in treatment—treating small to large volumes, oxidation and reduction of pollutants directly or indirectly, low energy use in comparison with thermal processes, and control of the reactions with a certain level of selectivity. Some of the disadvantages include stability of the electrodes, electrodes fouling, mass transfer limitations due to the size of electrode area, and reactions that are dependent on the conducting medium, electrolyte.

4.6   Adsorption and Ion Exchange

Waste pollutants can be removed from the waste stream by preferential accumulation of the pollutants at the surface of a solid phase or adsorbent. Adsorption is one of the more widely applied technologies for the treatment of industrial wastewater. Ion exchange is a form of adsorption whereby an ion in the solid phase is replaced by another ion in a solution in contact with the solid. Since replacement takes place at the interface, this process may be classified as adsorption. Adsorption is used to remove a wide range of pollutants, from synthetic organic chemicals such as pesticides and petroleum hydrocarbons to inorganic compounds such as heavy metals and anions such as perchlorate. Common adsorbents used for wastewater are activated carbon and synthetic ion exchange resins. Other adsorbents include activated alumina, forage sponge, and clays.

Adsorption isotherms are typically used to describe the equilibrium relationship between the bulk aqueous phase activity (concentration) of the adsorbate and the amount adsorbed on the interface at a given temperature. The more common models used are empirical models such as the linear model and the Freundlich model. First principles models such as the Langmuir model are also used.

Activated carbon comes in two forms: granular activated carbon (GAC) (average diameters range from 0.42 mm to 2.38 mm) and powdered activated carbon (PAC) (average diameter about 44 microns). GAC and PAC are made from wood, peat, lignite, bituminous coal, and coconut shells. The surface of GAC consists of different functional groups such as OH-, COO- that can adsorb a range of different compounds.[2] Most GAC systems are used as a tertiary process and are preceded with solids removal and filtration to minimize fouling of the GAC.

Synthetic ion exchange materials are made of cross-linked polymer matrix with charged functional groups attached by covalent bonding. The base material is polystyrene and is cross-linked for structural stability with 3 to 8 percent divinylbenzene. Functional groups on the base material include sulfonate (SO_3^-) groups for strong acid ion exchange resins and quaternary amine ($N(CH_3)_3^+$) groups for strong basic ion exchange resins. Cation resins are regenerated with strong acids such as sulfuric acid, nitric acid, and hydrochloric acid, while anion resins are regenerated by a strong base such as sodium hydroxide.

The reactor configuration, typically used for carbon adsorption and ion exchange systems, is the *fixed-bed system*. Important design parameters for GAC and ion-exchange fixed-bed systems include the type of GAC or ion exchange resin used, surface loading rate, GAC or ion exchange resin usage rate in terms of mass of GAC or ion exchange resin used per volume of water treated, GAC or ion exchange resin depth or volume based on required breakthrough of a pollutant, and the empty bed contact time. Methods used in sizing of GAC fixed-bed systems include the use of pilot- and laboratory-scale column tests such as the rapid small-scale column tests (RSSCTs), the bed depth service time method (BDST) and the kinetic approach.[2,10,12] Typical GAC surface loading rates are between 2 and 30 m/hr, average GAC depth is 1 meter and the empty bed contact times are between 10 and 30 minutes. In the case of ion exchange systems, the volume of resin needed to treat a wastewater is based on pilot studies or estimated based on data provided by resin suppliers.[13] Typical surface-loading rates range from 5 to 60 m/hr, and the resin bed depth for co-current regeneration and counter-current packed bed systems are 1.2 meter and 2 meters, respectively.

5 BIOLOGICAL WASTE TREATMENT

Biological processes harness the metabolic abilities of microorganisms to degrade organic materials (both dissolved and suspended organics) into stable and simple end products, along with the production of energy for growth and maintenance and the synthesis of new microbial cells. Several factors impact the microbial processes. These factors may be broadly divided into substrate-related, micro-organisms-related, and environmental-related. Substrate-related factors include the impact of the physical-chemical properties (e.g., structure of organic compounds, solubility, sorption) and concentrations on the degradation of the compound. Microorganisms-related variables include selection and acclimatization of the microorganisms to the compound, presence of certain species of microorganisms, and the required enzyme systems. Environmental-related conditions include the presence of electron acceptors, pH of the wastewater, presence of nutrients, temperature, and total dissolved solids. Two of the important environmentally related conditions in wastewater are the availability of electron acceptors and nutrients. To facilitate the microbial oxidation-reduction reaction, oxygen is used as an electron acceptor for aerobic metabolism while nitrate, Mn(IV) and Fe(III), sulfate, and carbon dioxide are used for anaerobic metabolism. The electron acceptor that derives the maximum free energy by the microorganisms will be used first. In a closed environment with a fixed amount of oxygen, oxygen will be used up first by the dominant aerobic heterotrophs. When the oxygen is used up, denitrifiers will use nitrate as electron acceptors followed by sulfate reducers, fermenters, and finally methanogens. The two nutrients that are most likely to be limiting are nitrogen and phosphorus. Nutrients must be provided at a minimum level in order to sustain microbial growth. The approximate formula for a bacteria cell is $C_5H_7O_2NP_{0.074}$ indicating that the ratio of C:N for cell synthesis and as an energy source is approximately 10:1. In typical wastewater systems, a ratio of C:N:P of 100:10:1 is used. Micronutrients such as sulfur, potassium, calcium, magnesium, iron, cobalt, and molybdenum are also needed.

5.1 Treatment Processes

Biological processes can be divided into aerobic (oxygen as the electron acceptors) or anaerobic (nitrate, sulfate, Fe(III) and carbon dioxide as the electron acceptors). For each biological process using different electron acceptors, treatment technologies may be classified as suspended growth, fixed film (biofilm), or hybrid (combination of biofilm and suspended growth). Examples of aerobic-suspended growth technologies include the activated sludge process and the sequencing batch reactor, while anaerobic systems include the conventional high-rate anaerobic digesters. In recent years, membrane bioreactors have emerged as a very strong contender to replace the activated sludge system for suspended growth technologies. Examples of aerobic fixed-film technologies include trickling filters and rotating biological reactors, while anaerobic fixed-film technologies include anaerobic biofilters. Some of the newer fixed-film technologies include the biological aerated filters or the combined oxic-anoxic biological filters. Biological treatment is widely used for wastewater from pulp and paper, food processing, oil and gas, and petrochemical industries.

5.2 Activated Sludge Process

Suspended solids consisting of microbial cells and inert materials are kept in suspension in the aeration basin of the activated sludge process by continuous aeration of the basin with air or oxygen. The mixed liquor (suspended solids and treated effluent) are transferred to a sedimentation tank where suspended solids and the treated effluent are separated by gravity settling (see Figure 9). Part of the suspended solids at the bottom of the sedimentation (called *sludge*) is returned to the aeration tank to maintain a suspended solids concentration in the aeration tank. The rest of the sludge are wasted and treated accordingly. The efficiency of the activated sludge process and the quality of the final effluent are highly dependent on maintaining the active biomass in the aeration basin, the settling characteristics of microbial sludge produced, and the performance of the settling tanks.[14] Operational characteristics play an important role in avoiding poor settleability and sludge bulking.[15]

The hydraulic retention times (HRTs) of the activated sludge range from 4 hours to as high as 24 hours. With recirculation of sludge and wasting in the sedimentation tank, the suspended solids in the aeration basin are maintained at a concentration of 2,500 to 4,000 mg/L, while the sludge-retention times (SRTs) are between 10 and 30 days.

5.3 Sequencing Batch Reactors

Sequencing batch bioreactors (SBRs), as indicated by its name, operate in a batch mode with two or more bioreactors in a typical system.[16] Treatment in a SBR is accomplished by operating the reactor in a sequence of events within a cycle (see Figure 10). For a treatment plant with two SBRs, wastewater is

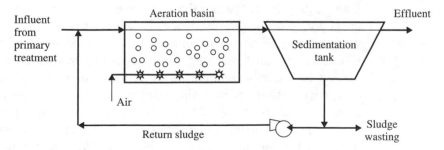

Figure 9 Schematic diagram of an activated sludge process.

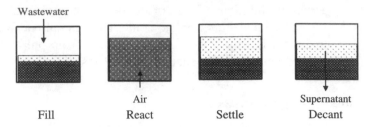

Figure 10   Different sequences within a cycle in a sequencing batch bioreactor.

added, in the *fill* sequence, to the first SBR that contains the biomass for the biochemical reactions from the previous cycle. When the first SBR is filled, the influent wastewater is diverted to second SBR. The wastewater in the first SBR is aerated and treatment of the wastewater occurs. This is called the *react* sequence. After the react sequence, the air is turned off and the biomass is allowed to settle in the *settle* sequence. The settled supernatant is then decanted in the *decant* sequence and discharged, leaving behind the biomass for the next cycle, where the sequences are repeated. Typically about a third of the reactor is decanted. The time period for a cycle varies from 4 hours to as much as 24 hours, depending on the wastewater. The biomass in the SBR is typically maintained between 2,000 and 4,000 mg/L.

The SBR is highly flexible in that the sequences can be manipulated for different redox conditions (anaerobic, anoxic, and oxic conditions), allowing for the treatment of organic compounds that are more amenable to anaerobic degradation followed by oxic degradation. A novel variation of the sequencing batch bioreactor is the hybrid system, called the sequencing batch biofilm bioreactor (SBBR). The SBBR has a support media that allows a biofilm to grow, along with maintaining suspended growth within the reactor.[17] This allows the SBBR to retain a higher concentration of biomass within the bioreactor, resulting in higher biochemical reactions.

5.4   Membrane Bioreactors

Poor settling of sludge and the requirement of large sedimentation tanks surface areas are some of the disadvantages with activated sludge systems, especially when space availability is limited. To overcome some of these disadvantages, separation of the suspended solids from the treated effluent may be accomplished by using membranes such as microfiltration and ultrafiltration instead of gravity settling, as in sedimentation tanks.[18] These reactors are called membrane bioreactors (MBR). Membranes can be installed either inside (submerged) or outside (sidestream) the reactors. Figure 11 shows the submerged configuration, which is the more common and cost-effective system used for membrane bioreactors.[19] Tubular membrane modules are usually installed for sidestream

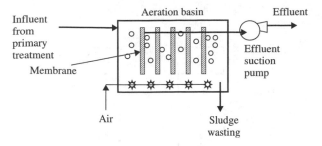

Figure 11 Schematic diagram of a membrane bioreactor.

configuration while plate-frame and hollow fiber modules are generally applied in submerged configuration. Sidestream tubular modules are typically used for treatment of harsh industrial wastewaters such as low pH or high pH wastewaters where ceramic membranes are needed.[20]

Some of the advantages of MBR over conventional activated sludge systems include excellent effluent quality, smaller plant size, lower sludge production, high operational flexibility, high decomposition rate of organics, better process reliability as well as microbial separation and odor control.[21-23] Disadvantages include fouling of the membranes, higher capital cost, and energy consumption, as well as higher aeration requirements than the activated sludge process.

MBR allows the control of SRT independently from HRT resulting in much longer SRTs (typically 30 to 50 days), as compared to activated sludge (10 to 30 days).[23] MBRs operate at high MLSS concentrations (4,000 to 10,000 mg/L) and at low HRTs (4 hours) resulting in a smaller bioreactor volume that can be as small as one-fourth of the size of an activated sludge plant. Sludge produced in MBRs may be as low as 0.22 kg MLSS/kg BOD_5 at 50 days SRT as compared to 0.7 to 1 kg MLSS/kg BOD_5 at 10 to 20 days SRT for activated sludge.[24,25]

5.5 Biological Aerated Filters

Biological aerated filters (BAFs) have a support media for the growth of a biofilm and are fully submerged with air injected into the reactor.[26] BAFs are well suited as a secondary treatment process or as an add-on treatment process. BAFs are suitable for the upgrade of existing treatment process and can be built modularly.[27] BAFs are operated either in an upflow or downflow mode (Figure 12a and b). Downflow systems with countercurrent air flow have the advantage of efficient mass transfer of oxygen to the biofilm, while upflow systems with co-current air and wastewater flow can handle higher influent flowrates than downflow systems (Figure 12b). In upflow systems, odor problems are reduced because the wastewater is fed from the bottom. Upflow systems can be modified to include different electron acceptors zones. For example, an anoxic (nitrate as the electron acceptor) or anaerobic zone can be created at the bottom

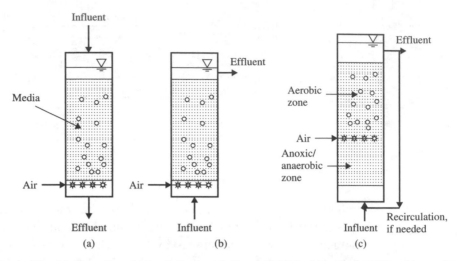

Figure 12 (a) Downflow biological aerated filters (BAFs), (b) upflow BAFs, (c) combined anoxic/oxic upflow BAFs.

of the BAF with an oxic (oxygen as the electron acceptor) zone in the upper half of the BAF by injecting air at a certain depth within the filter (see Figure 12c). For example, the Biostyr® technology marketed by Veolia Water has a combination of oxic-anoxic zones in a single reactor.[28] Media used for BAFs can be sunken type for downflow and upflow configurations, or floating type for upflow systems.[26] Materials used include proprietary materials such as 3- to 5-mm fired clay material, 2- to 4-mm polystyrene beads, a 60:40 mixture of poly-propylene, and calcium carbonate and common materials such as 5-mm diameter sand.[14,29]

High organic loadings of 2.5 kg BOD_5/m^3-day have been reported as compared to 0.06 kg BOD_5/m^3-day for activated sludge plants.[30] HRTs of BAF are typically in the range of 1 to 4 hours, although Pujol et al. reported that HRT as low as 10 minutes did not seemed to influence the reactor performance treating municipal wastewater. Depths of media may range from 1.6 m, to as much as 2.5 m.[31]

6 SUMMARY

The Clean Water Act and the National Pretreatment Program (40 CFR 403) regulate the discharge of wastewaters into bodies of water and into municipal wastewater treatment plants via the sewers. Compliance of these regulations is enforced by U.S. EPA and by state authorities. Industries can draw on the different physical, chemical, and biological treatment technologies to treat industrial wastewaters to stay in line with the regulations and to protect the environment from pollution. Selection of treatment technologies is dependent on the quantity and quality of the wastewater, the final effluent concentrations requirement, capital and operating costs, and on-site constraints such as land availability, labor availability, and labor skill levels. A thorough understanding of the physical principles, the chemical or biological reactions within each treatment process, is essential in the selection of the right treatment process for an industrial wastewater. By combining a series of treatment processes, most wastewaters can be treated to levels where it can be safely discharged. In areas where water is scare, many industries are reusing treated wastewater for utility use or for noncontact water such as flushing of toilets. As more and more industries explore different ways of implementing sustainability in their operations, conservation and reuse of water will become an important factor.

REFERENCES

1. P. L. McCarthy, "Removal of Organic Substances from Water by Air Stripping in Control of Organic Substances," in *Water and Wastewater*, B. B. Berger (ed.), EPA-600/8–83–011, Office of Research and Development, 1983.

2. J. C. Crittenden, R. R. Trussell, D. W. Hand, K. J. Howe, and G. Tchobanoglous, "Water Treatment: Principles and Design," 2nd ed., John Wiley, Hoboken, NJ, 2005.

3. J. Officer, J. A. Ostrowski, and P. J. Woollard, "The Design and Operation of Conventional and Novel Flotation Systems on a Number of Impounded Water Types: Particle Removal from Reservoirs and Other Surface Waters," *Water Sci. & Technol., Water Supply*, **1**(1), 63–69, (2000).

4. U.S. Army, "Oil/Water Separator Installation and Maintenance: Lessons Learned," U.S. Army Center for Public Works Technical Note, Aberdeen, Maryland, 31 October 1996.

5. J. Hoigne and H. Bader, "The Role of Hydroxyl Radical Reactions in Ozonation Processes in Aqueous Solutions," *Water Res*, **10**(377) (1976).

6. W. M. Copa and W. B. Gitchel, "Wet Air Oxidation," in *Standard Handbook of Hazardous Waste Treatment and Disposal*, H. M. Freeman (ed.), McGraw-Hill, New York, 1988.

7. E. F. Gloyna and L. Li, "Progress in Supercritical Water Oxidation: Research and Development," Fifth Int'l. Chem. Oxidation Symposium and Principles and Practices Workshop, Nashville, Tennessee, 1995.

8. K. Juttner, U. Galla, and H. Schmieder, and "Electrochemical Approaches to Environmental Problems in the Process Industry," *Electrochimica Acta*, **45**, 2575–2594 (2000).

9. "Electrochemical Process Oxidizes PCBs," *Chemical Engineering Progress*, **96**(12), 17 (2000).

10. R. L. Droste, *Theory and Practice of Water and Wastewater Treatment*, John Wiley, New York, 1997.

11. M. Modell, "Supercritical Water Oxidation," in *Standard Handbook of Hazardous Waste Treatment and Disposal*, H. M. Freeman (ed.), McGraw-Hill, New York, 1989.

12. T. D. Reynolds and P.A. Richards, *Unit Operations and Processes in Environmental Engineering*, PWS Publishing Company, Boston, 1996.

13. Dow, "Dowex Marathon C, Ion Exchange Resins: Engineering Information," Dow Chemical Company, Midland, Michigan, 2002.

14. Metcalf and Eddy, *Wastewater Engineering, Treatment and Reuse*, 4th edition. McGraw Hill Inc., New York, 2003.

15. L.D. Benefield and C.W. Randall, *Biological Process Design for Wastewater Treatment*, Prentice-Hall, Englewood Cliffs, NJ, 1980.

16. G. Demoulin, A. Rudiger, M. C. Goronszy, "Cyclic Activated Sludge Technology—Recent Operating Experience," *Water Science and Technology*, **43**(3), 331–337 (2001).

17. R. S. Protzman, P. H. Lee, S. K. Ong, and T. B. Moorman, "Treatment of Formulated Atrazine Rinsate by *Agrobacterium radiobacter* Strain J14a in a Sequencing Batch Biofilm Reactor," *Water Research*, **33**(6), 1399–1404 (1999).

18. T. Stephenson, S. Judd, B. Jefferson, and K. Brindle, *Membrane Bioreactors for Wastewater Treatment*, IWA Publishing, London, 2000.

19. M. D. Knoblock, P. M. Sutton, and P. N. Mishra, "Lessons Learned from Operation of Membrane Bioreactors," CD-ROM Proceedings of WEFTEC 71st Annual Conference and Exposition, Orlando, Florida, October 3–7, 1998.

20. C. B. Ersu and S. K. Ong, "Operating Characteristics and Treatment Performance of a Membrane Bioreactor Using Tubular Ceramic Membrane," IWA Environmental Biotechnology: Advancement on Water and Wastewater Applications in the Tropics, Kuala Lumpur, Malaysia, December 9–10, 2003.

21. B. Zhang, K. Yamamoto, S. Ohgaki, and N. Kamiko, "Floc Size Distribution and Bacterial Activities in Membrane Separation Activated Sludge Processes for Small-scale Wastewater Treatment/Reclamation," *Water Science and Technology*, **35**(6), 37–44 (1997).

22. G. Crawford, D. Thompson, J. Lozier, G. Daigger, and E. Fleischer, "Membrane Bioreactors-A Designer's Perspective," CD-ROM Proceedings of WEFTEC 73rd Annual Conference and Exposition, October 14–18, 2000, Anaheim, California.

23. C. Visvanathan, R. Ben Aim, and K. Parameshwaran, "Membrane Separation Bioreactors for Wastewater Treatment," *Critical Reviews in Environmental Science and Technology*, **30**(1), 1–48 (2000).

24. K. Takeuchi, O. Futamura, and R. Kojima, *Integrated Type Membrane Separation Activated Sludge Process for Small Scale Sewage Treatment Plants*. Ebara Infilco Ltd., Tokyo, 1990.

25. Hsu M. and T. E. Wilson, "Activated Sludge Treatment of Municipal Wastewater—USA Practice," in *Activated Sludge Process Design and Control: Theory and Practice*, Volume 1, W. W. Eckenfelder and P. Grau (eds.), Technomic Publishing Co., Inc, 1992.

26. L. G. Mendoza-Espinosa and T. Stephenson, "A Review of Biological Aerated Filters (BAFs) for Wastewater Treatment," *Environ. Engineering Sci.*, **16**(3):201–216 (1999).

27. W. S. M'Coy, "Biological Aerated Filters: A New Alternative," *Water Environment and Technology*, **42**, 39–42 (February 1997).

28. V. R. Borregaard, "Experience with Nutrient Removal in a Fixed-Film System at Full-Scale Wastewater Treatment Plants," *Wat. Sci. Tech.*, **36**(1), 129–137 (1997).

29. J. H. Ha, S. K. Ong, and R. Surampalli, "Nitrification and Denitrification in Partially Aerated Biological Aerated Filter (BAF) with Dual Size Sand Media," IWA Specialty Conference, Wastewater Reclamation and Reuse for Sustainability, Jeju, Korea, November 8–11, 2005.

30. A. J. Smith, J. J. Quinn, and P. J. Hardy, "The Development of an Aerated Filter Package Plant," in 1st Int. Conf. Advances in Water Treat. and Environ. Man., Lyon, France, June 27–29, 1990.

31. R. Pujol, H. Lemmel, and M. Gousailles, "A Keypoint of Nitrification in an Upflow Biofiltration Reactor," *Wat. Sci. Tech.*, **38**(3), 43–49 (1998).

32. U.S. EPA, *Development Document for Effluent Limitations Guidelines and Standards for the Metal Finishing Point Source Category*, EPA 440/1-83-091, Environmental Protection Agency, Washington, DC, 1983.

33. U.S. EPA, *Advanced Photochemical Oxidation Processes*, EPA l6251R-981004, Office of Research and Development, Environmental Protection Agency, Washington, DC, 1998.

CHAPTER 9

THE ENVIRONMENTAL IMPACTS OF PACKAGING

Eva Pongrácz
University of Oulu, Finland Department of Process and Environmental Engineering

1 INTRODUCTION 238

2 FUNCTIONS OF PACKAGING 239
  2.1 Protection Function 239
  2.2 Distribution Function 240
  2.3 Household Function 240
  2.4 Intermediate Function 240
  2.5 Advertising Function 241
  2.6 Image-component Function 241
  2.7 Value-forming Function 241
  2.8 Waste-Reduction Function 241

3 PACKAGING MATERIALS 242
  3.1 Paper/board 242
  3.2 Glass 243
  3.3 Steel 243
  3.4 Aluminum 244
  3.5 Plastics 245
  3.6 Composites 245
  3.7 Degradable Plastics 247
  3.8 Wood 248

4 CONSUMPTION OF PACKAGING MATERIALS 248

5 ENERGY USE 250

6 ROLE OF PACKAGING IN POLLUTION 251
  6.1 Litter 251
  6.2 Water Pollution 252
  6.3 Air Pollution 252
  6.4 Solid Wastes 253

7 ENVIRONMENTAL ASSESSMENT OF PACKAGING MATERIALS 253

8 RECOVERY OF POSTCONSUMER PACKAGING 255
  8.1 Waste Packaging in Municipal Solid Waste 255
  8.2 Packaging Waste Reduction 256
  8.3 Choice of Waste Management Options 257
  8.4 Recycling of Packaging 258
  8.5 Recycling of Plastic Packages 259
  8.6 Feedstock Recycling 260
  8.7 Comparison of Mechanical and Feedstock Recycling 262
  8.8 Glass Recycling 263
  8.9 Steel Recycling 264
  8.10 Recycling Aluminum Packages 264
  8.11 Recycling Multimaterial Packages 264
  8.12 Safety of Recycled Materials in Packages 265
  8.13 Disposal of Biodegradable Polymers 266
  8.14 Used Packages as a Source of Energy 267
  8.15 Recovery of Wooden Packaging 268
  8.16 Construction Materials from Waste 268

9 PACKAGING SYSTEMS 268

10 LIFE-CYCLE ASSESSMENT OF PACKAGING SYSTEMS 272

11 CONCLUSIONS 273

1 INTRODUCTION

The need for packaging and the development of packaging was caused by the fact that the production and the consumption took place at separate places and times, and the produced goods had to be distributed and transported. Packaging became a connecting link between production and consumption, and the importance of this link is growing in urbanized societies. More than 150,000 people are being added to urban population in developing countries every day. In the mid-twentieth century, only one-third of the world's population was urban. The prediction is that by 2025, two-third of the world's people will live in cities. This means that more people will live in cities than occupied the whole planet in the 1980s.[1] In such level of urbanization, distribution of goods, especially food, is crucial, and the role of packaging is enormous.

Since the 1970s, when litter was of significant concern, packaging has often been associated with wasteful behavior. This is partly due to the fact that packaging wastes are a very visible part of environmental problems. The negative image of packaging does not, however, translate into consumer hostility at point of sale. Products and not packages are bought. The package is not noticed during purchase, transport, and use of the product—in fact, it is not noticed until the minute the product is consumed and the package had fulfilled its function and turns into waste. At that minute, the package is already seen as an environmental burden, wasting resources. Those concerned about the state of the environment can take part in reducing this burden through packaging recovery programs.

Packaging has positive and negative impacts on the environment. The negative impacts include resources use and the effects of packaging-related wastes and emissions. The positive impact is that packaging consumer goods facilitates their distribution, and thus makes it possible to obtain goods otherwise not

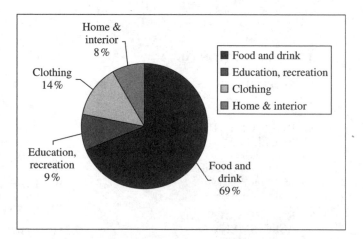

Figure 1 Packaging use by type of goods. (From Ref. 2.)

accessible. Environmentally conscious packaging enables satisfying human needs in an effective way.

In developed countries, food packaging represents more than two-thirds of all packaging.[2] Due to this reason, this chapter is mainly concerned with food packaging. Figure 1 illustrates the breakdown of packaging types.

2 FUNCTIONS OF PACKAGING

Although the package is tailor-made for the product, all products are made for consumers. Until the middle of the twentieth century, groceries were a meeting point: People talked there, discussed with the shopkeeper, asked information about the products. In modern supermarkets the function of the shopkeeper is taken by packaging. Modern packaging is an expressive form of the consumer lifestyle, giving character to the product. Geiger listed the functions of packaging as follows: protection, distribution, household, intermediate, advertisement, image-component and value forming functions.[3] In addition, packaging, especially food packaging, can have an important waste reduction function.[4]

2.1 Protection Function

Today, an important function of packaging is protection and preservation. This was not always the case. Until the early nineteenth century, food was preserved by salting, smoking, and drying. The situation changed, however, during the Napoleonic Wars. While on the campaign trail, French troops suffered from scurvy and starvation. Napoleon issued a challenge for a better way to preserve food, and in 1809 Nicolas Appert discovered that cooked foods could be kept from spoiling if the air were eliminated. Appert preserved food by boiling it and packing it in jars, thus creating the process of canning.[5]

In the 1860s, Louis Pasteur discovered why removing air preserves food. Air carries living organisms, including mold and bacteria. By heating liquids such as milk and beer to at least 70°C for 15 seconds, these harmful organisms can be killed.[6] Today, food spoilage in Western Europe is less than 3 percent for processed food and 10 to 15 percent for fresh food. In lesser-developed countries where packaging is minimal, food spoilage can reach 50 percent.[7]

The protective function is more and more important in the present trend of increasing urbanization. For example, in Finland, the Finnish Association of Packaging Technology and Research concluded that future packaging trends do not depend on materials on hand but on more important factors such as product protection and distribution. Recently, the use of active packaging became widespread, in which packaging is combined with the use of means that assure the preservation of the product, such as protective gas, oxygen removal, and so on. *Smart* packages are also more common: The packaging includes an indicator for additional safety, with which one can follow the state of the product, such as temperature, leaking, spoilage, and so on.[8]

2.2 Distribution Function

Packaging helps loading, collection, and transport of the product. Distribution of bulk and liquid products is virtually impossible without packaging. Protective packaging such as bubble wrap or foam peanuts ensures safe journey. Corrugated paperboard and polystyrene foam hold expensive electronics equipment securely in their cartons and cushion them against falls, shifts, and bumps. Prior to loading onto ships, trucks, or planes, these cartons are stacked on pallets and wrapped with a sheet of self-clinging stretch wrap. This very strong, yet thin, film stabilizes the load, keeping it from shifting and falling. Fewer falls mean reduced damage and breakage, keeping both waste and related disposal costs to a minimum.

Fragmentation of consumer markets is considered one of the major challenges of the future of packaging. Packages should function both in the traditional and new channels of distribution, such as via the Internet. In the latter case, the package must protect the product delivered in the same transport package together with other products that may require different storage temperatures.[9]

2.3 Household Function

Some packages directly enhance consumption or further preparation of the product. Probably, the most famous example is the *TV dinner*, which allowed meals to go from the freezer to the oven to the table. Later, the metal tray was replaced by a plastic one, permitting even faster food preparation in the microwave oven. Many packages make it easy for us to use the products they contain (e.g., squeezable bottles, reclosable liquid board containers, plastic bottles with handles, and pop-up dispenser tops).

A change in social trends has been noted in Finland. Although home-cooked meals still hold value, few women can afford the time it involves on a daily basis. Because of the growth in jobs held by women in the professional sector, after a busy working day, convenience calls for pre-prepared food. However, the social aspect of the family spending time together is still significant.[8] Interviews conducted in Finland led to the conclusion that consumers anticipate growth of take-away dining in the future, which will increase the demand for convenience packages in which the food can be delivered, heated, and served by the consumers.[9]

2.4 Intermediate Function

The intermediate function of packaging is very important in modern marketing. The product is offering itself, and promotes the meeting with the purchaser. The package takes over the role of the sellers, helps to make a favorable impression, aids identification, and stimulates purchase.

2.5   Advertising Function

Today, it is universally acknowledged that packaging decisions can have a significant impact on sales.[10] A visually pleasing package attracts attention, which is important in an increasingly competitive environment. Generally speaking, a new, more appealing, and/or visually effective packaging system is unlikely to immediately change the well-established shopping habits of people who do not buy the brand. Instead, the impact is more subtle: A new design may drive nonusers to take a second look at the brand, shift their perceptions somewhat, and perhaps lead them to consider it as an acceptable alternative. Above all, the package offers information about the content, the product itself. It is a message of the manufacturer to the consumer. Food packages contain preparation instructions; they also provide nutritional, dietary, and ingredient data. Packages of all types include safety and storage information, plus any necessary warnings.

2.6   Image-component Function

The brand, trademark, and other media elements are integral parts of a package, promoting the creation of an image. One of the most recognizable packages in the world is the Coca-Cola bottle; its shape is designed to be recognized even in the dark by touch.

2.7   Value-forming Function

From an economic point of view, packaging has a very important role in the sales process. Without packaging, many products, especially bulk products, cannot be sold to the customer. For instance, a barrel full of toothpaste would be very difficult to sell, but portioning it into squeezable tubes makes it possible to put on the market. Thus, packaging creates value for the toothpaste. The role of packaging as a marketing tool will be strengthened in the future.[9]

2.8   Waste-Reduction Function

Packaging reduces waste in two important ways. First, it keeps food from spoiling and having to be discarded. In the United Kingdom, the proportion of food that is unfit for consumption before it reaches the consumer is 2 percent, whereas in developing countries, where packaging is not as widespread, this loss can be in excess of 40 percent.[11] Second, packaging permits foods to be processed more efficiently. For example, 50 years ago, people went to a butcher for chicken. For every 1,000 chickens sold, the butcher threw away 750 kg of feathers, viscera, and other waste products. Today, chicken producers ship the edible parts to market and process the rest into byproducts such as animal feed and fertilizer. It takes only about 7.7 kg of packaging to ship those 1,000 chickens to grocery stores. That 7.7 kg of packaging permits the 750 kg of waste to be used efficiently rather than merely thrown away.[12] As paper, metal, and glass packaging increase, food waste decreases. Increases in plastics packaging create the greatest reductions in food waste.[13] Overall, for every 1 percent increase of packaging, food waste decreases by about 1.6 percent.[14]

It is important to note that while increasing the product-per-package size, one can save on the packaging material, as well as on the unit price of the product for the consumer. This solution is not always generally applicable. Recent trends in Finland point at the increasing number of one- and two-member families that prefer smaller packages.[9] Buying a large package also has the risk that the product will not be consumed within warranty time and thus will be disposed of. Although one ought to aim at overall waste reduction, packaging material cannot be saved to the detriment of product spoilage and discard.

3   PACKAGING MATERIALS

The first packages served as containers, and their principal function was to hold food and water. They were probably taken directly from nature, such as leaves and shells. Later, containers were fashioned from natural materials: wooden logs, woven plant fibers, pouches made from animal skins. The next containers developed by early societies were clay pots, which date back to 6000 B.C. The first known pottery is from Syria, Mesopotamia, and Egypt. Besides being functional, clay bowls, vases, and other vessels were an artistic medium that today provide important clues regarding the culture and values of ancient peoples. Although no longer a significant packaging medium, clay still continues to have a major artistic value.[15]

Today, a wide range of materials are used for packaging applications, including metal, glass, wood, paper or pulp-based materials, plastics, ceramics, or a combination of more than one materials as composites. They are applied in three broad categories of packaging:

1. *Primary packaging,* which creates sales unit and is normally in contact with the goods
2. *Secondary packaging,* or collection packaging such as cardboard boxes, wooden crates, or plastic containers used to carry quantities of primary packaged goods.
3. *Tertiary packaging,* or transport packaging that is used to assist freight transport of large quantities of goods, such as wooden pallets and plastic shrink-wrap

3.1   Paper/board

Paper manufacturing uses cellulose fibers that form bonds with each other. Carton boxes are very effective and versatile packaging media and provide protection against contamination and breakage. It is easy to print on, collect into secondary packages, and pile on shelves at the point of sale. After use, carton is 100 percent recyclable and is often used as raw material for the manufacture of packaging papers and boards. Corrugated board is made by combining several layers or paper, with the inner layers called fluting. The cardboard box is a very versatile and widely used packaging medium. It is the most broadly used material in secondary packaging.

Proper management of forests can guarantee a continued supply of wood for paper and other purposes. Most of the trees used to make paper are trees planted explicitly for manufacturing paper. Thus, less paper usage means fewer trees planted by commercial harvesters. Moreover, harvesting and planting trees may have other environmental benefits. Trees consume large amounts of carbon dioxide. For example, U.S. forests could be consuming as much carbon dioxide as the United States emits, if they were growing forests. Mature forest ecosystems made up of combination of growing trees and dead material, give off as much carbon dioxide as they consume.[14]

3.2 Glass

Glass continues to be an important packaging material. It was first used in Egypt and Babylon as long ago as 2500 B.C. when it was formed into jewelry and small containers. The major event in the history of glass was the discovery of blow molding. Around the first century A.D., Syrian artisans found that molten glass could be blown into different shapes, sizes, and thicknesses. This eventually led to the mass production and wide availability of all types of glass containers.[16]

Glass is manufactured by fusion at very high temperatures (up $1650°C$) of naturally occurring minerals such as sand (SiO_2), soda ash ($NaCO_3$), and limestone ($CaCO_3$).[17] On the one hand, cullet melts more readily, and the melting does not significantly degrade the materials. No physical difference can be measured between virgin and recycled grades. This alone makes glass recycling sensible. On the other hand, supply of sand is plentiful. 27.72 percent of the Earth's crust is made up of Si, the second in quantity after oxygen. Soda ash is rare and expensive, and is mainly produced from NaCl. Na as an element makes up 2.83 percent of the Earth's crust, and also is largely present in ocean matter. Limestone is abundant, and relatively inexpensive. Ca makes up 3.83 percent of the Earth's crust. Approximately 70 percent of total glass consumption is used for packaging purposes.[18]

3.3 Steel

Although metals such as copper, iron, and tin began coming of age at the same time as clay pottery, it is only in more modern times that they began to play a unique role in packaging. In many cases, metal containers proved to be stronger and far more durable than other materials.[19]

Steel is smelted from naturally occurring iron ore at around 1400°C. Iron alone is in abundant supply, makes up 5 percent of the Earth's crust. Iron is scarcely known in a pure condition, but is used in impure form, containing carbon. If low carbon concentration is required, steel must be purified. The carbon content of the steel is burnt out at temperatures of around 650°C. Alloys for special applications require further processing and adding minerals such as chromium, nickel, tungsten, vanadium, and titanium for enhancing the physical and/or chemical properties.[20] For packaging purposes, the *tinplate* is used: a cold-reduced low-carbon sheet protected by coating on both sides with a very thin layer of tin.[18] The British Navy began using tin cans widely in the early 1800s, and canned food began appearing in English shops by 1830.[19]

The *tin-free steel* is made corrosion-resistant by a very thin coating of chromium phosphate, chromium or chromium oxide, or aluminium. Steel use for packaging purposes makes up around 5 percent of the total world steel consumption. Tin and chrome are in rather short supply; tin makes up 0.4 percent, and chromium 0.01 percent of the Earth's crust, but because amounts used are so small, supply does not appear likely to be a problem in the near future.[18]

3.4 Aluminum

Tin and steel cans became widely accepted during World War II. This rising demand also led to rising costs of tin plate, causing can producers to look for an economical replacement. Aluminum filled this need and, in 1959, the Adolph Coors Company became the first American brewer to package beer in an aluminum can.[21]

Different alloys and gauges of aluminum foil are used for different packaging applications, with most alloys including up to around a 3 percent mix of iron, silicon, and manganese, with tiny amounts of copper occasionally added for extra strength. The thinnest foil used for wrapping chocolates may be only 6 microns thick, with household wrapping and cooking foil between 11 and 18 microns, foil for packaging lids between about 30 and 40 microns, and foil for containers generally between 40 and 90 microns.[22]

Aluminum itself is plentiful, makes up 8.13 percent of the Earth's crust—but never as free metal but as silicates, from which the extraction is expensive. Commercial production of aluminum is from bauxite. Aluminum from bauxite is smelted in electric arc-furnaces on temperatures of around 800°C. The overall energy consumption of aluminum manufacture is very energy demanding. However, the main aluminum smelting plants are in countries such as Norway, Canada, and Scotland, where renewable resources are used for energy production (water power). Again, near-term material shortages seem unlikely, mostly due to the ready recyclability of aluminum. About 25 percent of total aluminum consumption is used for packaging purposes.[18]

3.5 Plastics

Plastics are macromolecular polymeric materials. The majority of plastics in packaging are thermoplastic organic polymers, that is in the main chain have only carbon-carbon bonds, such as polyolefines, polystyrene (PS), polyvinyl-chloride (PVC), but there are semiorganic polymers, as polyamide (PA), and polyesters (PE). Plastics for packaging are in the form of foils (up to 0.2 mm thick), and sheets (above 1 mm). Foils are used for packages with flexible wall as bags, and sheets are used for rigid wall packages.[23] The five largest volume polymers used in packaging are polyethylene, polypropylene, polystyrene, polyvinyl chloride (PVC), and polyethylene terephthalate (PET). Packaging is the major use for polyethylene and polypropylene. High-density polyethylene (HDPE) is used in applications such as containers, milk and detergent bottles, bags, and industrial wrapping. Low-density polyethylene (LDPE) is used for pallet and agricultural film, bags, coatings, and containers. Polypropylene is employed in film, crates, and microwavable containers. Polystyrene finds use in jewel cases, trays, and foam insulation, while PET is used in bottles, film, and other food-packaging applications.[24] Plastics are also increasingly used in secondary and transport packaging; re-usable plastics boxes and trays are replacing single-use cardboard and wooden boxes.

The role of plastics in packaging is substantial. Plastics represent 20 percent by weight of all packaging materials and are used to package 53 percent of all goods. In comparison, glass, which also represents 20 percent of all materials, packages only 10 percent of all goods.[2] Plastics, for the most part, are based on petroleum and natural gas, but plastics' production accounts only for about 2 to 4 percent of overall consumption of oil and natural gas.[18] The packaging industry is one of the major users of plastics; however, plastic packaging often accounts for just 1 to 5 percent of the product's overall weight. In Western Europe, about 37 percent of plastics are used for packaging purposes. Figure 2 illustrates the plastic consumption by industrial sector in Western Europe in 2003.[25]

Plastics have a negative image due to their fossil content. A comparison between plastics and gasoline based on their crude oil equivalent reveals that the average per-capita plastics consumption in Western Europe equals approximately 32 liters of gasoline. In the United States, an equivalent amount of gasoline used for the production of all plastics for packaging would equal a mere 19 days of automotive travel.[26] This suggests that only a 5 percent improvement in gasoline mileage would offset the total amount of energy required for the production of plastics into packaging markets. This would appear to be a relatively small improvement to enjoy the benefits of plastics.

3.6 Composites

Composites are a combination of materials used for enhancing the content protection. Two or more separate layers materials are joined, most frequently paper

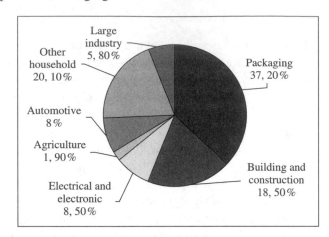

Figure 2 Plastic consumption by industrial sector in Western Europe in 2003. (From Ref. 25.)

or board, and aluminum foil or plastics. The use of combinations has advantages from technological and economic point of view. Often their use is the only technologically feasible solution. Flexible and semi-rigid-wall packaging materials are produced by the following methods: coating, laminating, and co-extrusion. The most commonly used combined packaging materials follow.[23]

Paper–plastic Composites
Paper–plastic composites are the most commonly used combination. The paper gives rigidity, and the plastic gives low permeability and heat sealability. All the paper–plastic combinations can be used with board; the most commons are the cardboard–PE and the board–PP combinations.

Cellophane Composites
To reduce the water absorption of cellophane, and to improve its resistance properties and sealability, the most frequently used method is lacquering. Cellophane can also be combined with several plastics by extrusioning or laminating.

Plastic–plastic Composites
The concept of modern multilayer packaging means that there is a minimum use of plastic material, because various characteristics can be combined into one thin packaging film. The most widely used plastic for combination is PE. The PA–PE combination processed by coextrusion has a good gas resistance. It can be deep drawn, and its fat resistance and flexibility are suitable, too. These combinations are very widely used for vacuum packages. In addition, the PET–PE combination is heat resistant, and the PE–PP is sterilizable. As an illustration of the effectiveness of plastic composites, Williams presents multilayer plastic

film used for sausage packaging (about 0.1 mm in thickness) containing a layer of polyethylene (an excellent barrier to moisture) and a layer of polyamide (an excellent barrier to oxygen).[7] If polyamide alone was used, the film would need to be at least five times thicker to provide the same barrier to moisture. If the film was pure polyethylene, it would need to be 100 times thicker to provide the same barrier to oxygen. What is more, the multi-layer film offers excellent puncture and abrasion resistance, as well as heat sealability.

Composites add substantial savings in materials and energy, considerably lower costs, and much less packaging waste. Saving fuel also means lower emissions during transportation. Williams's estimate for Germany indicated that if the 32,000 tons of multilayer packaging used in 1991 should be substituted by other materials, 71,000 tons of paper, 100,000 tons of glass, 110,000 tons of steel, and 9,000 tons of aluminum would be needed, a total of 293,000 tons altogether.[7] Not only it is nine times more packaging weight, but four and a half times more energy would be needed to produce the packaging, and the cost of packaging would increase three times. Going a step further, even assuming 90 percent collection rate, 90 percent of that quantity to be sorted and 95 percent of sorted material recycled would leave 67,000 tons of waste for disposal by other means. Even if certain amounts of the substitute materials would be combusted—such as paper—there would be 36 percent lower energy recovery than with plastics. Williams concluded that multilayer plastic packaging minimizes the quantity of waste destined to landfill. It uses less energy to produce, its energy content can be efficiently recovered, and it is a cost-effective solution.

3.7 Degradable Plastics

In nature, all organisms re-enter the carbon cycle by degradation into basic elements that serve as a foundation for development and continues sustainment of life. This same logic leads to the development of degradable plastics: To design and engineer strong, lightweight, useful disposable plastics that can break down under environmental conditions in waste disposal systems to products that can be utilized by the ecosystem (carbon cycle).[27] One contribution to a more sustainable recovery of plastic waste might be the use of compostable plastics.[28] The American Society for Testing and Materials (ASTM) provides a definition of degradable plastics:

- *Degradable plastics* are plastic materials that undergo bond scission in the backbone of a polymer through chemical, biological, and/or physical forces in the environment at a rate that is reasonably accelerated, as compared to a control, and that leads to fragmentation or disintegration of the plastics.
- *Biodegradable plastics* are those degradable plastics, where primary mechanism of degradation is through the action of micro-organisms such as bacteria, fungi, algae, yeasts.

- *Photodegradable plastics* are those degradable plastics where primary mechanism of degradation is through the action of sunlight.
- *Biodegradation of plastics* is conversion of all constituents of a plastic or hybrid material containing plastics to carbon dioxide, inorganic salts, microbial cellular components, and miscellaneous byproducts characteristically formed from natural materials.

There are specialized applications where biodegradable materials have an edge, such as in conjunction with organic waste. According to Reske,[28] compostable plastics and packaging are ready for the market. In many applications in the food sector (especially for fruit and vegetables), increasing amounts are being used in a number of EU countries and worldwide. Items that could help avoid floating marine litter would be invaluable.

3.8 Wood

Wood as packaging material is largely used for transport packaging, in the form of crates and pallets. Pallets are a universal and critical part of product transportation. Forty percent of all hardwood lumber produced in the United States is reported to have been made into solid wood packaging. The pallet industry uses approximately 4.4 billion board feet of hardwood lumber and 2.1 billion board feet of softwood lumber for the production of 400 to 500 million solid wood pallets annually.[29] While the amount of new wood pallets manufactured increases slightly, in the same time the percentage of hardwood used is reduced and the recovery of pallets increases.[29]

4 CONSUMPTION OF PACKAGING MATERIALS

The average household buys goods packed in 190 kg of packaging, using 7 GJ energy each year. Packaging is typically 9 percent of the weight of the packaged product.[2] Table 1 summarizes the package weight to product weight percentage for some consumer goods.[30]

The most effective packages, those that contribute only 1 to 10 percent of the packed product's overall weight, are paper, plastics, or composites. From 11 to 20 percent, the fairly effective packages include plastic and aluminium packages. In the category of 21 to 40 percent, the less-effective packages, we have mainly large volume, light weight products (cereal flakes), liquid goods in more sophisticated, rigid-wall plastic containers (roll-on deodorant, dishwasher detergent), and tin-canned goods. From 40 percent up, the "ineffective" packages include glass packages and extremely low-specific-weight goods, such as deodorant spray, and goods portioned into extremely small and light quantities, such as tea, seasoning, and pills. This would indicate that a move toward more effective packaging options requires the use of flexible wall packages, plastic-plastic, or paper-plastic combinations, avoids low-specific-weight products and goods in extremely small

Table 1 Percentage of the Package's Weight, Compared to the Packed Product

Package Weight to Product Weight Ratios (%)	
1–10%	
1	500 g of pasta in PE bag
	61 g bar of chocolate in plastic wrapper
2.7	1 l milk in paper+PE box
3	1 l soft drink in PET bottle
3.3	1 kg of coffee in brick pack
3.5	1/2 kg of meet on foam tray
4	0.33 l soft drink or beer in aluminium can
	1 l ice-cream in HDPE box
5	fruit juice in aseptic box
	250 g of cold cuts vacuum packed
5.3	2 dl yogurt in plastic cup
5.29	bag of potato chips
6.6	400 g of margarine in plastic tub
	150 g of cold cuts vacuum packed
6.7	1 l ketchup in plastic squeeze-bottle
7.4	10 eggs in pulp tray
9	bar of soap in paper box
9.5	fabric softener in HDPE bottle
11–20%	
11.9	85 g cat food in aluminum pouch
12.4	1/2 l oil in plastic bottle
13.4	500 g of canned food
18.5	2 dl of shampoo in plastic bottle
21–30%	
23	400 g cereals in PP bag and paper box
25	150 g of canned food
31–40%	
34	deo roll-on in HDPE bottle
40	150 g cereals in PP bag and paper box
49	1 l dishwasher liquid
41–60%	
53	0.3 l glass bottle of beer or soft drink
56	deodorant in spray bottle
57	150 g jam in glass jar

(*continued overleaf*)

Table 1   (*continued*)

61–100%	
68	0.5 l salad dressing in glass bottle
74.5	100 tablets in PS bottle and carton box
80	0.5 l oil in glass bottle

>100%	
160	a box of 25 tea bags
588	tablets in blister package and paper box
611	20 g of seasoning in glass bottle

Note: From Ref. 30. Copyright The Cygnus Group. All rights reserved.

Table 2   Availability of Packaging Raw Materials

Packaging Material	Raw Material	Fossil Resource	Renewable Resource	Overall Resource
Paper/board	Wood, natural fibers	Nil	All	Very abundant
	Auxiliary chemicals	All	Nil	
Metals:				
Iron	Iron ore, scrap iron	About half	About half*	Limited
Tin chromium	Tin and chrome ores	Nearly all	Insignificant*	Severely limited
Aluminium	Aluminium ore Scrap	Majority (but plentiful)	Minority* but growing	Moderately limited
Glass	Sand, soda	Majority (but abundant)	Minority* but growing	Abundant
Plastics	Crude oil (now)	Almost all	Little	Moderately limited
	Biomass (wood sugar)	Nil	All	Very abundant
	Auxiliary materials, e.g. N, Cl, S, O	Some, but abundant	Some	Very small factor, no limitation

Note: Based on Ref. 31. *Recycling

portions, and uses only refillable glass.[30] The availability of all packaging materials is summarized in Table 2.[31]

5   ENERGY USE

Packaging materials use energy in their manufacture and distribution, and contribute to the energy required for transporting products. Energy input is required

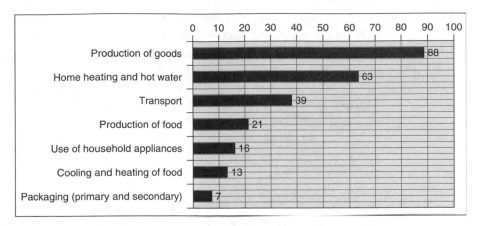

Figure 3   Energy consumption of household activities (GJ/household/year). (From Ref. 2.)

in several ways to produce and distribute packages. First, energy is used in converting the basic raw materials into packaging raw materials. Second, energy is used to convert the packaging materials into packages. Generally, all stages take place in different geographical locations. Recycling systems add to this energy demand, with further transportation need and processing of discarded packages. Notwithstanding, studies show that the household activities with the highest environmental impact are:[2]

- Production of food and goods
- Home heating and hot water
- Transport
- Use of household appliances.

As Figure 3 illustrates, the energy consumption of packaging is relatively small. From the different packaging materials, aluminium and glass manufacture consume the most energy. Aluminium manufacture uses large amounts of electrical energy in refining the metal from ore. Smelting of one ore batch of around 800 kg takes 3 to 4 hours. Glass manufacture, apart of being an energy-intensive, high-temperature process, also contributes to large transportation costs due to its heavy weight. The practice of refilling is also very energy intensive due to high transportation distances from numerous shops to refill centers. Plastics use primarily petroleum and natural gas, both for the energy needed in manufacturing and for the content of the material itself. It is estimated that around 2 to 4 percent of all petroleum consumption is used for plastics manufacture.[18]

6   ROLE OF PACKAGING IN POLLUTION

6.1   Litter

Litter constitutes only a minor part of total wastes, but it is of widespread concern. It is an unpleasant sight, constitutes a hazard to many animals, and is a possible health hazard to humans. Litter is often equated with packaging. Packaging materials (glass and plastic bottles, cans, paper cups, paper and plastic wrappings) are indeed the main constituents of litter. Excluding unofficial dumps, the proportion of packaging is usually a quarter to half by weight, but because of the low bulk density, packaging is often the majority by volume. Packaging litter constituted 11.91% of all litter in Ireland in 2006 and is the third largest component after cigarette and food related litter.[32]

The effect of plastics litter on the marine environment is also of particular concern. It originates from both land and sea sources, and the debris is of three types: fishing gear, such as nylon lines, buoys and nets; packaging bands, straps, and synthetic ropes; and general litter, such as bags, bottles, and plastic sheeting.[33] The UN Group of experts on the Scientific Aspects of Marine Pollution (GESAMP) concluded the following: chemical contamination and litter can be observed from the poles to the tropics and from beaches to abyssal depths—in short, throughout the whole length, breadth, and depth of the world ocean. The Annex V of the International Convention for the Prevention of Pollution from Ships (MARPOL), which came into force in December 1988, makes it illegal for vessels of the 131 ratifying nations to dispose "into the sea ... all plastics, including but not limited to synthetic ropes, synthetic fishing nets, and plastic garbage bags."[34]

6.2 Water Pollution

Pollution arises from wastewater discharge of some packaging material manufacturing or related activities. One of the basic water-polluting activities is paper production, releasing biological oxygen demand (BOD), chemical oxygen demand (COD), volatile suspended solids (VSS), and total suspended solids (TSS). In addition, the manufacture of miscellaneous materials used in packaging, such as adhesives, coatings, and inks is a source of hydrocarbon pollution. The discharge of cooling water from electricity generation in turn causes thermal pollution. Subjects of concern are also accidental emissions during production, or processing of packaging materials, especially the drainage of fire-fighting activities during accidental fires.

Finally, water pollution arises from landfill leachates, although the causes of leachates are, rather, the remains of products on the packages. Historical packaging can also be the source of organic plasticizers for PVC, or lead and cadmium from pigments.[18]

6.3 Air Pollution

The main source of air pollution is the packaging material manufacturing process. Some of the emissions, such as vinyl chloride, CFC, and hexane can arise from accidental fires, or waste-incineration activities. Direct packaging-related emissions arise from landfill sites, as a consequence of decomposition of wood and paper, releasing CO_2, and methane. In addition, CO_2 emission arises from glass and steel manufacture.

Packaging-related sources of pollution are also electricity generation (CO_2, SO_2, NO_x emissions) and transportation-related emissions (e.g., CO_2, SO_2, NO_x, dust, hydrocarbons). It is increasingly important to take into account the transportation-related emissions, especially when considering reuse, or recovery.[18]

6.4 Solid Wastes

Packaging-related solid wastes arise already at extraction and processing of raw materials. These wastes often end up in landfill sites. Further preconsumer and postconsumer wastes have to be distinguished. The general public is conceiving of only the postconsumer solid wastes, although that is only a part of all packaging-related wastes. Most of the preconsumer packaging waste of packaging material, or package manufacture is, however, recycled in house. The nonrecyclable part of preconsumer packaging wastes is disposed.

Recovery, and in particular recycling of postconsumer packaging wastes, does not stop further generation of wastes. First, not all the collected material is recycled and, second, the product made from the recycled material will end up being a waste sooner or later as well. As for incineration of postconsumer packages, it may mean a volume decrease of 20 to 40 percent.

An indirect packaging-related solid waste source is slag for producing the electricity that was consumed by packaging activity.[18]

7 ENVIRONMENTAL ASSESSMENT OF PACKAGING MATERIALS

The Danish minister of environment in 1988 announced that within a few years the manufacture and use of polyvinyl chloride (PVC) products had to be reduced as much as technically and economically possible due to their environmental impacts of production, use, and disposal. This preventive environmental policy was mainly based on the emission of hydrogen chloride and dioxins from waste incineration. A study of the technical, economic, and environmental consequences of a substitution was initiated by the National Agency of Environmental Protection. The goal was to collect background data for the upcoming negotiations between the environmental authorities and PVC-industry and manufacturers of PVC products in Denmark. The environmental assessment focused on PVC and 11 alternative materials, such as polyethylene (PE), polypropylene (PP), polyethylene terephtalate (PET), polystyrene (PS), polyurethane (PUR), synthetic rubbers (EPDM, CR and SBR), paper, impregnated wood, and aluminum.[35]

The assessment of each material was conducted in three steps. First, a screening of the life cycle for the potentially most severe impacts of the material was accomplished by consulting experts in material-, health-, and environmental sciences, and a chemical profile, including four to five chemicals or chemical groups

Table 3 Comparison of PVC with Alternative Packaging Materials

Material	Impacts
PVC	• Potential severe impact areas of exposure to the carcinogenic vinyl chloride monomer in the work environment and the discharge of dioxins in wastewater. • Exposure to vinylchloride, chlorine, or hydrogen chloride, heavy metals, phosgene, and dioxins generated in accidents (e.g., fires), or in the production and use of PVC • Incineration of PVC-containing waste generates hydrogen chlorine, dioxins and heavy metals that are emitted to the atmosphere, or contaminate incinerator ashes or filter residuals.
EPDH (ethylene-propylene-diene)	• Use of halogen-based flame retardants in special products as well as possible exposure to neurotoxic n-hexane and carcinogenic benzene at production and processing. *PS*:
PS	• Production requires more energy, than the production of PVC. Some typical products are expanded with CFC or azodicarbonamide (sensitising agent) with severe external and work environmental impacts, respectively.
Impregnated wood	• Manufacturing involves high exposure to wood dust, expected to be carcinogenic, and accidental releases of tributyotin (wood preservatives) constitute a major risk to the aquatic environment.
Paper	• Production is dominated by sulphate-mass and, in some countries, chlorine-based bleaching resulting in waste water strained with oxygen-consuming pollutants and chloroorganics, for example dioxins.
Aluminum	• Production of virgin aluminium involves very high energy consumption, and the work environment includes severe potentials of exposures to carcinogenic polyaromatic hydrocarbons (PAH's). • Approximately only one fifth of the raw material ends up in the final product, thus the production results in major amounts of solid waste and sludge to be disposed of.
PUR	• Implies occupational exposure to highly toxic isocyanides in the production, processing, manufacturing and in fires. • PUR is commonly expanded with CFC. • Halogen-based flame retardants are frequently used in the production of PUR.
Synthetic rubbers, CR(chloroprene), SBR (styrene-butadiene):	• Involve carcinogenic substances in the work environment of production and processing (vulcanisation). • CR may generate hydrogen chloride and dioxins when incinerated or burned.

Note: From Ref. 35.

characterizing the material, was established. Second, data on the key consequences were collected and evaluated from readily available literature and interviews with experts from Danish Technological Institute, the industry, and environmental authorities. Finally, the evaluation of each material was used to develop an impact profile for the material as such, and for each of the alternative materials a comparison to PVC was made. The results of the study are summarized in Table 3. In summation, from the alternative materials evaluated, *PE, PP* and *PET* proved to be environmentally preferable to PVC.[35]

8   RECOVERY OF POSTCONSUMER PACKAGING

8.1   Waste Packaging in Municipal Solid Waste

Packaging is required to give protection to a product until the last bit of the product is consumed. This would mean that most packages, especially the reclosable ones such as plastic containers, glass jars, are still in perfect shape when empty. They could still continue to be used for the same purpose as designed: containment, protection, and use of the product. Discarded waste packages are thus not necessarily useless; they just are not used anymore. The problem is not in the package itself, but in the possibilities. If there are no containers for separate waste collection, and there is no need for constantly rising amount of butter boxes, jam jars, and no possibility to burn part of the waste, then the only choice is to send the packaging waste to landfill, regardless of the environmentally consciousness of the consumer.[26] Some thin, lightweight packs may not be worth collecting for recycling because too much energy would be needed to collect and clean them. But they have environmental advantages in other ways, such as allowing more goods and less packaging to be packed in fewer trucks thus reducing transport pollution.[2]

Packaging is often cited as one of the reasons of rising amount of municipal wastes. In the United States the amount of municipal waste increased five times as quickly as the population over the period 1920 to 1970.[36] It is, however, not due to packaging only. The reasons of the growing amount of municipal waste are rising level of affluence, advent of build-up obsolescence, demand for convenience products, cheaper consumer products, changing patterns of taste and consumption, and, in part, the proliferation of packaging.[36]

Generally, it is estimated that packaging constitute one third of household waste. The other two components of high percentage are biogenic material at 30 and newsprint, 20 percent.[37] In the United States, the trend is similar. By volume, packaging constitutes up to 30 percent of household waste; by weight, about one-third is biogenic material and one-fifth is newsprint.[33] The importance of distinction between classification by weight or volume can be shown with the following examples: In Austria, packaging constitutes 30 percent of household waste by weight, and 50 percent by volume.[38] Table 4 illustrates the amount of packaging waste (PW) related to municipal solid waste (MSW).[39]

In the United Kingdom in the 1975 to 1995 period, although the volume of discarded packaging materials in the domestic waste bins has risen, but the weight remained approximately the same.[40] Most probably, this is due to lightweighting

Table 4   Packaging Waste Generation in Selected Countries and Communities

	Packaging Waste (million tonnes)	PW/MSW (%)	PW per capita (kg)
OECD	140.0	33	181
EEC	50.5	49	154
USA	56.8	27	210
Japan	20.0	41	163
United Kingdom	7.7	44	134
France	10.0	59	181
Germany	10.0	49	181
Italy	12.0	68	188

Note: From Ref. 39. Reprinted with Permission from Elsevier.

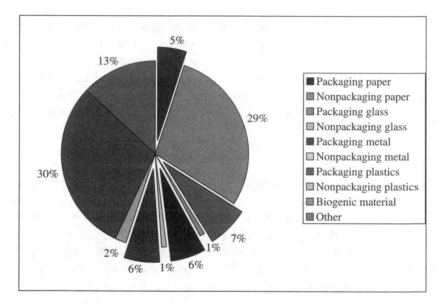

Figure 4   Composition of garbage in domestic wastebaskets in UK. (From Ref. 40.)

of packages, and the widespread use of plastics. Figure 4 outlines the composition of garbage in domestic wastebaskets in the United Kingdom.[40]

8.2   Packaging Waste Reduction

The way of reducing packaging wastes is lightweighting of packages. It combines the commercial benefit of lower unit cost with the improved resource efficiency. It is a result of improved packaging material and package manufacture, which allows the use of lighter, thinner-walled packages. An important way of lightweighting is also the product innovation as the introduction of concentrated, and dried products. There are, however, limits to lightweighting of individual packages, which could overweigh the benefits:[41]

- Increased amounts of secondary and transport packages, are needed.
- Thinner packages are more fragile and may result higher wastage.
- Inadequate product protection may lead to greater spoilage.

Reuse of packages also contributes to waste reduction. Refillable bottles are the best-known examples of reusable packages, but not the only ones. Refillable bottles are returned into the bottling plant in reusable crates; reusable crates are also traditional in bakery industry. Another application of reuse is the so-called *refill pack*. The consumer buys a sturdier reclosable container once, and further on purchases the product in lighter refill poach. With no need for an opening and reclosing device, the refill packages can be reduced to the minimum needed to protect and contain the product. It is a lightweight, space-efficient system that minimizes distribution costs and transport pollution while giving the consumer all the benefits of a durable and convenient container to use at home, the refill pack has many advantages.[42] Refill packs are attractive for the consumer for their lower price. Refill-pack systems also instill brand loyalty, a considerable marketing benefit for the company.

8.3 Choice of Waste Management Options

Nations are considering restrictions on packaging and controls on products in order to reduce solid-waste generation rates. Local and regional governments are requiring wastes to be separated for recycling, and some have even established mandatory recycling targets. Secondary and tertiary packaging materials are normally in larger quantities and have less material variation. Thus, they are relatively easier to collect and sort by wholesalers or retailers for recycling or reuse purposes. Primary packaging materials are not only more dispersed into households, they are also largely mixed, contaminated, and often damaged. Thus, they pose problems in recycling or reuse of the materials.[43]

Previously considered as a local issue, it is now clear that solid waste management has international and global implications. The European Community has been criticized for setting rigid recycling percentages for packaging materials. Isoaho argues that regulation based on material, product or source classifications are very difficult to manage, especially if they are too detailed at international level.[44] Conditions in the countries are different, citizens react in a different way. The international political supervision should remain to decide policies, to create general strategies, and to agree to standards of environmental effects and management quality. This implies focusing on energy and material policy, supervision instruments, and management environments rather than on, for example, recycling percentages or single products. Within countries there are different waste management regions, and within these, there are different collection areas characterized by waste generators, their density, and the specific waste stream volume.

8.4   Recycling of Packaging

The concept of recycling to conserve resources is based on the assumption that a recycling requires fewer raw materials and less energy, and generates fewer emissions into the environment, than manufacturing new material. However, for recycling to be environmentally beneficial, the effects of the collection, transportation, and reprocessing operations must be less harmful than those resulting from the extraction and processing of the virgin raw material that the recycled product replaces.

Germany has long been regarded as the most advanced country in Europe in packaging recycling. The Law on Waste Management (Abfallgesetz) passed in 1986 laid the foundations for later German packaging waste-management strategies. The law gave the environmental ministry extensive rights, and resulted in the packaging Directive (Verpackungs-verorderung) in 1991. The ordinance obliges the producers and retailers to take back and dispose of the packaging waste in an environmentally sensible way. In order to take care of the packaging waste, which now had to be dealt with separately, the Duales System Deutschland GmbH (DSD) was set up. The concept behind the DSD is that the organization gives various packaging materials the right to bear the "green dot" and thus be recycled by the disposal network set up by the DSD. Companies that want their packages to bear the green dot must first pay a per-package fee to the organizations. DSD has a rather controversial and difficult history, attacked by the public, environmental, and trade organizations, political parties, and the media. In its early years it was on the edge of bankruptcy. It has received a lot of bad press and political criticism for exporting collected waste abroad to countries such as Indonesia and China instead of recycling them in Germany. Hanisch quotes that, at that time, Germany faced a severe landfill shortage, with packaging waste amounting to a significant percentage—30 percent by weight and 50 percent by volume—of the nation's total municipal waste stream.[45] However, according to Rathje and Murphy, the claim that packaging waste is a major constituent in landfills is simply a myth.[46] Over a period of five years, a U.S. "Garbage Project" excavated 14 tons of waste from nine municipal landfills. This project sought to address the claim that fast-food packaging and polystyrene foam were the major elements of American trash. They found that out of the 14 tons of excavated waste, 1 percent was polystyrene foam, and less than 0.5 percent was fast-food packaging.[46] Similarly, it was calculated that, in Finland, packaging waste constitutes approximately 1.5 percent of all landfilled waste.[47] Viewed in such a context, packaging waste does not appear to be overflowing landfills.

The European Parliament and Council Directive 94/62/EC on Packaging and Packaging Waste ("Packaging Directive") first came into force at the end of

1994 and has both environmental and single market objectives. The Packaging Directive aims to harmonize the management of packaging waste in the EU and tackle the impact that packaging and packaging waste have on the environment. Although the primary objective is to increase the recovery and recycling of packaging waste in a consistent way in all member states of the EU (so as to avoid barriers to trade), priority is also given to reducing the amount of packaging used and the reuse of packaging. The Packaging Directive sets member states mandatory recovery and recycling targets, the first of which were to be met in 2001. A revised Packaging Directive (2004/12/EC) was published in February 2004. It sets new recovery and recycling targets, as a percentage of all packaging waste to be met by December 31, 2008, illustrated in Table 5.

8.5 Recycling of Plastic Packages

Plastics packaging wastes present a number of challenges in terms of recovery due to the composition and diversity of the plastics used and the fact that mixed waste is often dirty or contaminated. First, the two different ways of recycling need to be distinguished: *mechanical recycling* and *feedstock recycling*. At mechanical recycling, the plastic waste is used as a secondary raw material to replace primary (virgin) plastics. From collected bottles, detergent, and fertilizer bottles, water pipes can be manufactured by mechanical recycling. Waste plastic films can be recycled into waste bags, or cable coatings. In Western Europe good progress has been made, and most countries have increased their recycling rates. However, challenges still remain for many countries to meet the minimum recycling target of 22.5 percent, set by the European Packaging Directive. As a whole, mechanical recycling of post-user plastic packaging waste increased by 12.6 percent in Western Europe in 2002—in the meantime, packaging waste increased by just 3.5 percent. Consequently, the recycling rate went up from 20.5 percent in 2001 to 22.4 percent in 2002. In terms of the total recovery (recycling

Table 5 Revised Targets of the Packaging Directive (2004/12/EC)

The new targets are:	
Minimum recovery	60%
Recycling	55–80%

Minimum material-specific targets are:	
Glass	60%
Paper/board	60%
Metals	50%
Plastics	22.5%
Wood	15%

and energy recovery) of plastic packaging waste in Europe, comparing recovery in 2001 and 2002, the rate increased from 49.4 percent to 52.5 percent as a result of an increase in mechanical recycling and countries adopting best waste collection practices.[25]

There are several limits to plastic packaging recycling. Most restrictive are the technical limits. Due to the aging of the material and pollutants such as additives, colors, and dirt, recycled postconsumer plastics can never completely replace virgin material.[48] Citizens are generally eager to recycle plastics; however, in Finland, for example, local authorities perceived plastics recycling as problematical due to difficulties in separating different types of plastics.[49] A survey of developmental needs in waste management by the Technological Research Centre of Finland concluded that the realisation of plastics recycling is not meaningful when plastic waste is collected from municipalities.[50] Due to the restrictive technical limits, mechanical recycling is not the major route in packaging plastics waste management, Plinke and Kaempf recommend feedstock recycling such hydrogenation, pyrolysis, gasification, or others.[48]

8.6   Feedstock Recycling

The expression *feedstock recycling* is used for methods when the waste plastic's energy content is used by other methods than simple combustion, also referred to as *tertiary polymer recycling*.[51] These processes are not recycling by the classical understanding of the word. Since plastics are generally high-caloric-value products ranging from approximately 18,000 to 38,000 kcal/kg, using them for their energy alone or for related chemical production could be an alternative option.[52] For example, one could take the view that the crude oil content of the plastic is temporarily used by the plastic to serve as a package. After its function as a package has been served, the fossil energy could be used. Feedstock recycling includes the following methods:

Use in Blast Furnaces

A potential use of plastic waste is in blast furnaces as a reducing agent to withdraw oxygen from the iron ore, substituting heavy oil currently used. Since the use of solid plastic waste involves significant additional investments, the main interest is in using plastic oil obtained from the fluidized-bed pyrolysis process. The plastic oil does not contain sulphur, so its use involves process-technical advantages, as sulphur variations in fuel oil can be regulated with the aid of plastic oil. Plastic oil may also substitute for heavy fuel oil, either as such, or mixed with fuel oil, if the waste plastic does not contain chlorine. Plastic types produced from polyethylene and polypropylene were found to be best suited for the production of plastic oil. In autumn 1999, blast furnace tests were carried out in Raahe Steel factory in Finland, and plastics were melted successfully in heavy fuel oil, without any additives.[53]

Thermolysis

Thermolysis is performed at a temperature lower than 500°C and in the absence of oxygen.[51] Compared to incineration, *thermolysis* is considered as a viable alternative to treat MSW, especially in regions with a low population density. At the end of thermolysis, the waste will have lost approximately 60 percent of its weight. As opposed to incineration, thermolysis does not produce slag. There remains only a mixture of carbonaceous solid fuel, metals, and minerals. The thermolysis process can be called thermolytic sorting, since it isolates the combustible organic compounds from the noncombustible ones (water, minerals, metals), and only the combustible ones are burned.[54]

Pyrolysis

Pyrolysis for the simultaneous generation of oils and gases can be convenient to obtain hydrocarbons and even recover crude petrochemicals, or to generate energy from waste plastics.[55] *Pyrolysis* involves heating of a feed in an inert atmosphere at a temperature ranging from 500° to 800°C, to produce three forms of energy: gas, liquid, or charcoal. Pyrolysis is an extremely versatile process, and the reaction products can be controlled by means of the type of process and the operating conditions. The main purpose is to convert biomass and waste into high-energy condensable *pyroligneous liquid*, which is much easier to manage than bulky waste. Pyrolysis is an endothermic (heat-absorbing) reaction. While at higher temperatures the gas yield increases, char yield is maximized at low heating temperatures.[56] Pyrolysis of high-PVC solid waste in a fluidized bed at low temperature gives a chlorine-free fuel for a fluidized-bed combustor (FBC), plus concentrated HCl. The process has thermal efficiency of approximately 36 percent, depending on the pyrolysis temperature and the PVC content. Hydrochloride recovery can be above 90 percent at a pyrolysis temperature of 310°C.[57]

Gasification

Gasification is, technically, a compromise between combustion and pyrolysis: It proceeds in reaction with air, oxygen, or steam at temperatures in the range of 700° to 1,000°C. It can be considered to be a partial oxidation of carbonaceous material leading, predominantly, to a mixture of carbon monoxide and hydrogen (rather than carbon dioxide and water produced by direct combustion), known as *synthesis gas* or *syngas*, due to its application in a variety of chemical syntheses.[56] These gases contain *chemical energy* that can be tapped as required. The advantage of this technology, over straightforward combustion, is that the lower bed temperatures employed in the process give good chances that problematical elements such as potassium, sodium, and chlorine can be retained in the ash.[58]

Hydrogenation

Hydrogenation, usually in the presence of catalysts, is the final method of feedstock recycling considered. In the process, the polymers are cracked in a hydrogen atmosphere at a temperature in the area of 400°C, and at a pressure of 300 bar. Compared to treatment in the absence of hydrogen, *hydrogenation* leads to the formation of highly saturated products, avoiding the presence of olefins in the liquid fractions, which favors their use as fuels. Moreover, hydrogen promotes the removal of hetero-atoms (Cl, N, and S) that may be present in the polymeric wastes.[56] The end product is a synthetic crude oil, which can then be used as a raw material by the petrochemical industry. Hydrogenation suffers from several drawbacks, mainly the cost of hydrogen and the need to operate under high pressures.

8.7   Comparison of Mechanical and Feedstock Recycling

A study by the Association of Plastics Manufacturers in Europe (APME) assessed the environmental impacts of mechanical and feedstock recycling and energy recovery of waste plastics. It was compared in terms of consumption of resources and environmental emission pollution potential. The criteria of "consumption of energetically exploitable resources" and "contribution to the greenhouse effect" lead to the following order of preference for feedstock recycling and energy recovery processes:[59]

- Use as feedstock in blast furnaces
- Thermolysis to petrochemical products
- Fluidized-bed combustion
- Hydrogenation, together with vacuum residue oils
- Incineration in domestic waste incinerators
- Fixed-bed gasification, together with lignite
- Gasification together with lignite in the fluidized-bed

The first three processes reduce the contribution to the greenhouse effect in comparison to landfilling. All these processes reduce the eutrophication and acidification potential in comparison to landfill. The overall volume of waste produced was found least in the waste incineration. In summary, from an ecological point of view and on the basis of the comparative analysis of feedstock recycling and energy recovery, APME recommends the following recovery processes:[59]

- Use as reducing agents in blast furnaces
- Thermolysis to petrochemical products
- Fluidized-bed combustion

Mechanical recycling processes have ecological advantages over feedstock and energy recovery processes, if *virgin* plastic is substituted in a ratio of 1:1.[59] With this prerequisite, mechanical recycling processes reduce the consumption of resources and emissions in comparison to feedstock recycling and energy recovery processes. However, because of the aging of the material and presence of pollutants such as additives, colors, and dirt, recycled postconsumer plastics can never completely replace virgin material. As a consequence of the restrictive technical limits, Plinke and Kaempf believe that physical recycling will not be the major route in plastics waste management, and other routes must be applied.[48] APME concurs, and asserts that if considerably less than 1 kg of virgin plastic is substituted by 1 kg of waste plastic, mechanical recycling processes no longer have an advantage over feedstock recycling and energy recovery processes.[59]

Notwithstanding, feedstock recycling is not widely used in Europe, and the amount of plastics recycled by tertiary method has been rather low and stagnant, while mechanical recycling has steadily grown since 1991.[60]

8.8 Glass Recycling

While the plastics industry is faced with a jumbled collection of mixed plastics types, which is their responsibility to sort, glass has been collected for a number of years and sorted by color, with very low levels of contamination. Despite both those factors, and despite the fact that governments everywhere are proposing measures to further increase glass recycling, the amounts of glass collected are giving rise to considerable problems. The EC Packaging Directive, and world-wide various national initiatives, require significant increases in the tonnage of packaging collected. Glass is recycled to save raw materials and energy and to reduce waste. The glass industry across Europe has always referred to the fact that an additional 10 percent of glass scrap (cullet) results in 2 percent energy reduction. The amount of energy used per kilo to make bottles dropped 11.8 percent between 1986 and 1990, which can be attributed to the increased recycling rate (11 to 26.3 percent) and general improvements in furnace technology. The glass companies, however, may not always get the full benefit of energy saving. The cullet may cause problems in the furnace and one of the possible solutions involves using that energy saving.[61] The glass industry has claimed that every tonne of cullet used in the manufacture of new glass saves the equivalent of 30 gallons of oil. However, the experience of the British Glass Recycling Company is that the extra handling involved with more cullet going into the furnace outweighs the energy savings.[62] Glass manufacturers whose production processes were designed for using virgin raw materials of a specified and predicable quality cannot easily replace those virgin raw materials with contaminated and nonstandard secondary materials. They will have to make modifications to their facilities. Contamination from paper, plastics, and the original contents of the jars and bottles does not offer any real problems. The cleaning processes for cullet use no water and very little energy. Aluminum and tin-plate caps have been more of a problem. The major problem, however, is caused by ceramics. They can escape detection, be broken up like glass, but then do not melt in the furnace. Also, the type of glass used for oven-safe dishes is chemically quite different from container glass and is not compatible with it.[61]

In Finland, glass is still perceived as *the* symbol of recycling; however, authorities find glass recycling the most problematic, due to its relatively low retail value, contrasted with high collection, transport, and treatment costs.[49] The raw materials for glass are cheap and rather plentiful. Processing facilities are sparse requiring long distance transport of collected cullet. Moreover there is an over-supply throughout Europe and North America, causing quantities of glass to be landfilled. In many countries there is more green glass collected than can be used for reprocessing, because of lack of market demand. For this reason, research continues to explore alternative uses for excess cullet, such as glass in asphalt, bricks and other glass matrix composites.

8.9   Steel Recycling

Purchased scrap as industrial scrap or obsolete scrap makes up approximately 26 percent input in steel production. Use of postconsumer packaging material is problematic because of the contaminants. Organic residues burn out, producing fumes, and the tin coating is difficult to remove if dissolved, but it can be removed with electrolytic, or alkaline method. These methods are very costly, and due to low percentage of steel used in packaging, the income of tin is too low to make the process profitable. Tin left in the steel, up to 0.1 percent, does strengthen the steel, but above that amount it makes the steel more brittle. The major problem, however, is lead from solder. It has a low solubility, basically causes no problem in steel production, but penetrates through furnace bricks and may cause steel breakout. Moreover, it is toxic so it must be removed from the dust.[17]

8.10   Recycling Aluminum Packages

Aluminum is one of the success stories of recycling. The major fact in aluminum recycling is that approximately 95 percent of the energy needed to produce virgin aluminum is saved and 97% less water pollution is created by using reclaimed aluminum rather than producing new metal from ore.[63]

Recycling one kilogram of aluminium also saves up to 8 kilograms of bauxite, four kilograms of chemical products and 14 kilowatt hours of electricity. The recycling rate for aluminium cans is already above 90% in some countries such as Brazil and Japan, the European average is 52%, Norway being the champion with 93%. The average can coming out of a store is re-melted and back on the shelf within 6-8 weeks.[64] In some places, similarly to PET and glass bottles, aluminum cans have a deposit. Cans are dropped into re-vending machines, which spit out a receipt you can use in the shop. The deposit/refund system is generally welcome by shops, since it increases consumer goodwill.

8.11   Recycling Multimaterial Packages

A criticism against multimaterial packages is that they cannot be recycled, and monomaterial packages should be preferred. However, there are hardly any monomaterial packages on the market today. Glass packages have plastic and/or paper labels and nonglass closures; metal containers have polymer coatings on the inside to prevent product-package interaction, and are either lithographed on the outside or have a paper label. Even the traditional tinplate is a multimaterial package of steel coated with tin. Many plastic packages have labels made from paper or plastic, the latter frequently being a different polymer than the package itself. Most packages are thus multimaterial.

The most widely recognized multimaterial package is the aseptic package. *Aseptic packaging* means that a sterilized product (e.g., fruit-juice concentrates) is packed in sterilized conditions. This results in a considerably increased shelf life. The aseptic beverage carton consists of three materials: a central core of paperboard (typically 80 percent by weight) coated on the outside by a thin layer of polyethylene and on the inside by two layers of polyethylene with a very thin layer of aluminium foil in between. This use of multiple layers, also called *lamination*, makes the best use of the resources required to produce the carton by optimizing the physical properties of each material. No monomaterial could give the same performance achieved by these three materials in combination. This type of packaging saves energy, because a truck carrying filled cartons contains 95 percent product and 5 percent package.[65]

The aseptic beverage carton can be incinerated. The calorific value is 20.5 MJ/kg, approximately half of fuel oil. The carton not only releases a lot of energy during combustion, but it also burns cleanly. The thin layers of polyethylene become water vapor and carbon dioxide when burnt. If aseptic cartons are incinerated, the aluminium foil becomes aluminium oxide, a compound that occurs naturally in the Earth's crust. Beverage cartons can also be recycled by several different ways, including compressing them into chipboard and separating the different component materials to produce other products. In the latter case, the paper fibers are repulped in a hydrapulper. After 20 to 30 minutes, the paper fibers become separated from the polyethylene, and the aluminium foil remains trapped between the polyethylene layers. The separated fibers can be used to manufacture writing paper and household tissue. The remaining components can be individually recycled into raw materials, or used as a clean source of energy.[65] One of the biggest challenges for effective recycling of the beverage cartons, however, is efficient collection and separation.

8.12 Safety of Recycled Materials in Packages

A safety issue arises in conjunction with the use of recycled materials that come into contact with foods and beverages. The concern here is whether the original containers were used to store poisonous materials.

The consumption of PET for mainly packaging purposes has reached an amount of over 2.5 million tonnes per year in Western Europe.[25] PET is not limited to bottles and jars; various types of disposable cups and trays are popular as well. It is, therefore, not surprising that the reintegration of recycled PET into the manufacture of packaging is of much interest. However, the chronic exposure to even small amounts of toxic contamination can be hazardous for the consumer. Since PET is considered to be an effective barrier against contaminants, there was a proposal to encapsulate postconsumer PET into unused PET. The resulting multilayer PET product is recyclable and can be reintegrated into the manufacture of packaging in the same way. In the case of soft-drink bottles, postconsumer recycled PET is *coinjected* as a middle layer between two unused PET layers, the one inside preventing the beverage from direct contact with the recycled material. Also in the case of cups made of *coextruded* PET film, the postconsumer recycled material is again brought back into the loop of recycling. The effectiveness of such *functional barrier layer* was tested by varying its thickness.[66]

The use of recovered fibers in food packaging was also reported.[67] The paper from recovery operation arrives at the recycling mill in bales. The source of these bales varies from high-quality virgin fibers, such as envelope and boxboard clippings, to lower-quality materials including old newspaper, used paper, corrugated board from grocery stores, and paper of various types collected in curbside collection programs. Fibers from high-quality sources mixed with some from other origins would not cause adulteration of packaged food. First, the process of resuspending the fibers in water will accomplish some cleaning even without the addition of chemicals to enhance this action. The agitation and high water-to-fiber ratio present in direct entry processes would be expected to reduce many types of contamination. Second, the short fibers, commonly called *fines*, are expected to hold more than their share of contaminants, due to the higher specific surface area. This effect is observed, for example when PCBs are present in recovered fiber. Since the loss rate of these fines will exceed the loss rate of the longer fibers (the system design is likely to accomplish this segregation), unwanted contaminants will be selectively removed from the final paperboard product. Taken together, the two removal processes should provide significant reductions in the levels of unwanted substances in the finished material. Third, any remaining unwanted substances seem unlikely to migrate to food in quantities sufficient to pose unacceptable levels of risk of adverse health effects. The risk of adverse health effects due to exposure to any substance or mixture rests on the level of the exposure. It was concluded that the levels of the few substances likely to be found in recycled board are unlikely to result in migration of these substances to food at harmful levels.[67]

8.13 Disposal of Biodegradable Polymers

For biodegradable polymers, organic recycling is the most desirable choice. However, for this, a recovery system for organic waste must exist. As a prerequisite for coming on the market, biodegradable polymers may neither leave behind harmful materials in the compost nor impair the organic recycling process. The most crucial problem of composting, however, is the market of the compost. The experience of the European Investment Bank is that compost derived from municipal waste has difficulties in finding markets, in particular when there is a lack of separate waste collection, because of the risk of heavy metal contamination.[68]

8.14 Used Packages as a Source of Energy

Dry combustible fraction of household waste can be utilized for energy production. Those used packages, which cannot be recycled in a practical way but are dry and combustible, could be separated and used as fuel in existing boilers. Weight for weight, plastics and paper contain more energy than coal. This makes plastic and paper waste an invaluable fuel source, helping other materials to burn and also reducing fossil-fuel consumption. Even more resources are saved when energy is harnessed to provide heat and power.[7]

A research project in Finland investigated the use of different types of packaging waste as a secondary fuel in a circulating fluidized bed (CFB) boiler.[69] The effect of limestone addition and the role of the sulphur-to-chlorine ratio in the fuel on the emissions were also investigated. Emissions from the co-combustion of packaging waste were compared to the reference peat-coal combustion. The main primary fuel was peat and coal at a ratio of 55/45, which also was used as a reference when evaluating emission data. The shredded waste was cocombusted at a rate of 10 to 20 percent of the thermal feed. Four different waste materials were tested, two representing clean postindustrial and two dirty postconsumer combustible wastes, as follows:[69]

1. The liquid packaging board (LPB) consisted of unprinted polyethylene-coated board cut-out waste from the production of milk cartons.
2. The mixed board and flexible packaging material (MB/FP) consisted solely of printed production waste. The main components were cardboard, paper, plastics, metallized foil, and laminated aluminum foil.
3. Refuse derived fuel (RDF) consisted of shredded fuel derived from municipal solid waste. It was processed at Stormossen mechanical waste treatment plant at Vaasa. The pretreatment of RDF consisted of shredding, separating the organic fraction by a mechanical classifier, crushing, separating the magnetic metals with a magnet, and secondary crushing.
4. The mixed plastic waste (MP) was collected from Helsinki in public, separate-bin, municipal collection. All fuels were handled outdoors in bulk with a frontloader, and consequently were wet and contained sand.

The result of this investigation showed that the combustible fraction of waste materials, mainly consisting of used packaging, can be safely utilized as co-fuel in modern power plants, as up to 20 percent of the thermal feed with fossil fuels. In addition, local utilization would save energy in transportation.[69]

8.15 Recovery of Wooden Packaging

Wooden pallets are often used only once and disposed of afterward. Approximately 171 million of the 400 to 500 million pallets are recovered and used as fuel.[70] The new packaging waste directive of EU (2004/12/EU), prescribing recycling targets for packaging materials also includes wooden packaging and its recovery. For example, in Finland about 90 percent of wooden packaging is recovered and used as fuel.[71]

8.16 Construction Materials from Waste

A number of studies have been carried out to assess the feasibility of manufacturing constructional materials from rubbish to avoid landfill disposal. An example of these is the *neutralysis process*, which combines pulverized MSW with clay to form pellets, which are calcined to produce a light aggregate material.[73] In addition, several reports are available on the use of the ash of incinerated MSW as a concrete aggregate.[73,74]

9 PACKAGING SYSTEMS

Packaging systems can be defined a set of operations that fulfill the function of creating sales units of the product.[26] Figure 5 illustrates the packaging operation, and Figure 6 depicts a schematic of the packaging system.

Identifying the boundaries of the packaging system is especially important when considering life-cycle studies. The following can be pointed out:

- *The packaging operation is a crucial part of the packaging system.* Raw-material acquisition reaches over the product's system: During the acquisition of the raw materials, such secondary materials may appear that are not needed for the given system. For example, woodchips at tree cutting for paper production is not brought into the products system, but is used for another purpose.

- *The product's manufacturing or processing is partly included in the packaging system.* The product's properties may be modified (e.g., portioning) to ease packaging, or eventually embedded in the packaging operation as, for example, vacuum treatment of ground coffee.

- *Waste treatment is reaching into product's system, since the product system also creates wastes.* All the effects of the waste treatment are part of the given product's or package's system.

- *The impacts of secondary material processing may be part of the package or the secondary product system.* It is a matter of allocation.

The choice of the packaging operation is a complex decision, and it is based on the following main groups of factors (shown in Figure 7):[26]

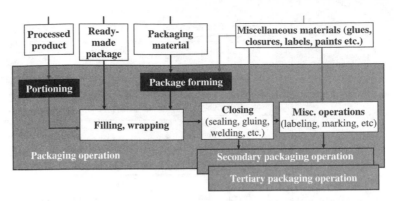

Figure 5 The steps of the packaging operation. (From Ref. 26.)

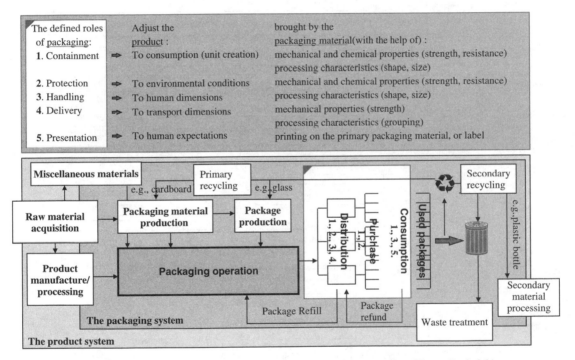

Figure 6 The packaging system as part of the product system. (From Ref. 26.)

1.a *Quality protection:* This involves protecting the product from the environmental, such as physical effects: damaging the unity of the product by force, pressure, or chemical effects, as damages by moisture, gases.

1.b *Loss prevention:* This involves protecting the product from losses, by chemical (evaporation, sublimation), or physical way (leaking, scattering), or pilferage in the shop.

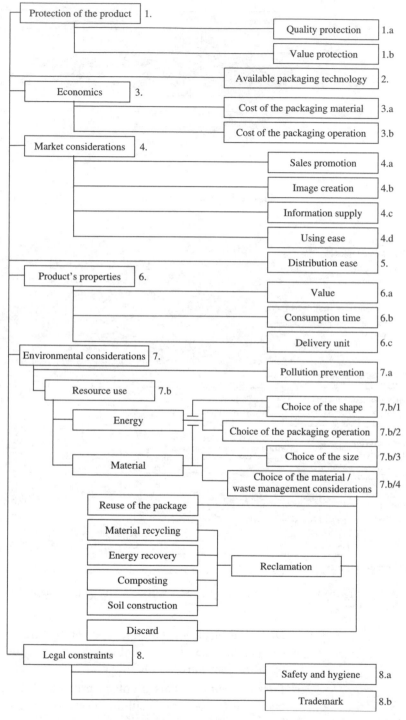

Figure 7   Factors influencing the choice of a package. (From Ref. 26.)

2 *Available packaging technology:* All the technologies used for packaging goods. This does not necessarily mean the best available technology.

3.a *Cost of the packaging material:* Cost will be allocated to the product's price.

3.b *Cost of the packaging operation:* Cost includes the cost of the packaging machinery, and also the cost of running the technology (handling, maintenance, amortization, energy consumption).

4.a *Sales promotion:* The advertisement value of the package, or certain property of the package that makes the product more attractive for the consumer is a consideration.

4.b *Image creation:* The package should create a positive image of the product or company. This is generally an aesthetics issue but could also be a property of the package, such as its strength or reliability.

4.c *Information supply:* Information is printed on the package about the product, or package, such as wastes management instructions.

4.d *Using ease:* The package should promote the use and consumption of the product.

5. *Distribution ease:* The package's function should help the product to get to the consumer.

6.a *Value:* The price of the product.

6.b *Life-cycle of the product:* This indicates for how long and within what conditions the packed product will be kept before purchased, and used, once opened (consumption time).

6.c *Delivery unit:* The amount (weight, volume, pieces) of the product in the package.

7.a *Pollution prevention:* The package's function to prevent the pollution or disturbance of the environment by the product, especially when the product is harmful to the environment.

7.b1 *Choice of the shape:* The shape should contain a unit of product while using the least amount of packaging material, with high operational performance.

7.b2 *Choice of the packaging operation:* There are *resource use considerations.* The choice is between the least energy-consuming operation, or a high operational performance ensuring low percentage of waste.

7.b3 *Choice of the size:* The packaging material per product ratio should be minimized. This choice will also consider resource use.

7.b4 *Choice of the packaging material: (Waste management considerations)* Choice of the material, with considerations of the package's possible reuse or recovery (material recycling, composting, energy recovery, soil construction, etc.) after the product has been consumed.

Generally, the package's properties are adjusted to the packed product, but in rare cases the product's properties are changed (e.g., avoid unnecessary protruding parts) with the purpose of creating a more practical package shape. The motives are primarily economic; simpler shapes are easier and more effective to collect into secondary (retail) packages. Another motive is the environmental concern: simpler shapes use fewer materials, and effective collection to secondary packages also entails reduced resource exploitation. Waste management options are considered already at package design; design for reuse, recovery, or eventually disposal, are important tools of environmental marketing.

10   LIFE-CYCLE ASSESSMENT OF PACKAGING SYSTEMS

LCA has its roots as far back as the early 1960s. That time, resource and environmental profile analyses (REPA) were done, with the goal to predict how changes in population would affect the world's total mineral and energy resources. In 1972, the office of solid waste at USEPA initiated regulatory activity for the packaging industries in the United States. The environmental effects of packages, especially of beverage containers, were of particular interest in the 1970s, due to litter problems. The Commission of the European Communities introduced its Liquid Food Container directive in 1985. Finally, in 1995, a standard methodology for LCA for packaging was developed for the European Community. Based on this methodology, an LCA study was done for compact detergent-packaging systems. The study showed that from an environmental point of view, it is advantageous to use refill packages compared to master packages.[75]

Many life-cycle assessments (LCA) and life-cycle inventories (LCI) have been carried out on packages and packaging materials since the 1970s. Most of those, which were analyzing whole packaging systems, were made to compare beverage packaging systems. The main question was whether single-use plastic or refillable glass bottles should be used. Generally, glass and returnable bottles are conceived as environmentally more acceptable. However, the long transportation distances, ineffective returns of empty bottles, and the simple fact that for the lighter weight of plastic bottles, a fully loaded truck transports 1.867 times more water in polyvinyl chloride (PVC) bottles than in glass bottles, shed a different light on plastic bottles.[76] For example the study promoted by Lox concluded that one-way plastic bottles can be environmentally preferable to returnable glass because of the role of transportation. The fossil fuel use and the consequent contribution to global warming are higher for glass, even if it is assumed that it is refilled 20 times.

A thorough study on Finnish beverage systems (refillable bottles, recovered steel cans, and single-uses PET bottles) was also done. The study stated that environmental quality does not correlate with the sole energy intensity. Topology and operational relations have a decisive effect in the assessed environmental quality of the studied system. In Finland, there is no clear winner in the comparison between different packaging options. Assessed environmental quality depends strongly on how the system is assumed to be put together and operate. When whole packaging systems are assessed, and the results may affect the material flows in the society, it is not enough to assess only ecological consequences. It is probable that if there is a working system of a beverage packaging system, either returnable glass, single-use plastic or recycled aluminum, to change it may involve environmental effects that may surpass the expected benefits of the change.[77] In Finland, packaging of brewery products and soft drinks is predominantly based on refillable bottles (glass and plastic, respectively). Return and reuse systems of bottles are comprehensive and effective, and thus the degree of reuse is high—about 80 percent.[78]

There are also several points in an LCA that can significantly change the result of a study. These are, among others: the functional unit, system boundaries (geographical, natural as well as life cycle), data quality, and allocation. A traditional problem in LCA is how to deal with processes or groups of processes with more than one input and/or output, and how to deal with the use of recycled material in another product than the original. A crucial problem of evaluation and interpretation of the inventory results is that they depend on social and political preferences rather than on technical development.

11 CONCLUSIONS

The concern about the effect of packaging on the environment derives from their relatively high percentage in the household waste. This, however, indicates rather the level of consumption than excessive packaging. Packages are made to deliver the product to the consumer; hence, they cannot be viewed separately, either from the product or from consumption. Increasing amounts of packages in the waste stream only indicates increasing consumption.

Packaging is strongly influenced by social desires, political preferences, and regulatory and economical effects. In addition, packaging is not only a product—a package—but a system, and the package itself cannot be separated from its content. Consequently, an environmental assessment cannot mean only ecological impact analysis; neither can the judgment of environmental friendliness be based solely on the type of packaging material. Plastics have perhaps the most negative image, albeit of being lightweight and sturdy, thus giving high protection value with low environmental impact. It is especially true for composites, which combine several materials for better protection, while using minimal amount of the individual materials.

The use and reuse of glass packaging is a state-specific question. The raw material is in plentiful supply and there are no technical barriers to its reuse or recycling. If there is a well-working system of glass reuse, changing it may involve significant expenditures, as well as environmental impacts. While in several countries reusable glasses are not efficient due to large transportation distances, in other countries, with a good infrastructure, reuse is preferred. Another form of packaging reuse, the refill packages, are environmentally and economically favoured and their increment would be preferred. Reducing package weight and making products more concentrated also lead to better resource use.

Reclamation of packaging wastes through recycling is strongly promoted by legislative bodies in the European Union as way of reducing the environmental impacts of packaging. From an ecological point of view, mechanical recycling processes have ecological advantages over feedstock and energy recovery processes if virgin plastic is substituted in a ratio of near 1:1. On the basis of the comparative analysis of feedstock recycling and energy recovery, the following recovery processes can be recommended: use as reducing agents in blast furnaces, thermolysis to petrochemical products, and fluidized-bed combustion.

Finally it can be asserted that while packaging plays an important role in achieving a sustained development, its most important actors are the consumers themselves. No regulation can be as effective as a well informed, environmentally conscious, ethical public.

REFERENCES

1. World Resources Institute. World Resources 1996–1997. New York: Oxford University Press (1996).
2. J. M. Kooijman, *Environmental Impact of Packaging: Performance in the Household*, INCPEN, London, 2000.
3. S. Geiger, "The Role of Packaging in Increasing Sales," *Journal of Marketing*, **2** (1973).
4. INCPEN, INCPEN brief on packagings contribution to waste minimisation. Waste Minimisation Workshop 27/03/2002, Westminster Central Hall, London, 2002.
5. B. Barbour, *The Complete Food Preservation Book*, David McKay Co., New York, 1978.
6. J. H. Alexander. "Solid Waste in Perspective." In *First Annual Packaging and Government Seminar*, The Packaging Institute (May, 1997).
7. V. Williams, "Plastic Packaging for Food: The Ideal Solution for Consumers, Industry and the Environment," *R'95 Recovery, Recycling, Re-integration*. Proc. Cong. pp. II. 49–56 (1995).
8. A. Leppänen-Turkula, T. Meristö, & T. Järvi-Kääriäinen, "Pakkaus 2020, Tulevaisuuden visioita Suomen pakkausalalle" ("Packaging 2020, Future Scenarios for the Finnish Packaging Branch"), *Pakkausteknologia—PTR ry (Association of Packaging Technology and Research) PTR Report*, **48** 60 (2000).
9. V. Korhonen, "Elintarvikepalkkaamisen nykytila ja tulevaisuuden näkymiä" ("The Present State and Future Trends of Food Packaging"), *Pakkausteknologia—PTR ry (Association of Packaging Technology and Research) PTR Report*, **46**, 61 (2000).
10. S. Young, "Packaging Design, Consumer Research, and Business Strategy: The March toward Accountability," *Design Management Journal* (Fall 2002).
11. G. Cybulska, *Waste Management in the Food Industry: An Overview*, Campden and Chorleywood Food Research Association Group, UK, 2000.

12. INCPEN, *Packaging Saves Waste*, The Industry Council for Packaging and the Environment, London, 1987.

13. H. Alter "The Origins of Municipal Solid Waste: The Relations between Residues from Packaging Materials and Food." In: *Waste Management and Research*, Vol. 7, pp 103–114 (1989).

14. L. Scarlett, "A Consumer's Guide to Environmental Myths and Realities," in *NCPA Progressive Environmentalism, Trade & Aid Resource Book*, National Center for Policy Analysis. USA, 1995.

15. E. Cooper, *A History of Pottery*, St. Martin's Press. New York, 1972.

16. F. Kumper and K. G. Beyer, *Glass: A World History*, New York Graphics Society. Greenwich. CT, 1966.

17. W. Baumgart, "Glass." In: *Process Mineralogy of Ceramic Materials*, W. Baumgart, A. C. Dunham and A. C. Amstutz (eds.), 1984.

18. S. E. M. Selke, *Packaging and the Environment: Alternatives, Trends and Solutions*, Technomic Publishing Co., Lancaster, PA, 1990.

19. P. Knauth, *The Metalsmiths*, Time-Life Books. New York, 1974.

20. E. T. Turkdogan, Fundamentals of Steelmaking. Institute of Materials, Cambridge: The University Press, 1996.

21. Coors Brewing Company, http://www.coors.com/factsheets/EnvironmentFactSheet.pdf, Last accessed 5.1.2007.

22. Aluminum Packaging Recycling Organization (Alupro) http://www.alupro.org.uk/ last accessed 5.1.2007 American Plastics Council (APC), *Waste Prevention—Is Recycling Enough?* Transcript of a Washington Policy Forum sponsored by APC and Harper's magazine.

23. O. Rockstroh, *Handbuch der industriellen verpackung*, München: Moderne Industrie, 1972.

24. R. Beswick and A. G. Dunn, "Plastics in Packaging—Western Europe and North America," Market Report, 2002.

25. Plastics*Europe*, "An analysis of plastics consumption and recovery in Europe, 2002 & 2003," 2004.

26. E. Pongrácz, "The Environmental Effects of Packaging," Licentiate thesis. Tampere University of Technology, Department of Environmental Technology, Institute of Environmental Engineering and Biotechnology, Tampere, Finland, 1998.

27. J. Reini, *Biodegradable Plastics—Options for the Future*, TEKES, Helsinki, 1992.

28. J. Reske, "Beauty of Bioplastics. Compostable Packaging Sees a Bright Future Ahead," *Waste Management World* (March–April 2005).

29. J. Bejune, R. Bush, P. Arman, B. Hansen and D. Cumbo, "Pallet Industry Relying More on Recovered Wood Material," *Pallet Enterprise* 22(10), pp. 20–27 (2002).

30. R. Lilienfeld, *A Study of Packaging Efficiency as It Relates to Waste Prevention*, The Cygnus Group, Ann Arbor MI, March 1995.

31. L. L. Katan, Packaging, Environment and Recycle, 1987.

32. *The Litter Monitoring Body*. July 2006 results http://www.litter.ie. Last accessed 5.1.2007.

33. UNEP—Earthwatch, *Chemical Pollution: A Global Overview*, A Joint Publication of the International Register of Potentially Toxic Materials and the Global Environment Monitoring System's Monitoring and Assessment Research Centre, 1992.

34. International Maritime Organization, International Convention for the Prevention of Marine Pollution from Ships, 1973 as modified by the protocol of 1978 relating thereto (MARPOL 73178), http://www.imo.org, last accessed 5.1.2007.

35. K. Christiansen, L. Hoffmann, L. E. Hansen, A. Schmidt, A. A. Jensen, K. Pommer, and A. Grove, "Environmental Assessment of PVC and Selected Alternative Materials," in *UNEP/IEO (United Nations Environmental Programme, Industry and Environmental Office) Cleaner Production Programme; Working Group on Policies, Strategies and Instruments to Promote Cleaner Production* (Seminar), Trolleholm, Sweden. pp. 112–115, 1991.

36. M. Gandy (ed.), *The Recycling and the Politics of Urban Waste*, Earthscan. London, 1994.

37. M. Herron, "Packaging and the Environment," *UNEP/IEO (United Nations Environmental Programme/Industry and Environmental Office) Cleaner Production Programme; Working Group on Policies, Strategies and Instruments to Promote Cleaner Production* (Seminar), Trolleholm, Sweden. pp. 30–35, 1996.

38. J. Mayr, "Packaging and the Environment in Austria," *UNEP/IEO (United Nations Environmental Programme/Industry and Environmental Office) Cleaner Production Programme; Working Group on Policies, Strategies and Instruments to Promote Cleaner Production* (Seminar), Trolleholm, Sweden. pp. 53–64, 1991.

39. O. Arrango and C. Bertucci, "Municipal Solid Waste: A Quantitative Overview," in *The Management of Municipal Solid Waste in Europe*, A. Q. Curzio, L. Prospetti, and R. Zobali (eds.), Elsevier, Amsterdam, 1994, pp. 16–38.

40. A. Porteous (ed.), *Packaging Waste Management Opportunities*, Pira International Report, Surrey, 1995.

41. M. Flood, "Over-Packaging, or Over-Consumption," in *Warmer Bulletin*, No. 37, pp. 16–17 (1993).

42. D. Perchard "Impact of Legislation—Europe." In: Murphy, Mary. (ed.). *Packaging and the Environment*. Surrey: Pira International. pp. 31–45 (1992).

43. G. Davis and J. H. Song, Biodegradable Packaging Based on Raw Materials from Crops and Their Impact on Waste Management," *Industrial Crops and Products. An International Journal*, **23**(2), 147–161 (2006).

44. S. Isoaho "Recycling Policy and Strategy." In: *Report for ISWA Working Group Meeting*, Madrid, Spain, September 29–30, 1994. pp. 9–17 (1995).

45. C. Hanisch, "Is Extended Producer Responsibility Effective?" *Environmental Science and Technology*, **34**(7), 170A–175A (2000).

46. W. Rathje and C. Murphy, "Five Major Myths about Garbage, and Why They're Wrong," *Smithsonian*, **23**(4), 113–122 (1992).

47. E. Pongrácz, "Re-defining the Concepts of Waste and Waste Management: Evolving the Theory of Waste Management," Doctoral dissertation. University of Oulu. Department of Process and Environmental Engineering, Oulu, Finland, 2002.

48. E. Plinke and K. Kaempf, "Ecological, Economical, and Technical Limits to Recycling of Packaging Materials," in *R'95 Recovery, Recycling, Re-integration*, Proc. Cong. pp. I.179–180, 1995.

49. S. and K. Vrpilainen "A Present Day Realization of Waste. Management by Finnish Local Authorities." *Ymparisto ja terveys* 31 (7–9): 14–23 (2000).

50. J. Anhava, E. Ekholm, E. Ikäheimo, K. Koskela, M. Kirvi, and M. Walavaara, *Jätehuollon ja materiaalikierrätyksen teknologiat ja niiden kehittämistarpeet* (*Technologies of Waste*

Management and Material Recycling; and Their Development Needs), TEKES, Teknolo-giakatsaus 102/2001. Helsinki, Finland, 2001.

51. N. Horvat and F. T. T. Ng, "Tertiary Polymer Recycling: Study of Polyethylene Ther-molysis as a First Step to Synthetic Diesel Fuel," *Fuel*, **78**(4), 459–470.

52. A. L. Bisio and N. C. Merrieam, "Technologies for Polymer Recovery/recycling and Potential for Energy Savings," in *How to Manage Plastics Waste: Technology and Market Opportunities*, Ch. 3. A. L. Bisio and M. Xanthos (eds.), Münich, Germany, Carl Hanser Verlag, 1994, pp. 15–31.

53. V. Arpiainen, J. Rant, L. Hietanen, E. Leppämäki, and K. Sipilä, *Muovin pyrolyysin esiselvitys (Prestudy of Pyrolysis Technology for Plastic Waste)*, Report ENE 1/34/99, VTT Energy, Espoo, Finland, 1999.

54. C. Braekman-Danheux A. D'haeyere, A. Fontana, and P. Laurent, "Upgrading of Waste Derived Solid Fuel by Steam Gasification," *Fuel* **77**(1–2), 55–59 (1998).

55. N. Kiran, E. Ekinci, and C. E. Snape, "Recycling of Plastic Wastes via Pyrolysis," *Resources, Conservation and Recycling*, **29**(4), 273–283.

56. J. Aguado and D. Serrano, "Feedstock Recycling of Plastic Wastes," Royal Society of Chemistry Clean Technology Monographs. Cambridge, UK, 1999.

57. R. Zevenhoven, E. P. Axelsen, and M. Hupa, "Pyrolysis of Waste-derived Fuel Mixtures Containing PVC," *Fuel*, **81**(4), 507–510 (2002).

58. R. Vesterinen Gasification of Waste Preserved Wood Impregnated with Toxic Inorganic and/or Organic Chemicals. Gasification tests with impregnated waste wood at the 5 MW Jalasjärvi gasification plant. Espoo: VTT, 1995. 38 p. + app. 35 p. (VTT Publications 244). (1995).

59. APME (Association of Plastics Manufacturers in Europe) Life Cycle Analysis of Recy-cling and Recovery of Household Plastics Waste Packaging Materials. Summary Report (1995).

60. N. Mayne, "Plastics Recycling Around the World," *Waste Management World* (July–August 2003), pp. 69–75.

61. WARMER, "Glass Recycling: The British Industry Viewpoint," *Warmer Bulletin*, No. 40, 14–15 (1994).

62. O. Tickell, "The Future for Glass Recycling in the U.K.," *Warmer Bulletin*, No. 39, pp. 14–15 (1993).

63. Can Manufacturers Institute, Environmental Issues, Recycling Fun Facts, http://www.cancentral.com/, last accessed 5.1.2007.

64. International Aluminium Institute, Aluminium Recycling Facts, http://www.wend-aluminium.org/environment/recycling, last accessed 5.1.2007.

65. G. L. Robertson, "Multimaterial Package Design: Resource Optimization for a Sustainable Environment," *R'95 Recovery, Recycling, Re-integration*, Proc. Cong. pp. II. 141–146, 1995.

66. Z. Harmati, E. M. Moset, P. O. Boll and F. R. Lang. "How Effective Does an Unused PET-Barrier-Layer Protect Food Against Contamination" in: *R as Recovery Recycling, Reintegration*, Proc. Cong. pp II 179–180, 1993.

67. B. F. Vincent, "The Use of Recovered Fiber in Food Packaging," *Tappi Journal*, **78**(8), 49–52, (1995).

68. E. Greppi, "EIB and the Financing of Municipal Solid Waste Disposal," in *The Management of Municipal Solid Waste in Europe. Economic, Technological and Environmental Perspectives*, A.Q. Curzio, L. Prosperetti and R. Zoboli (eds.), Elsevier Science B.V., The Netherlands, 1994.

69. H. Manninen, M. Frankenhaeuser, T. Järvi-Kääriäinen, and A. Leppänen, *Used Packaging as a Source of Energy*. PTR—Association of Packaging Technology and Research. Report No. 37, 1994.

70. T. M. Smith, M. Reichenbach, S-A Molina–Murillo, and R. Smith, *Potential Effect of International Phytosanitary Standards on Use of Wood Packaging Material*, Research report. University of Minnesota, Department of Bio-based Products, 2004.

71. A. Leppänen-Turkula, "The new packaging directive increases recycling targets" (in Finnish), *Kehittyvä Elintarvike*, **4** (2004).

72. A. Krol, K. White, and B. Hodgson, "Production of Lightweight Aggregate from Wastes: The Neutralysis Process," Proc. International Conference on Environmental Implications of Construction with Waste Materials, Maastricht, The Netherlands, November 10–14, 1991.

73. T. Mangialardi, "Sintering of MSW Fly Ash for Reuse as a Concrete Aggregate," *Journal of Hazardous Materials*, **87**(1–3), 225–239 (2001).

74. K. Nishida, Y. Nagayoshi, H. Ota, and H. Nagasawa, "Melting and Stone Production Using MSW Incinerated Ash," *Waste Management*, **21**(5), 443–449 (2001).

75. R. G. Hunt and W. E. Franklin, "LCA—How it Came About—Personal Reflections on the Origin and the Development of LCA in the USA," *The International Journal of Life Cycle Assessment*, **1**(1), pp. 4–7 (1996).

76. F. Lox (ed.), *Waste Management—Life Cycle Analysis of Packaging*, Final Report. Study Realised by the Consortium Vrije Universiteit Brussel, Vlaamse Instelling voor Technologisch Onderzoek, Belgian Packaging Institute, for the European Commission, DG XI/A/4, 1994.

77. H. Mälkki, S. Hakala, Y. Virtanen, and A. Leppänen, *Life Cycle Assessment of Environmental Impacts of Finnish Beverage Packaging Systems*. PTR—Association of Packaging Technology and Research. Report No. 43, 1995.

78. PYR (The Environmental Register of Packaging, Finland), "Recovery statistics," http://www.pyr.fi/. Last accessed 5.1.2007.

CHAPTER 10

AQUEOUS PROCESSING FOR ENVIRONMENTAL PROTECTION

Fiona M. Doyle
University of California, Department of Materials Science and Engineering, Berkeley, California

Gretchen Lapidus-Lavine
Depto. Ingeniería de Procesos e Hidráulica, Universidad Autónoma Metropolitana-Iztapalapa, México

1	INTRODUCTION	279
2	ENVIRONMENTAL PROTECTION	280
	2.1 Issues and Challenges	280
	2.2 Potential and Limitations of Aqueous Processing	281
3	UNIT OPERATIONS IN AQUEOUS PROCESSING	281
	3.1 Precipitation	282
	3.2 Reductive Precipitation	286
	3.3 Electrolysis	288
	3.4 Electrokinetic Remediation	288
	3.5 Adsorption	289
	3.6 Biological Methods	291
	3.7 Solvent Extraction and Ion Exchange	291
	3.8 Membrane Processes	293
	3.9 Foam Separation Processes	294
4	CONCLUSIONS	296

1 INTRODUCTION

Aqueous processing for environmental compliance and protection encompasses water-based processes used to assist in complying with waste discharge and management regulations, along with water-based processes that lower the overall environmental impact of a particular operation. Two significant areas within this overall category are the *aqueous treatment of wastes* and *alternative, aqueous processes*. Solution processes are now usefully being applied to the treatment of obsolete electronic components, spent catalysts, and reactive process residues.[1–3] They are also used to recover metals from effluents generated in many manufacturing activities, including the manufacture of integrated circuits and printed circuit boards.[4–6] Aqueous processing is even used in recycling water itself.[7]

Aqueous processing is already being embraced for environmental protection, because of its numerous advantages. Moreover, it is highly likely that aqueous processing will truly dominate the menu of possible technological approaches

279

during the next few decades. Before discussing why aqueous processing offers such potential, and how much scope there is for aqueous processing for environmental protection, it is worth clarifying the term *environmental protection* in the context of this chapter.

2 ENVIRONMENTAL PROTECTION

In the broader sense, *environmental protection* includes regulatory policy and behavior modification, as well as technical activities. The technical issues are ideally referred to as *environmental engineering*. However, this term has been appropriated for activities that react to problems, such as contaminated air and water. Clearly, the future will depend on proactive technologies that allow sustainable development and recycling of the Earth's resources, with minimal impact to the overall system of the planet. *Green manufacturing* and *soft processing* are other terms used for this approach. It is here that aqueous processing has so much to offer.

2.1 Issues and Challenges

One of the greatest challenges in developing technologies for environmental protection is to adopt a rigorous systems approach that allows the benefits and shortcomings of new processes to be compared quantitatively with those of existing processes. Much is unknown regarding the complex interplay between engineered processes and the natural environment, and the effects on organisms and ecosystems of long-term exposure to specific chemicals may only emerge after decades. Superimposed on this lack of information is the inherent difficulty in comparing dissimilar effects. Is it better to raise the level of trace metals in the groundwater below a leach residue storage pond, or to emit particulates into the air? When considering cleaning up contaminated sites, the environmental impact of generating energy, whether in internal combustion engines, in hydroelectric dams, or even by solar panels, is rarely considered. Yet appreciable energy is needed for large-scale projects, such as mines and mineral-processing facilities. In the United States, contaminants are often taken from contaminated sites to licensed management facilities that are specifically designed to contain the contaminant of concern. There are complex sociopolitical issues with the siting of such facilities, along with concerns over accidental release of the contaminant during transport. Finally, inadequacies are emerging in the cost/benefit analyses done when considering new technology for environmental protection.

With current energy, commodity, and land prices, there are many instances where byproduct recovery from wastes, or recycling of complex scrap materials, appears less economic than discarding the waste or scrap. However, such analyses seldom consider *future* costs associated with waste disposal. In the United States, companies and their successors are now being required to spend tens, sometimes hundreds, of millions of dollars to remediate environmental damage caused by waste-management practices that were perfectly legal when used decades ago. Had these future costs, and the associated legal costs, been anticipated when waste-management practices were being developed, waste minimization, byproduct recovery, and recycling would have appeared more economic. This is particularly true when considering wastes that are classed as hazardous because of trace concentrations of metallic contaminants. It is constructive to recast these problems as new opportunities for materials processing, with the eventual objective of converting the components of the waste to marketable byproducts. Although the market value of such byproducts is seldom high enough to cover all processing costs, waste treatment circumvents the cost of managing hazardous solid wastes and sludges.

2.2   Potential and Limitations of Aqueous Processing

Given the current and future importance of environmental protection, there is clearly an urgent need for new or improved technologies for recycling, and for waste treatment and minimization. There are undoubtedly many wastes for which physical separations and pyrometallurgical processes will be the most suitable. Nevertheless, aqueous processing offers enormous potential that is relatively untapped at present. Some of this underutilization stems from the fact that aqueous processing is a relative newcomer to the arena of minerals and materials processing. In addition, however, the boundary conditions of the environmental protection problem are changing. Aqueous processing offers the potential of minimal air pollution and use of hazardous chemicals, modest energy utilization and unprecedented flexibility and suitability for local, small-scale operations. Aqueous processing is suitable for both improving the environmental impact of existing materials production operations (environmentally protective aqueous processing) and for application well outside of the traditional materials processing arena (aqueous processing for environmental protection). However, it is essential to recognize the limitations of traditional water-based processes. Many *do* cause air pollution, are energy intensive, and use hazardous materials. Traditional aqueous-phase processes will, in general, require modification or replacement before they are optimal for environmental protection.

Bearing this in mind, this chapter summarizes the unit operations employed in aqueous processing, emphasizing their underlying principles. The goal is to provide the information needed to modify these unit operations for different types of solutions, and to achieve appropriate water-quality objectives.

3   UNIT OPERATIONS IN AQUEOUS PROCESSING

Many different solid, liquid, and airborne waste materials from a similarly large number of industries, such as manufacturing, electronics, batteries, pigments and paints, catalysts, polymers, printing and photocopying, fuels and lubricants, and dentistry and pharmaceuticals contain toxic materials. Improper disposal of any of these can produce contaminated drainage. Surface water and groundwater can become contaminated by metals from geological sources, particularly if the disturbances from construction and mining activities have accelerated the natural weathering processes.[8] In many cases, the aqueous solutions contain some metals at levels high enough to be deleterious to aquatic organisms, to enter the food chain, or to have an adverse effect on wildlife living close to contaminated bodies of water.

The detoxification of dilute solutions is particularly challenging. To treat such solutions, it is necessary to devise some way of removing contaminant species (physically or chemically) from a large volume of solution. With the exception of H^+ and OH^- ions that neutralize basic or acidic solutions, any chemical reagents added to the solution should also be removed, so as not to introduce additional contaminants. The reagent and energy costs incurred should be as low as possible, and any reagents used should be recycled to the greatest extent possible.

Many commercial aqueous phase processes could be modified to treat such solutions. This chapter discusses precipitation, reductive precipitation, electrolysis, electrokinetic remediation, adsorption, solvent extraction and ion exchange, biological processes, membrane processes, and foam separations. Specifically, it focuses on their advantages, shortcomings, and possible modifications that might broaden their applicability.

3.1   Precipitation

Chemical precipitation techniques are selected according to the solubility products of different compounds, the cost of reagents, the impact of residual dissolved species on water quality, and the marketability or disposal of the resulting solid phase.

Theoretical Considerations

Thermodynamics dictates that the maximum solubility of a salt in a liquid solution is given by the product of the activities of the uncomplexed ions, raised to the appropriate power dictated by the stoichiometry of the salt. For a salt $M_m N_n$:

$$K_{sp} = (a_M)^m (a_N)^n = (\gamma_M [M^{n+}])^m (\gamma_N [N^{m-}])^n \tag{1}$$

where K_{sp} = Solubility product of salt $M_m N_n$
$\quad a_i$ = activity of ion i in the solution
$\quad \gamma_i$ = activity coefficient of i in the solution
$\quad [i]$ = concentration of i in the solution in moles per liter

In dilute aqueous solutions (up to approximately 0.001M), ionic activity coefficients are essentially equal to unity; therefore equation (1) reduces to

$$K_{sp} = [M^{n+}]^m [N^{m-}]^n \tag{2}$$

In most cases, the contaminant ion whose concentrations must be lowered is cationic, which is represented by M^{n+}. The cationic concentration is lowered by adding N^{m-} from an external source and/or lowering the value of K_{sp} by adjusting the solution temperature. Composite plots of solubilities, such as those compiled by Monhemius, can be exceedingly useful when selecting precipitation methods appropriate for individual solutions.[9]

The species referred to in equations (1) and (2) are the uncomplexed ions, and not the total concentrations of the contaminant or counter-ions. This becomes important when, as in most industrial waste streams, there are other anions that may complex M^{n+}. The total concentration of soluble M is that of the uncomplexed species plus all of the M present in complexed form:

$$[M]_T = [M^{n+}] + \sum_x \sum_y X \lfloor M_x A_y^{(xn-yk)+} \rfloor \tag{3}$$

where $[M]_T$ = total concentration of soluble M
$[M_x A_y^{(xn-yk)+}]$ = concentration of the complex $M_x A_y^{(xn-yk)+}$
A^{k-} = anionic species that complexes with M^{n+}
x is the variable describing the stoichiometric amount of the cationic species in complexes; the summation is done for all values of x and y.

The concentration of each complexed M species may be calculated, using the free anion or ligand concentration and the stability constant of the complex:

$$[M_x A_y^{(xn-yk)+}] = K_{xy}[M^{n+}]^x [A^{k-}]^y \tag{4}$$

where $[A^{k-}]$ = concentration of uncomplexed anion
K_{xy} = stability constant for the complex $M_x A_y^{(xn-yk)+}$

Relevant stability constants may be found in databases such as those published by the National Institute of Standards and Technology.[10] The approach is exemplified below for precipitation of a hydroxide in the absence and presence of complexing anions.

Example 1   Chromium(III) may be precipitated from tannery wastewater with sodium hydroxide, forming $Cr(OH)_3$, whose solubility product is 6.3×10^{-31}.[11] Tannery wastewater, initially at pH 3.5, has the following average composition:[12]

Chromium, as Cr(III)	3,950 ppm
Sulfate, as SO_4^{2-}	3,525 ppm
Chloride, as Cl^-	22,070 ppm

Calculate the minimum pH that would be required to lower the chromium concentration to 1 ppm, a) Considering only the chromium and b) considering complexation of the chromium by the chloride and sulfate ions.

Solution: Chromium(III) forms several soluble complexes with hydroxide, sulfate, and chloride compounds, which increase its overall solubility: $CrOH^{2+}$,

$Cr(OH)_2{}^+$, $Cr_2(OH)_2{}^{4+}$, $Cr_3(OH)_4{}^{5+}$, $Cr_4(OH)_6{}^{6+}$, $CrSO_4{}^+$, $CrOHSO_4$, and $CrCl^{2+}$. Therefore, the total solubility of Cr(III) in solution is given by the following species balance:

$$[Cr(III)]_T = [Cr^{3+}] + [CrOH^{2+}] + [Cr(OH)_2{}^+] + 2[Cr_2(OH)_2{}^{4+}]$$
$$+ 3[Cr_3(OH)_4{}^{5+}] + 4[Cr_4(OH)_6{}^{6+}] + [CrSO_4{}^+] + [Cr(OH)SO_4] + [CrCl^{2+}]$$

The concentration of each soluble complex is related to the concentration of free (uncomplexed) Cr^{3+} and that of the complexing ligands, through these stability constants:[10]

$$\frac{[Cr(OH)^{2+}]}{[Cr^{3+}][OH^-]} = 2.82 \times 10^9 \qquad \frac{[Cr(SO_4)^+]}{[Cr^{3+}][SO_4{}^{2-}]} = 3.98 \times 10^2$$

$$\frac{[Cr(OH)_2{}^+]}{[Cr^{3+}][OH^-]^2} = 1.26 \times 10^{17} \qquad \frac{[Cr(OH)(SO_4{}^{2-})]}{[Cr^{3+}][OH^-][SO_4{}^{2-}]} = 4.47 \times 10^4$$

$$\frac{[Cr_2(OH)_2{}^{4+}]}{[Cr^{3+}]^2[OH^-]^2} = 1.0 \times 10^{24} \qquad \frac{[CrCl^{2+}]}{[Cr^{3+}][Cl^-]} = 0.1$$

$$\frac{[Cr_3(OH)_4{}^{5+}]}{[Cr^{3+}]^3[OH^-]^4} = 1.0 \times 10^{37}$$

$$\frac{[Cr_4(OH)_6{}^{6+}]}{[Cr^{3+}]^4[OH^-]^6} = 1.58 \times 10^{80}$$

Each equation is solved for the complex of interest and substituted into the species balance equation, along with the values for the chloride and sulfate concentrations. The equation is then solved for the pH that gives 1.923×10^{-5} moles per liter (1 ppm) for the total soluble chromium ($[Cr(III)]_T$). From inspection of the stability complexes, it is clear that hydroxide complexes Cr(III) much more strongly than does either sulfate or chloride. Accordingly, the pH needed to attain 1 ppm Cr(III) is 8.12, regardless of whether one considers complexation by sulfate and chloride; at this pH the dominant soluble complex is $Cr_4(OH)_6^{6+}$. Had complexation by hydroxide not been considered, the minimum pH would have been erroneously calculated as 5.51.

Neutralization of Acid and Precipitation of Metal Hydroxides

Neutralization and precipitation is the technique used most widely for treating metal-contaminated acid-rock drainage, metal ion spills, along with some plating and pickling wastes.[13] Hydrated lime is usually used as the neutralizing agent, because of its low cost, ready availability in many areas, and efficacy.[14] However, lime yields voluminous carbonate/hydroxide sludges. Furthermore, gypsum is produced when treating sulfate-containing solutions. The hydroxides of the toxic metals are generally dispersed in the resulting sludges, and may be present in sufficient quantity to render the entire residue hazardous from a regulatory standpoint. Clearly, this is an area that might benefit from more thoughtful design and rigorous process control. Recycle of sludge to the precipitation stage has been used to increase the particle size, and hence improve the settling and filtering properties of the resulting precipitates.[15,16] By staging the neutralization process, it should be possible to neutralize the bulk of the acid at low pH, with little precipitation of metals, thereby producing a voluminous but environmentally benign residue, and then raise the pH in a separate reactor to produce a low-volume residue in which the metals are concentrated. NaOH, Na_2CO_3, MgO, and other alkaline materials have been tested as neutralizing agents, but tend to give poor settling residues, in addition to being costly.[17] Recent attention to the problem of alkaline cement kiln dusts raises the intriguing possibility of using these wastes to treat acidic effluents.[18]

Precipitation of Metal Sulfides

The precipitation of metal sulfides has very strong potential for detoxifying aqueous solutions, stemming from the very low solubility products of the sulfides of many toxic metals.[19,20] Metal sulfides might be floated from the solution being treated,[21,22] or might be separated by settling or filtration.[23] Such sulfides could be fed to smelters, thereby obviating the need for waste management. Sulfide could be introduced from various sources, such as hydrogen sulfide gas, sodium sulfide, or a combination of these. Hydrogen sulfide gas is particularly attractive, because it can be generated by anaerobic sulfate-reducing bacteria.

Separate bioreactors allow treatment of solutions containing higher concentrations of metal ions than can be tolerated by most sulfate-reducing bacteria, and separation of different metal ions in the effluent.[24] Up to 5000 m^3/d of groundwater contaminated by zinc, cadmium, and excessive sulfate at the Budelco zinc refinery in the Netherlands has been treated using bacterially generated sulfide.[25] All sulfide precipitation systems suffer the potential disadvantage of relatively high residual levels of sulfide species, due to strong hydrolysis of the S^{2-} ion. This requires an additional oxidation stage, either biotic or abiotic, to remove sulfide species from solution prior to discharge.

Precipitation of Fe(III) compounds

Many trace contaminants co-precipitate with Fe(III), behavior that can be problematic in many hydrometallurgical processes. However, such behavior is useful in removing arsenate from solutions, and has been tested for a broader range of contaminants.[26,27] There are unresolved questions on the long-term stability of such residues under ambient conditions.[28,29] The theoretical basis behind cationic precipitation with Fe(III) or Al(III) is essentially the same as the previous case—the formation of insoluble compounds, which have specific solubility products (K_{sp}).[30,31]

3.2 Reductive Precipitation

The phenomenon of *cementation* was known to Chinese alchemists in the second century B.C. Copper was produced commercially from contaminated mine waters by cementation with metallic iron in China in 1086 A.D..[32] This process, ideally represented as

$$Fe^o + Cu^{2+} \rightarrow Fe^{2+} + Cu^o \tag{5}$$

later found broader application in Europe—for example, at Rio Tinto in Spain. Kennecott developed a conical precipitation reactor that offered significantly better yields than the launders traditionally used for this reaction.[33] The subsequent development of solvent extraction reagents specific to copper caused solvent extraction-electrowinning to supplant cementation for commercial production of copper, because of its superiority in terms of cost, product yield, and purity. However, cementation has found fairly widespread use for detoxifying water contaminated by copper.

Theoretical Considerations

The copper cementation reaction consists of two half-cell reactions: one oxidizing (loses electrons) and the other reducing (gains electrons):

$$
\begin{aligned}
Fe^o &\rightarrow Fe^{2+} + 2e^- & E_h^o &= -0.440V \\
Cu^{2+} + 2e^- &\rightarrow Cu^o & E_h^o &= 0.337V
\end{aligned}
\tag{6}
$$

Each has a characteristic reduction potential, E_h^o, reported for the standard state, with reference to the standard hydrogen reaction. When combined, the two half-cell potentials give the electrochemical cell potential, ΔE^o. Thermodynamics dictates that the cementation reaction will proceed as long as the following relation is valid:

$$\frac{zF\Delta E^o}{RT} > \ln \prod \frac{a_{products}}{a_{reactants}} = \ln \frac{a_{Fe^{2+}} a_{Cu^o}}{a_{Fe^o} a_{Cu^{2+}}} \tag{7}$$

where, z = number of electrons transferred = 2
   F = Faraday's constant (96,480 coulombs/mole)
   R = the gas constant (8.314 J/mol K)
   T = absolute temperature
   a_i = activity of species i

Since the activities of pure solids are unity, equation (7) reduces to

$$\frac{zF\Delta E^o}{RT} > \ln \frac{a_{Fe^{2+}}}{a_{Cu^{2+}}} \tag{8}$$

For most contaminated solutions, the concentrations of metal ions are sufficiently dilute that $\gamma_i \cong 1$; hence the activities of the ions can be approximated by their concentrations, yielding:

$$\frac{zF\Delta E^o}{RT} > \ln\frac{[Fe^{2+}]}{[Cu^{2+}]} \tag{9}$$

The concentration of a species i is given by its initial value, modified by the amount reacted or produced. This material balance may be expressed as:

$$[i] = [i]^o \pm \text{moles produced or consumed} = [i]^o \pm \varepsilon$$

where $[i]^o$ = initial concentration of i

ε = moles reacted or consumed

Example 2   Copper is to be removed from a waste solution from processing integrated circuits containing 0.2 M $CuSO_4$ and 10^{-3} M Fe^{2+}. What is the maximum amount of copper that can be removed from this solution using metallic iron?

Solution: The reaction of copper ions with the iron will proceed to equilibrium as long as there is enough metallic iron in the system. In this case, thermodynamics predicts that

$$\frac{zF\Delta E^o}{RT} = \ln\frac{[Fe^{2+}]}{[Cu^{2+}]}$$

Substituting for the values of the variables and constants ($z = 2$, F $= 9.65 \times 10^4$ J/mole-V, $\Delta E^o = 0.337 - (-0.440) = 0.777$ V, R $= 8.314$ J/mole K and T $= 298$ K), the ratio of ionic concentrations becomes 1.9×10^{26}.

In order to calculate the equilibrium concentrations of the ions, the concentrations of the ions are replaced with their respective material balances:

$$1.9 \times 10^{26} = \frac{[Fe^{2+}]}{[Cu^{2+}]} = \frac{[Fe^{2+}]^o + \varepsilon}{[Cu^{2+}]^o - \varepsilon} = \frac{10^{-3} + \varepsilon}{0.2 - \varepsilon}$$

Solving for ε, the equilibrium concentration of Cu^{2+} is extremely small. Hence the maximum amount of copper that can be removed is essentially 0.2 mol/L. Of course, at equilibrium, the concentration of ferrous ions becomes 0.2M, which may be unacceptable for some applications.

Unfortunately, the copper product requires extensive refining before being marketable. Moreover, scrap iron cannot displace zinc, cadmium, or many other trace contaminants from solution. Zinc is more versatile than scrap iron, having a lower reduction potential. However, although zinc is used extensively in hydrometallurgical processes to cement copper, nickel, and cobalt from zinc leach solutions before electrolysis[34,35] and to recover gold and silver in the Merrill-Crowe process,[36] the Zn^{2+} ions introduced into solution would be unacceptable for environmental applications, even if the process were economically attractive.

Metals such as copper, nickel, and cobalt can be precipitated from aqueous solutions by hydrogen reduction. However this requires buffering (ammoniacal solutions are used in the classical Sherritt-Gordon process) and high hydrogen pressures, which would be difficult to attain for most environmental applications.[37]

Zero-valent iron (ZVI) barriers are a specific application of reductive precipitation that recently have been utilized for removing low concentrations of some halogenated organic compounds from waste or run-off streams.[38,39] Interestingly enough, ZVI, which functions at ambient temperature and near neutral pH values, has been found to decompose chlorinated compounds by a reductive mechanism in the absence of air and by an oxidative one in its presence.[40] In the latter case, the ZVI actually catalyzes the formation of the hydroxyl radical, a very potent oxidant. Attention has been placed recently on the use of nano-sized ZVI for in-situ remediation of soils and groundwater.[41]

3.3   Electrolysis

Electrolysis is more versatile than chemical reduction as a means of removing most toxic ions from aqueous solutions in a metallic form. Many metals with negative standard reduction potentials can be electrowon from acidic solutions with high current efficiencies, because of the high polarization of hydrogen discharge on many metals. The half-cell reactions occurring during electrolysis are generally

$$
\begin{aligned}
\text{Cathode:} \quad & M^{n+} + ne^- \rightarrow M^o \\
\text{Anode:} \quad & 2H_2O \rightarrow O_2 + 4H^+ + 4e^-
\end{aligned}
\tag{10}
$$

Hence, it can be seen that the metal ions are replaced by hydrogen ions. Although no extraneous metallic species are introduced into solution, in some circumstances the acid may require neutralization prior to discharge.

Probably the three greatest challenges in using electrolysis for dilute mixed solutions are to achieve acceptable mass transport around the cathode, to prevent excessive discharge of hydrogen, and to prevent side reactions that further reduce current efficiencies (e.g., the recycle of Fe^{3+}/Fe^{2+} between the anode and cathode). A large number of novel cells with separate anode and cathode compartments, moving and particulate beds, enhanced stirring, and so on have been designed to address these problems, and are the subject of ongoing developmental work.[42–45] As the energy efficiency of these is improved, electrolysis is likely to become much more widely used for environmental applications.

3.4   Electrokinetic Remediation

Electroosmosis and *electrophoresis* are electrokinetic methods that are being considered for contaminated soil remediation.[46–50] The principle behind these techniques is the relative movement of components of the aqueous-solid system generated by the application of an electrical field. In electroosmosis, the field induces flow of solution through a porous bed with surface charge, while in electrophoresis an applied field induces motion of charged particles dispersed in a solution. In both cases, these methods are used for soil with very low permeability because their action is not affected by external forced convection. However, only electroosmosis is effective to treat neutral species.

3.5   Adsorption

Adsorption is the preferential partition of a solute at a solid-liquid interface. Partition is driven by a coulombic or chemical affinity between the high-energy substrate surface compounds or functional groups and the sorbate (substance that is adsorbed). The strength and selectivity of the surface-sorbate interactions are generally determined by the chemical reactions that occur between the two. One example is the bond formed between a specific surface group and a cation with which it forms a sparingly soluble salt. In this case, the adsorption is comparable to the precipitation reaction and can be treated quantitatively in the same manner as solubility products. Since the reactions take place at the substrate-solution interface, the adsorption capacity is proportional to the substrate surface area and the functional group density.

Theoretical Considerations

Most adsorption takes place in a layer adjacent to the solid-solution interface. If there are negligible interactions between species adsorbed on adjacent surface sites, adsorption can be described quantitatively by a Langmuir type isotherm:

$$\theta = \frac{C_{\text{ads}}}{N_{\text{s}}} = \frac{kC_i}{1 + kC_i} \tag{11}$$

where θ = fractional surface coverage or adsorption sites occupied
$\quad C_{\text{ads}}$ = moles of i adsorbed per gram of substrate
$\quad N_{\text{s}}$ = moles of adsorption sites per gram of substrate
$\qquad$ (adsorption capacity)
$\quad k$ = adsorption constant = f (nature of affinity, temperature)
$\quad C_i$ = concentration of sorbate i in the solution

The magnitude of k is determined by the adsorption strength or chemical affinity between the substrate and sorbate. The value of N_{s} is often affected by the solution pH, since H^+ and OH^- compete with species i for adsorption sites; when this happens, the number of sites available for the adsorption of i is reduced. The term C_i may also be affected by the solution pH or complexing agents; a complexed species may be favored over an uncomplexed one, or vice versa. This is seen for the adsorption of metallic species on clays, where hydroxycomplexes of the metal ion are favored.

Equation (11) implies that adsorption is reversible, which is generally true. The adsorbed contaminant can, in many cases, be desorbed from the substrate through a change in solution conditions (elution), which shift the equilibrium state of the isotherm; this is accomplished commonly by a change in temperature, solution pH or addition of a complexing agent, which changes the concentration or affinity of species i with the surface functional groups. In this manner, the substrate is regenerated and may be reused. However, the elution process produces another solution, hopefully more concentrated, that must be treated by another method to recover the contaminant.

Adsorption onto Oxides and Clay Minerals
The adsorption of heavy metals onto amorphous or crystalline forms of iron oxide and clays occurs in nature and is phenomenologically related to the binding of contaminant to the superficial ferric and/or aluminium ions.[51,52] Although, this behavior explains the concentration of metal contaminants in soils, it does not constitute a viable method for trapping low concentrations of contaminants from aqueous streams because of its limited adsorption capacity.

Adsorption on Carbon
Carbon is used extensively in water treatment for removing organic contaminants, such as phenols and related compounds. Carbon is also used widely for sorbing gold from cyanide leach solutions.[53] However, carbon and many other carbonaceous materials are capable of sorbing a wide range of other metals. Activated carbons from various sources have been reported to be effective at removing Cr(VI), Pb, and Hg from dilute solutions, and less effective for Mo, Cu, Pb, Ni, Zn.[54,55] Sorption is pH and temperature dependent,[56] and the behavior and mechanisms are undoubtedly highly substrate-specific.[57-58] It is quite likely that application of the engineering expertise now developed for adsorption of gold on carbon could be usefully directed toward detoxification of water.

The ability of low-rank coals to adsorb many metallic cations from aqueous solutions is well recognized.[59,60] Bituminous coal has been used to remove cadmium and mercury[62] from water, the latter after various chemical pre-treatments.[61] Coal-derived humic acids show similar sorption properties.[63] If coals possessed appropriate ion exchange properties, they could be treated by combustion, because of their low cost and high calorific value, although care would have to be taken to minimize emissions of potentially hazardous air pollutants during combustion. The metals would report to the ash, which might be stable, or require leaching for stabilization. Unfortunately, our own studies on oxidized bituminous coal revealed a low sorptive capacity for metals, which would limit its commercial utility.[64,65]

3.6 Biological Methods

There has been extensive interest in biological methods for sorbing toxic metals from dilute solutions.[66–70] Numerous organisms are capable of taking up appreciable levels of metals from solutions.[71] These organisms include bacteria, algae,[72] fungi, yeasts, and a large number of aquatic plants,[73–76] including *Typha* and *Sphagnum*.[77] Many engineered effluent treatment systems use these organisms.[78–81] Other biomass, including activated sludge,[82] treated sewage sludge, bark,[83–85] rice husks, and lignite, have sorptive properties, resulting from the metal-complexing ability of common functional groups within the biological tissue. Biopolymers derived from algae and shellfish are similarly effective. In many cases, these methods are very promising.[86–88] However, most biological processes only concentrate the toxic metals from a dilute solution into a more concentrated biomass (although some organisms cause metals to accumulate into extracellular metabolic products).[86–89] Generally, the metal loadings in the biomass become great enough to prevent further metal uptake. Some systems are amenable to desorption of the metals, whereas many would require incineration.[90]

The use of biological processes to remove trace metals from effluents and other solutions,[91–93] and to replace solvents and etchants that are becoming increasingly problematic, are additional areas where adaptation of commercial materials processes could have a huge influence in a larger societal context. Many microorganisms have been reported to accumulate metal ions within their structure. However, this subject is quite extensive and deserves a chapter in itself.

3.7 Solvent Extraction and Ion Exchange

Similar to adsorption, solvent extraction and ion exchange involve a reaction between the contaminant species and a reactant or functional group in another phase (solvent extraction) or on a substrate surface (ion exchange). Solvent extraction has three attractive features that have led to widespread adoption in chemical engineering and materials processes:

1. It is selective (and is therefore useful for separations).
2. It can be operated continuously in closed loops with modest reagent make-up requirements.
3. It generates a marketable product.

Solvent extraction processes have been devised for treating etchants, removing impurities from bleed streams in primary metallurgical processes, and treating similar solutions.[94] Solvent extraction has also been tested for treating mine waters.[95] Unfortunately, the liquid extractants and diluents invariably have finite solubilities in water, which may be high enough to deteriorate water quality.

Moreover, the loss of reagents to a large volume of dilute solution can be exceedingly costly. Thus, this technique has not found widespread application for treating dilute effluents or contaminated solutions prior to discharge.

The finite solubility of solvent extraction reagents in aqueous streams is not generally problematic in closed loops, because losses are limited. However, ion exchange has significant advantages for environmental applications.[96] The most obvious technical advantage is the fact that because the functional groups are grafted onto the matrix of a solid substrate, there is a much lower likelihood of them being lost into the solution being treated, and contaminating said solution. A less obvious advantage is the fact that ion exchange is inherently more robust, inasmuch as it presents fewer health and safety concerns when not being overseen by skilled personnel. Ion exchange materials are useful for treating dilute solutions because they allow metals to be removed without releasing the extractant to the solution. A limited amount of work has investigated the use of zeolites for detoxifying effluents contaminated by metal ions.[97] However, there is much more interest in organic ion exchangers. They have been used for many years for water softening and deionization, and have been investigated for treating acid-rock drainage,[98–100] but researchers have yet to find commercial application for removing toxic metal ions from waste streams. Chelating ion exchange resins of various types allow solutions containing both cationic and anionic impurities to be treated,[101–103] which would be even more useful for water treatment than in most materials separations. Selective elution, allowing immediate separation of metals, would be a further option when using such resins. Another approach for achieving selectivity for many contaminant ions over alkali and alkaline earth metals uses chelation by dissolved organic anions, followed by sorption onto anion-exchange resins.[104]

Theoretical Considerations

Ion exchange is similar to adsorption in that both involve affinity between a species i in solution and functional groups on a substrate surface; ion exchange may even be considered as a particular case of adsorption, where coulombic affinity is prevalent in a completely reversible process. This reversibility is promoted by competition between two ions for the same adsorption sites. In common water-softening systems, calcium ions replace sodium ions in the ion exchange resin, according to the following reaction:

$$Ca^{2+} + 2Na - | \rightarrow | - Ca - | + 2Na^+ \qquad (12)$$

where $-|$ and $|-$ represent surface function groups and $Na - |$ is a surface site occupied by a sodium ion. For dilute solutions, the equilibrium constant for equation (12) may be represented as

$$K_{IE} = \frac{[Na^+]^2[| - Ca - |]}{[Ca^{2+}][| - Na]^2} \qquad (13)$$

Let θ = fraction of surface sites occupied by calcium. If it is assumed that all surface sites are occupied either by calcium or sodium, and further that because of the difference in valence that one calcium site can accommodate two sodiums, then the surface coverage by sodium is $2(1 - \theta)$. Hence:

$$K_{IE} = \frac{[Na^+]^2\theta}{[Ca^{2+}](2(1 - \theta))^2} \tag{14}$$

The ion exchange resin can clearly be regenerated by passing a concentrated solution of sodium ions through the resin, reversing the reaction.

The major drawbacks with ion exchange technology, in its current state of development, are engineering problems, rather than chemical. The packed-bed columns conventionally used for ion exchange are inherently rather slow, at best only semicontinuous systems, and are sensitive to particulate matter, which plugs the columns and fouls the surface of the resin beads. Operational modifications, such as reciprocating flow ion exchange,[105] can improve performance and reduce the size of beds needed for a given separation, but are not continuous. The continuous countercurrent ion exchange system uses very short hydraulic pulses to move resin through loading and elution zones in a closed-loop contactor, and is reported to tolerate small quantities of particulate in the feed.[106] Continuous ion exchange processes have addressed some of the problems with conventional ion exchange, including resin-in-pulp methods.[107] However, breakage and attrition of the resins generate fine particles that pass through the screens used to separate them, leading to substantial losses of resin. Dramatic recent improvements in the mechanical properties of resins have not completely eliminated these problems.

3.8 Membrane Processes

Membrane processes have seen little use in commercial materials processes but are likely to be increasingly important for environmental applications. Here, new materials developments are rapidly overcoming problems that have hitherto restricted widespread adoption. Ideally, a membrane could halt all fluxes besides the one desired, while withstanding operating conditions such as pressure, thermal, osmotic pressure and activity gradients, and abrasion forces, without being susceptible to fouling by fines. In practice, such ideal behavior has not been achieved, and compromises are needed.

Membrane processes, such as reverse osmosis and dialysis, are already used for certain effluent treatment applications and desalination.[108] They can be operated continuously, and they allow recovery of the dissolved values. Membrane processes have benefited enormously from recent advances in membrane materials that can withstand high-pressure gradients and harsh chemical environments. Ultrafiltration of macromolecular complexes of metal ions, which shows more chemical specificity, was more promising for detoxifying effluents,[109] and the work is still ongoing. Particulate matter, which is present in many environmental and processing solutions, can dramatically reduce the permeability of membranes. This is still a serious limitation to their widespread use. Liquid membranes are much more experimental for either hydrometallurgical or environmental applications.[110,111] Although supported liquid membranes have been used for limited applications, they suffer from poor stability and limited service life, due to loss of the organic solvent from the pores of the support.[112–114] Now that environmental constraints are so different, this work is worth revisiting.

Theoretical Considerations

The criteria that dictate which substance will pass through the membrane are both thermodynamic and steric. The former imposes the condition that a concentration or fugacity gradient of the substance must exist between the solutions on either side of the membrane for flow to occur. The desired substance must also be small enough to enter and exit the pores in a timely manner. In filtration and ultrafiltration, certain substances are excluded because they do not fit through the pores—naturally in the former case and selectively combined with macromolecules in the second. In dialysis, the presence and absence of concentration gradients of the different components between the contaminated stream and the receiving solution primarily dictate which substance will be transported. The selection in reverse osmosis uses both criteria; for the water, increased pressure on the feed solution side induces the water to flow even against the concentration gradient by causing a fugacity gradient, while the flow of salts is restricted by transport issues. In all cases, it is important to evaluate the driving force or restrictions for the passage of each of the solution components.

Fluidized- and spouted-bed electrowinning, which promise to greatly improve the space-time yields of electrowinning, as well as allowing electrowinning to be performed on more dilute streams, such as process effluents, would be most effective if the membranes could be ideally selective, permitting the transport of just one type of ion, while also presenting a low Ohmic resistance.[115,116] Membrane processes could be invaluable for treating and recycling process water[117,120] or for treating bleed streams in metallurgical processes. Environmental gains could also be reaped from advanced batteries and fuel cells, the ability to detoxify a wide range of process effluents, and biomimetic separations in product manufacture, such as continuously removing fermentation products from a fermentation process producing pharmaceutical products or commodity bioproducts.

3.9   Foam Separation Processes

Ion flotation is a process in which metal ions are removed from dilute, aqueous solutions by interaction with dissolved, surface active organic collectors. The metal-collector complexes or ion pairs adsorb preferentially at solution-air interfaces, and hence can be removed by sparging air through the solution and separating the resulting foam.[121] Precipitate flotation is a related technique in which an insoluble compound that is hydrophobic, or can be rendered so using appropriate collectors, is precipitated and removed by flotation. The results of numerous studies on this and related flotation separation processes have been reviewed in a series of publications.[122–125]. Certain flotation separation techniques have been used in analytical chemistry.[126] Others have reviewed single- and multielement separations achieved by ion flotation.[127] Suitable systems have been identified for most of the elements of the periodic table, and the effect of operating conditions on the performance of ion flotation operations has been discussed.[128–130] Ion flotation has been tested at the pilot scale for recovering gold from cyanide leach solutions.[131–132]

The removal of cationic species such as Cu^{2+}, Cd^{2+}, Pb^{2+}, and Cr(III) from 3×10^{-4} M aqueous solutions has been investigated, using sodium dodecylsulfate as a collector and small additions of ethanol as a frother.[133–134] High ionic strengths appear to interfere with ion flotation of cationic species, although much of this effect appears to be due to the low concentration of free multivalent cations at high ionic strengths.[135] Although this effect can be addressed by using cationic collectors for the anionic contaminant species, a significant excess of cationic collectors is generally needed for effective flotation of anions.[136] Of interest from an engineering and economic perspective is the fact that metals can be recovered, and dodecylsulfate can be regenerated from the contaminant-dodecylsulfate complexes/precipitates.[137] The ability to recover a metal byproduct, and regenerate the collector for recycle to effluent treatment, provides the basis of a continuous flow sheet for effluent treatment, as sketched schematically in Figure 1.

Theoretical Considerations

The Gibbs adsorption equation has been used to estimate the adsorption density of different metal dodecylsulfates, Γ_i, at air solution interfaces, from experimental

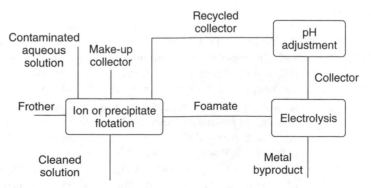

Figure 1 Schematic flowsheet for treatment of dilute solutions by ion or flotation, followed precipitate by electrolysis of the foamate.

measurements of surface tension, γ, as a function of the concentration, c_i:

$$f(c_i) = -\sum \Gamma_i RT \qquad (15)$$

where R is the gas constant and T the absolute temperature.[138] Metal ion removal rates predicted from these adsorption densities, superficial gas velocities, and estimated bubble areas correlated well with experimental measurements of metal removal rates. Above about 5×10^{-5} M, different metal dodecylsulfates exhibited different adsorption densities, in the order $Na^+ << Cu^{2+}$, $Cd^{2+} < Pb^{2+} < Cr^{3+}$, whereas at lower concentrations none exhibited significant surface activity, and hence should not be capable of further removal by adsorption. The lowest concentrations of metal ions achieved after prolonged gas sparging of metal-dodecylsulfate solutions correlated exceedingly well with these concentrations. The thermodynamic basis of the behavior suggests that collectors that can achieve specific water-quality objectives can be sought using our existing understanding of the thermodynamic behavior of surface active species.

4   CONCLUSIONS

There is a large and growing demand for good technology for removing contaminants from dilute solutions. Chemical precipitation methods are widely used, and could be improved significantly, drawing on existing engineering principles. Electrolysis and electrokinetic remediation are likely to become much more important with new developments in electrochemical engineering. Carbonaceous materials might gain broader acceptance as metal and organic sorbents. Biological methods are likely to increase in importance. Ion exchange and membrane processes are good at present, and novel configurations could further expand their applicability. Foam separations methods might be useful, although both the chemistry and engineering of these methods require further development.

REFERENCES

1. R. W. Gibson, D. J. Fray, J. G. Sunderland, and I. M. Dalrymple, "Recovery of Solder and Electronic Components from Printed Circuit Boards," in F. M. Doyle, G. Kelsall, and R. Woods (eds.), *Electrochemistry in Mineral and Metal Processing VI*, The Electrochemical Society, Pennington, NJ, 2003, pp. 346–354.
2. K. Koyama, M. Tanaka and J.-C. Lee, "Copper Recovery from Waste Printed Circuit Boards," in *Hydrometallurgy 2003, Proceedings of the 5th International Symposium*, C. A. Young, A. M. Alfantazi, C. G. Anderson, A. James, D. B. Dreisinger, B. Harris, (eds.), TMS, Warrendale, PA, 2003 pp. 1555–1563.
3. D. Pilone and G. H. Kelsall, "Metal Recovery from Electronic Scrap by Leaching and Electrowinning IV," in C. A. Young, A. M. Alfantazi, C. G. Anderson, A. James, D. B. Dreisinger, and B. Harris (eds.), *Hydrometallurgy 2003, Proceedings of the 5th International Symposium*, TMS, Warrendale, PA, 2003, pp. 1565–1575.

4. J. H. Huang and A. M. Alfantazi, "Electrowinning of Cobalt from a Sulfate-chloride Solution as an Option for Treatment of Industrial Effluents," in *Electrochemistry in Mineral and Metal Processing VI,* F. M. Doyle, G. Kelsall, and R. Woods (eds.), The Electrochemical Society, Pennington, NJ, 2003, pp. 336–345.

5. R. Ding, J. W. Evans and F. M. Doyle, "An Investigation of the Electrodeposition of Copper Relevant to the Removal of Dissolved Copper from Semiconductor Industry Waste Streams," in: *Electrochemistry in Mineral and Metal Processing VI, F. M. Doyle,* G. Kelsall, and R. Woods (eds.), The Electrochemical Society, Pennington, NJ, 2003, pp. 326–335.

6. W. Ewing, J. W. Evans, and F. M. Doyle, "The Effect of Plating Additives on the Recovery of Copper from Dilute Aqueous Solutions Using Chelating Resins," in *Hydrometallurgy 2003, Proceedings of the 5th International Symposium,* C. A. Young, A. M. Alfantazi, C. G. Anderson, A. James, D. B. Dreisinger, and B. Harris, (eds.), TMS, Warrendale, PA, 2003, pp. 753–762.

7. M. Brown, B. Barley, and H. Wood (eds.), *Minewater Treatment: Technology, Application and Policy*, IWA Publishing, 2003.

8. F. M. Doyle and A. H. Mirza, "Understanding the Mechanisms and Kinetics of Acid and Heavy Metals Release from Pyritic Wastes," in *Mining and Mineral Processing Wastes*, F. M. Doyle (ed.), SME, Littleton, CO, 1990, pp. 43–51.

9. A. J. Monhemius, "Precipitation Diagrams for Metal Hydroxides, Sulphides, Arsenates and Phosphates," *Trans. IMM*, **86**, C202–206.

10. NIST Standard Reference Database 46, Version 8.0, *NIST Critically Selected Stability Constants of Metal Complexes* (software), 2004.

11. J. A. Dean, *Lange's Handbook of Chemistry*, 13th ed., McGraw Hill, New York, 1985.

12. A. Esmaeili, A. Mesdaghinia, and R. Vazirinejad, "Chromium(III) Removal and Recovery from Tannery Wastewater by Precipitation Process," *American Journal of Applied Science*, **2**(10), 1471–1473 (2005).

13. W. F. Bialas and A. C. Middleton, "Mine Drainage Costly to Neutralize," *J. Water Poll. Control Fed.*, **49**, 2054 (1977).

14. D. J. Bosman, "Lime Treatment of Acid Mine Water and Associated Solids/liquid Separation," *Water Sci. & Tech.*, **15,** 71–84 (1983).

15. R. N. Kust, "Improved Technology for Precipitating Metal Hydroxides," in *Residues and Effluents: Processing and Environmental Considerations*, R. G. Reddy, W. P. Imrie, and P. B. Queneau (eds.), TMS, Warrendale, PA, 1992, pp. 793–800.

16. J. B. Pfeiffer, "The Treatment of Acid Mine Drainage—Two Years of Successful Operation," in *Extraction and Processing for the Treatment and Minimization of Wastes*, J. Hager, B. Hansen, W. Imrie, J. Pusateri, and V. Ramachandran (eds.), TMS, Warrendale, PA, 1994, pp. 715–726.

17. G. W. Heunisch, "Lime Substitutes for the Treatment of Acid Mine Drainage," *Mining Engineering*, **39**(1), 33–36 (1987).

18. United States Environmental Protection Agency, *Report to Congress on Cement Kiln Dust*, Office of Solid Waste, EPA, Washington DC, 1993.

19. A. Möller, A. Grahn, and U. Welander, "Precipitation of Heavy Metals from Landfill Leachates by Microbially Produced Sulphide," *Environmental Technology*, **25**(1), 69–77 (2004).

20. A. H. M. Veeken, S. de Vries, A. van der Mark, and W. H. Rulkens, "Selective Precipitation of Heavy Metals as Controlled by a Sulfide-Selective Electrode," *Sep. Sci. Technol.*, **38**(1), 1–19 (2003).

21. F. F. Aplan and J. W. Perez, "Ion and Precipitate Flotation of Heavy Metal Ions from Solution," in *Residues and Effluents: Processing and Environmental Considerations*, R. G. Reddy, W. P. Imrie, and P. B. Queneau (eds.), TMS, Warrendale, PA, 1992, p. 791.

22. C. C. Nesbitt and T. E. Davis, "Removal of Heavy Metals from Metallurgical Effluents by the Simultaneous Precipitation and Flotation of Metal Sulfides Using Column Cells," in *Extraction and Processing for the Treatment and Minimization of Wastes*, J. Hager, B. Hansen, W. Imrie, J. Pusateri, and V. Ramachandran (eds.), TMS, Warrendale, PA, 1994, pp. 331–342.

23. G. Klamp and D. Wanner, "Removal of Arsenic from Washing Acid by the Sachtleben-Lurgi Process," in *Residues and Effluents: Processing and Environmental Considerations*, R. G. Reddy, W. P. Imrie, and P. B. Queneau (eds.), TMS, Warrendale, PA, 1992, pp. 833–837.

24. R. W. Hammack, D. H. Dvorak, and H. M. Edenborn, "Selective Metal Recovery Using Biogenic Hydrogen Sulfide: Rio Tinto Mine, Nevada," in *Extraction and Processing for the Treatment and Minimization of Wastes*, J. Hager, B. Hansen, W. Imrie, J. Pusateri, and V. Ramachandran (eds.), TMS, Warrendale, PA, 1994, pp. 747–760.

25. A. L. de Vegt and C. J. N. Buisman, "Full-scale Biological Treatment of Heavy Metal Contaminated Groundwater," in *Treatment and Minimization of Heavy Metal-Containing Wastes*, J. P. Hager, B. Mishra, C. F. Davidson, and J. L. Litz (eds.), TMS, Warrendale, PA, 1995 pp. 69–79.

26. R. G. Robins, J. Y. Huang, T. Nishimura, and G. H. Khoe, "The Adsorption of Arsenate Ion by Ferric Hydroxide," in *Arsenic Metallurgy, Fundamentals and Application*, R. G. Reddy, J. L. Hendrix, and P. B. Queneau (eds.), TMS, Warrendale, PA., pp. 99–112 (1988).

27. K. Yang, M. Misra, and R. K. Mehta, "Removal of Heavy Metal Ions from Acid Mine Water by Ferrite Coprecipitation Process," in *Separation Processes: Heavy Metals, Ions and Minerals*, M. Misra (ed.), TMS, Warrendale, PA 1995, pp. 37–48.

28. R. G. Robins, "The Solubility of Metal Arsenates," *Metall. Transactions B*, **12B**, 103–109 (1981).

29. G. B. Harris, and S. Monette, "The Stability of Arsenic Bearing Residues," in *Arsenic Metallurgy, Fundamentals and Application*, R. G. Reddy, J. L. Hendrix, and P. B. Queneau (eds.), TMS, Warrendale, PA., 1988, pp. 469–488.

30. J. Farrell, J. Wang, P. O'Day, and M. Conklin, "Electrochemical and Spectroscopic Study of Arsenate Removal from Water Using Zero-Valent Iron Media," *Environmental Science and Technology*, **35**, 2026–2032 (2001).

31. G. B. Harris, "The Removal of Arsenic from Process Solutions: Theory and Industrial Practice," in *Hydrometallurgy 2003, Proceedings of the 5th International Symposium*, C. A. Young, A. M. Alfantazi, C. G. Anderson, A. James, D. B. Dreisinger, and B. Harris (eds.), TMS, Warrendale, PA, 2003, pp. 1889–1902.

32. T. N. Lung, "The History of Copper Cementation on Iron—the World's First Hydrometallurgical Process from Medieval China," *Hydrometallurgy*, **17**, 113–129 (1986).

33. H. R. Spedden, E. E. Malouf and J. D. Prater, "Cone Type Precipitators for Improved Copper Recovery," *Journal of Metals*, **18**(10), 1137 (1966).

34. A. J. Monhemius, "The Electrolytic Production of Zinc," in *Critical Reports on Applied Chemistry Volume 1: Topics in Non-ferrous Extractive Metallurgy*, A. R. Burkin, Blackwell Scientific Publications, Oxford, 1980, pp. 104–130.

35. R. D. Johnson, L. D. Palumbo and G. D. J. Smith, "Electrolytic Zinc at Port Pirie, South Australia," in *Zinc '83, 13th Annual Hydrometallurgical Meeting*, CIM, Edmonton, Alberta, Canada, paper 15, 1983.

36. R. Y. Wan, and J. D. Miller, "Research and Development Activities for the Recovery of Gold from Alkaline Cyanide Solutions," *Mineral Processing and Extractive Metallurgy Review*, **6**, 143–190 (1989).

37. A. R. Burkin, *The Chemistry of Hydrometallurgical Processes,* Van Nostrand, Princeton, NJ, 1966, Ch. 1.

38. R. Rangsivek and M. R. Jekel, "Removal of Dissolved Metals by Zero-valent Iron (ZVI): Kinetics, Equilibria, Processes and Implications for Stormwater Runoff Treatment," *Water Research*, **39**(17), 4153–4163 (2005).

39. G. Bartzas, K. Komnitsas, and I. Paspaliaris, "Laboratory Evaluation of Fe^0 Barriers to Treat Acidic Leachates," *Minerals Engineering*, **19**(5), 505–514 (2006).

40. C. Noradoun, M. D. Engelmann, M. McLaughlin, R. Hutcheson, K. Breen, A. Paszczynski and I. F. Cheng, "Destruction of Chlorinated Phenols by Dioxygen Activation under Aqueous Room Temperature and Pressure Conditions," *Industrial and Engineering Chemistry, Research*, **42**, 5024–5030 (2003).

41. W. Zhang and D. W. Elliot, "Applications of Iron Nanoparticles for Groundwater Remediation," *Remediation*, **16**(2), 7–21 (2006).

42. D. Pletcher and F. C. Walsh, "Water Purification, Effluent Treatment and Recycling of Industrial Process Streams," Ch.7, *Industrial Electrochemistry*, Chapman and Hall, London, 1990.

43. R. Kammel, "Electrorecovery of Metals from Dilute Process Solutions in Electroplating," in *Residues and Effluents—Processing and Environmental Considerations*, R. G. Reddy, W. P. Imrie, and P. B. Queneau (eds.), TMS, Warrendale, PA, 1991, pp. 777–786.

44. A. Espinola, L. F. Medina Oliveira, and F. L. Ayres, "Flow Electrolysis for Decontaminating Plate Industries Waste Waters," in *Extraction and Processing for the Treatment and Minimization of Wastes*, J. Hager, B. Hansen, W. Imrie, J. Pusateri, and V. Ramachandran (eds.), TMS. Warrendale, PA., 1994, pp. 369–376.

45. P. Aguirre, I. Gaballah, B. Fenouillet, S. Ivanaj, G. Lacoste, and R. Solozabal, "Recovery of Zinc, Copper and Nickel from Industrial Effluents Generated by the Electroplating Units Using the "3PE" Technology," in *Separation Processes: Heavy Metals, Ions and Minerals*, M. Misra (ed.), TMS, Warrendale, PA., 1995 pp. 257–268.

46. G. R. Eykholt and D. E. Daniel, "Impact of System Chemistry on Electroosmosis in Contaminated Soil," *Journal of Geotechnical Engineering*, **120**(5), 797–815 (1994).

47. L. M. Ottosen, H. K. Hansen, A. B. Ribeiro, and Villumsen, "A Removal of Cu, Pb, and Zn in an Applied Electric Field in Calcareous and Non-calcareous Soils," *J. Hazard. Mater.*, **85**(3), pp. 291–299 (2001).

48. A. Z. Al-Hamdan and K. R. Reddy, "Geochemical Reconnaissance of Heavy Metals in Kaolin after Electrokinetic Remediation," *Journal of Environmental Science and Health, Part A: Toxic/Hazardous Substances & Environmental Engineering*, **41**(1), 17–33 (2006).

49. R. E. Saichek and K. R. Reddy, "Effects of System Variables on Surfactant Enhanced Electrokinetic Removal of Polycyclic Aromatic Hydrocarbons from Clayey Soils," *Environ. Technol.*, **24**, 503–515 (2003).

50. M. M. Page, and C. L. Page, "Electroremediation of Contaminated Soils," *J. Envir. Engrg.*, **128**(3), pp. 208–219 (2002).

51. S. Dixit and J. G. Hering, "Comparison of Arsenic(V) and Arsenic(III) Sorption onto Iron Oxide Minerals: Implications for Arsenic Mobility," *Environmental Science and Technology*, **37**, 4182–4189 (2003).

52. A. Garca-Sanchez, A. Alastuey, and X. Querol, "Heavy Metal Adsorption by Different Minerals: Application to the Remediation of Polluted Soils," *The Science of the Total Environment*, **242**(1), 179–188 (1999).

53. Y. F. Jia, C. J. Steele, I. P. Hayward, and K. M. Thomas, "Mechanism of adsorption of gold and silver species on activated carbons, *Carbon*, **36**(10), pp. 1299–1308, (1998).

54. M. O. Corapcioglu and C. P. Huang, "The Adsorption of Heavy Metals onto Hydrous Activated Carbon," *Water Research*, **21**, pp. 1031–1044 (1987).

55. S. K. Srivasta, R. Tyagi and N. Pant, "Adsorption of Heavy Metal Ions on Carbonaceous Material Developed from the Waste Slurry Generated in Local Fertilizer Plants," *Water Research*, **23**, 1161–1165 (1989).

56. T. C. Tan and W. K. Teo, "Combined Effect of Carbon Dosage and Initial Adsorbate Concentration on the Adsorption Isotherm of Heavy Metals on Activated Carbon," *Water Research*, **21**, 1183–1188 (1987).

57. C. P. Huang and M. H. Wu, "The Removal of Chromium from Dilute Aqueous Solution by Activated Carbon," *Water Research*, **11**, 673–679 (1977).

58. H. Koshima and H. Onishi, "Adsorption of Metal Ions on Activated Carbon from Aqueous Solutions," *Talanta,* **33**, 391–395 (1986).

59. C. J. Lafferty, "The Use of Low Rank Brown coal as an Ion-Exchange Material," *Fuel*, 69, 84 (1990).

60. J. H. Kuhr, J. D. Robertson, C. J. Lafferty, A. S. Wong and N. D. Stalnaker, "Ion Exchange Properties of a Western Kentucky Low-rank coal," *Energy & Fuels*, **11**(2), pp. 323–326.

61. C. Venkobachar and A. K. Bhattacharya, "Cadmium Removal from Domestic Waters by a Low Cost Sorbent," *Water Supply*, **3**, 157–163 (1985).

62. M. P. Pandy and M. Chaudhuri, "Removal of Inorganic Mercury from Water by Bituminous Coal," *Water Research*, **16**, pp. 1113–1118 (1982).

63. E. Sebestová, B. Hemelíková and L. Minarík, in *Proc. 7th Int. Conf. on Coal Science*, K. H. Michaelian (ed.), Banff, Alberta, Canada, 1993, pp. 110–113.

64. D. L. Bodine and F. M. Doyle, "Removal of Cu2+, Cd2+ and Mn(VII) from Dilute, Aqueous Solutions by Oxidized Bituminous Coal," *Treatment and Minimization of Heavy Metal-Containing Wastes*, J. P. Hagar, B. Mishra, C. F. Davidson and J. L. Litz (eds.), TMS, Warrendale, PA., 1995, pp. 81–93.

65. F. M. Doyle and D. L. Bodine, "Treatment of Aqueous Streams Containing Strong Oxidants Using Bituminous Coal," in *Pollution Prevention for Process Engineering*,

P. E. Richardson and B. J. Scheiner (ed.), Engineering Foundation, AIChE, New York, 1996, pp. 103–110.

66. B. J. Scheiner, F. M. Doyle and S. K. Kawatra (eds.), *Biotechnology in Mineral and Metal Processing*, Society of Mining Engineers, Inc., Littleton, CO, 1989.

67. I. A. H. Schneider, J. Rubio and R. W. Smith, "Biosorption of Metals onto Plant Biomass: Exchange Adsorption or Surface Precipitation?" *Int. J. Miner. Process.* **62**, 111–120 (2001).

68. B. Volesky and Z. R. Holan, "Biosorption of Heavy Metals," *Biotechnol. Prog.*, **11**, 235–250 (1995).

69. R. H. S. F. Vieira, and B. Volesky, "Biosorption: A Solution to Pollution?" *International Microbiology*, **3**, 17–24 (2000).

70. B. Volesky, *Sorption and Biosorption*, BV-Sorbex, Inc., St. Lambert, Quebec, 2004, p. 326.

71. A. E. Torma and W. A. Apel, "Recovery of Metals from Dilute Effluent Streams by Biosorption Methods," in *Residues and Effluents—Processing and Environmental Considerations*, R. G. Reddy, W. P. Imrie, and P. B. Queneau (eds.), TMS, Warrendale, PA., 1991, pp. 735–746.

72. J. P. Gould, J. Bender, and P. Phillips, "Bioremediation of Metal-contaminated Water with Microbial Mats," *Pollution Prevention for Process Engineering*, P. E. Richardson and B. J. Scheiner (eds.), Engineering Foundation, AIChE, New York, 1996a, pp. 175–182.

73. L. C. Thompson and R. L. Gerteis, "New Technologies for Mining Waste Management—Biotreatment Processes for Cyanide, Nitrates and Heavy Metals," in *Mining and Mineral Processing Wastes*, F. M. Doyle (ed.), SME, Littleton, CO, 271–278 (1990).

74. I. A. H. Schneider, M. L. de Souza and J. Rubio, "Sorption of Copper Ions from Aqueous Solutions by Potamogeton Luscens Biomass," in *Separation Processes: Heavy Metals, Ions and Minerals*, M. Misra (ed.), TMS, Warrendale, PA, 1995, pp. 49–60.

75. I. A. H. Schneider, M. Misra and R. W. Smith, "Biosorption of Heavy Metals and Uranium from Dilute Solutions," in *Separation Processes: Heavy Metals, Ions and Minerals*, M. Misra (ed.), TMS, Warrendale, PA, 1995, pp. 81–89.

76. D. R. Crist, R. H. Crist, J. R. Martin, J. Chonko and H. Thuma, "Ion Exchange Reactions in Sorption of Metals by Peat Moss," *Pollution Prevention for Process Engineering*, P. E. Richardson and B. J. Scheiner (ed.), Engineering Foundation, AIChE, New York, 1996, pp. 193–199.

77. P. G. Bennett and T. H. Jeffers, "Removal of Metal Contaminants from a Waste Stream Using BIO-FIX Beads Containing Sphagnum Moss," in *Mining and Mineral Processing Wastes*, F. M. Doyle (ed.), SME, Littleton, CO, 1990, pp. 279–286.

78. D. A. Hammer, *Constructed Wetlands for Wastewater Treatment*, Lewis Publishers, Chelsea, MI, 1989.

79. N. Kuyucak and P. St-Germain, "Passive treatment methods for acid mine drainage," in *EPD Congress 1993*, J. P. Hager (ed.), TMS, Warrendale, PA, 1992 pp. 319–331.

80. J. A. Brierley and C. L. Brierley, "Reflections on and Considerations for Biotechnology in the Metals Extraction Industry," in *Hydrometallurgy: Fundamentals, Technology and Innovation*, J. B. Hiskey and G. W. Warren (eds.), SME, Littleton, CO, 1993, pp. 647–660.

81. J. J. Gusek and T. R. Wildeman, "New Developments in Passive Treatment of Acid-rock Drainage," *Pollution Prevention for Process Engineering*, P. E. Richardson and B. J. Scheiner (ed.), Engineering Foundation, AIChE, New York, 1996, pp. 29–44.

82. R. M. Sterritt, M. J. Brown and J. N. Lester, "Metal Removal by Adsorption and Precipitation in the Activated Sludge Process," *Environment. Pollut.*, **24**, 313–323 (1981).

83. I. Gaballah and Kilbertus, "Elimination of As, Hg and Zn from Synthetic Solutions and Industrial Effluents Using Modified bark," in *Separation Processes: Heavy Metals, Ions and Minerals*, M. Misra (ed.), TMS, Warrendale, PA, 1994, pp. 15–26.

84. I. Gaballah, D. Goy, G. Kilbertus, and J. Thauront, "A New Process for the Decontamination of Industrial Effluent Containing Cadmium Ions," in *EPD Congress 1994*, G. W. Warren (ed.), TMS, Warrendale, PA, 1994a pp. 43–52.

85. I. Gaballah, D. Goy, G. Kilbertus, B. Loubinoux, and J. Thauront, "Decontamination of Synthetic Solutions Containing Lead Ions Using Modified Barks," in *EPD Congress 1994*, G.W. Warren (ed.), TMS, Warrendale, PA, 1994b, pp. 33–42.

86. K. B. Guiseley, "Chemical and Physical Properties of Algal Polysaccharides Used for Cell Immobilization," *Enzyme Microbiol. Technol.*, **11**, 706–715 (1989).

87. S. Wang, M. Misra, R. G. Reddy, and J. C. Milbourne, "Selenium Removal from Solutions using Iota Chips," in *Residues and Effluents—Processing and Environmental Considerations*, R. G. Reddy, W. P. Imrie, and P. B. Queneau (eds.), TMS, Warrendale, PA, pp. 757–773 (1991).

88. B. Pesic, D. J. Oliver, R. Raman, and C. L. Lasko, "Application of Natural Polymers for Removal of Heavy Metals from Aqueous Solutions: Sorption of Copper by the Modified Chitosan," in *EPD Congress 1994*, G. W. Warren (ed.), TMS, Warrendale, PA, pp. 257–274 (1993).

89. J. P. Gould, P. Phillips, and J. Bender, "Field Demonstration of Manganese Removal from Mine Drainage Using a Novel Biotreatment Process," *Pollution Prevention for Process Engineering*, P. E. Richardson and B. J. Scheiner (ed.), Engineering Foundation, AIChE, New York, 1996b, pp. 121–133.

90. T. Sakaguchi, T. Tsuruta and A. Nakajima, "Removal of Uranium by Using Microorganisms Isolated from Uranium Mines," *Pollution Prevention for Process Engineering*, P. E. Richardson and B. J. Scheiner, Engineering Foundation, AIChE, New York, 1996, pp. 183–191.

91. E. B. McNew, J. M. Barnes, and A. E. Torma, "A Biosorption Approach to Removal of Trace Concentrations of Uranium and Heavy Metals from Aqueous Solutions," in *Emerging Process Technologies for a Cleaner Environment*, S. Chander (ed.) SME, Littleton, CO, 1992, pp. 191–196.

92. S. Nordwick, M. Zaluski, D. Bless and J. Trudnowski, "Development of SRB Treatment Systems for Acid Mine Drainage," in C. A. Young, A. M. Alfantazi, C. G. Anderson, A. James, D. B. Dreisinger, and B. Harris (eds.), *Hydrometallurgy 2003, Proceedings of the 5th International Symposium*, TMS, Warrendale, PA, 2003, pp.1837–1846.

93. B. Wahlquist, T. Pickett, J. Adams, and T. Maniatis, "Biological Water Treatment for Dissolved Metals and Other Organics," in *Hydrometallurgy 2003, Proceedings of the 5th International Symposium*, C. A. Young, A. M. Alfantazi, C. G. Anderson, A. James, D. B. Dreisinger, B. Harris (eds.), TMS, Warrendale, PA, 2003, pp. 1903–1912.

94. H. Reinhardt, "Some Problems in Metal Waste Recovery Using Solvent Extraction," *Hydrometallurgy* **81**, F1 (1981).

95. H. F. Svendsen and G. Thorsen, *Oslo Symp. on Ion Exchange and Solvent Extraction,* Section IV, 1982, p. 277.

96. W. H. Jay, "Application of Ion Exchange Polymers in Copper Cyanide and Acid Mine Drainage," in C. A. Young, A. M. Alfantazi, C. G. Anderson, A. James, D. B. Dreisinger, and B. Harris (eds.), *Hydrometallurgy 2003, Proceedings of the 5th International Symposium,* TMS, Warrendale, PA, 2003, pp. 717–728.

97. T. H. Eyde, "Using Zeolites in the Recovery of Heavy Metals from Mining Effluents," in *EPD Congress 1993,* J. P. Hagar (ed.), TMS, Warrendale, PA, 1992, pp. 383–392.

98. K. A. Prisbey, J. G. Williams and H. Lee, "Ion Exchange Recovery of Cobalt and Copper from Blackbird Mine drainage," *Research Technical Completion Report, Project A-067-IDA,* Office of Water Research and Technology, U.S. Department of the Interior, 1980.

99. J. Holmes and E. Dreusch, "Acid Mine Drainage Treatment by Ion Exchange," U.S. Environmental Protection Agency, Office of Research and Monitoring, EPA-R2–72-056, 1972.

100. R. C. Wilmoth, J. L. Kennedy, J. F. Hall, and C. W. Steuwe, "Removal of Trace Elements from Acid Mine Drainage," *Office of Research and Development, EPA-600/7–79–101,* 1979.

101. R. R. Grinstead, W. A. Nasutavicus, and R. M. Wheaton, "New Selective Ion Exchange Resins for Copper and Nickel," *International Symposium on Copper Extraction and Refining, Vol. II,* 105th Annual AIME Meeting, Las Vegas, pp. 1009–1024, 1976.

102. K. C. Jones and R. R. Grinstead, "Properties and Hydrometallurgical Applications of Two New Chelating Ion Exchange Resins," *Chemistry and Industry,* **8**, 637–641 (1977).

103. R. R. Grinstead, "Selective Absorption of Copper, Nickel, Cobalt and Other Transition Metal Ions from Sulfuric Acid Solutions with the Chelating Ion Exchange Resin XFS4195," *Hydrometallurgy,* **12**, pp. 387–400 (1984).

104. T. M. Harris, C. Binter, S. K. Gopidi and P. C. Allison, "Removal and Recovery of Heavy Metals from Wastewater by Chelation-assisted Anion Exchange," in *Extraction and Processing for the Treatment and Minimization of Wastes,* J. Hager, B. Hansen, W. Imrie, J. Pusateri and V. Ramachandran (eds.), TMS, Warrendale, PA, 1994, pp. 343–356.

105. A. F. Chinn, R. T. McAndrew, R. L. Hummel and J. E. Mouland, "Application of Short Bed Reciprocating Flow Ion Exchange to Copper/zinc Separation from Concentrated Leach Solutions," *Hydrometallurgy,* **30**, 431–444 (1992).

106. J. C. Milbourne and I. R. Higgins, "Recovery of uranium using continuous countercurrent ion exchange (CCIX^{TM})," in *Separation Processes: Heavy Metals, Ions and Minerals,* Ed. M. Misra, TMS, Warrendale, PA, pp. 3–14.

107. M. Streat, "Applications of Ion Exchange in Hydrometallurgy," *Hydrometallurgical Process Fundamentals,* R. Bautista (ed.), Plenum Press, New York, 1984, pp. 539–553.

108. R. D. Noble, "An Overview of Membrane Separations," *Sep. Sci. and Technol.,* **22**, 731–743 (1987).

109. H. Strathmann, "Selective Removal of Heavy Metals from Aqueous Solutions by Diafiltration of Macromolecular Complexes," *Sep. Sci. and Technol.,* **15**, 1135–1152 (1980).

110. N. N. Li, "Liquid Membrane Separation Process," *U.S. Patent* **3**, 696,028 (1972).

111. N. N. Li and H. Strathmann (eds.), *Separation Technology: Proceedings of the Engineering Foundation Conference* held at Schloss Elmau, Bavaria, West Germany, April 27–May 1, 1987.

112. P. R. Danesi, "Separation of Metal Species by Supported Liquid Membranes," *Sep. Sci. and Technol.*, **19**, 857–894 (1984–1985).

113. D. S. Flett and D. Pearson, "Role of Hollow Fibre Supported Liquid Membranes in Hydrometallurgy," in *Extraction Metallurgy '85*, Inst. Min. Metall., London, UK, 1985, pp. 1–21.

114. R. Chiarizia, E. P. Horwitz, P. G. Rickert and K. M. Hodgson, "Application of Supported Liquid Membranes for Removal of Uranium from Groundwater," *Sep. Sci. and Technol.*, **25**, 1571–1586 (1990).

115. V. Jiricny, A. Roy and J. W. Evans, "Copper Electrowinning Using Spouted Bed Electrodes: Part I. Experiments with Oxygen Evolution or Matte Oxidation at the Anode," *Metall. Mater. Trans.*, **B 33**, 669–676 (2002).

116. V. Jiricny, A. Roy and J. W. Evans, "Copper Electrowinning Using Spouted Bed Electrodes: Part II. Copper Electrowinning with Ferrous Ion Oxidation as the Anodic Reaction," *Metall. Mater. Trans.*, **B 33**, 677–683 (2002).

117. W. S. Ho and K. Poddar, "New Membrane Technology for Removal and Recovery of Chromium from Waste Waters," *Environ. Prog.*, **20**(1), 44–51 (2001).

118. F. Valenzuela, H. Aravena, C. Basualto, J. Sapag, and C. Tapia, "Separation of Cu(II) and Mo(IV) from Mine Waters Using Two Microporous Membrane Extraction Systems," *Sep. Sci. Technol.*, **35**(9), 1409–1421 (2000).

119. Y. Wang, Y. S. Thio, and F. M. Doyle, "Formation of Semi-permeable Polyamide Skin Layers on the Surface of Supported Liquid Membranes," *J. Membr. Sci.*, **147**, 109–116 (1998).

120. Y. Wang and F. M. Doyle, "Formation of Epoxy Skin Layers on the Surface of Supported Liquid Membranes Containing Polyamines," *J. Membr. Sci.*, **159**, 167–175 (1999).

121. F. Sebba, "Concentration by Ion Flotation," *Nature*, **104**, 1062–1063 (1959).

122. F. Sebba, *Ion Flotation*, Elsevier, Amsterdam, 1962.

123. B. L. Kerger, L. R. Snyder and C. Horvath, "Adsorptive Bubble Separation Process," in *An Introduction to Separation Science,* John Wiley, New York, 1973.

124. P. Somasundaran, "Separation Using Foaming Techniques," in *New Developments in Separation Techniques*, Marcel Dekker, New York, 1976.

125. A. N. Clarke and D. J. Wilson, *Foam Flotation: Theory and Applications*, Marcel Dekker, New York, 1983.

126. A. Mizuike, "Flotation," in *Enrichment Techniques for Inorganic Trace Analysis*, Chapter 10, Springer-Verlag, Heidelberg, 1983.

127. M. Caballero, R. Cela and J. A. Perez-Bustamante, "Analytical Applications of Some Flotation Techniques—a review," *Talanta*, **37**, 275–300 (1990).

128. G. A. Stalidis, K. A. Matis and A. I. Zouboulis, *Chimika Chronika*, **15**, 133–146 (1986).

129. S. P. Nicol, K. P. Galvin, and M. D. Engel, "Ion Flotation—Potential Applications to Mineral Processing," *Minerals Engineering*, **5**, 1259–1275 (1992).

130. F. M. Doyle, D. W. Fuerstenau, S. Duyvesteyn and K. Sreenivasarao, "The Use of Ion Flotation for Removing Trace Metals from Waste Water," *Proc. First Int. Conf. on*

Processing Materials for Properties, H. Henein, T. Oki (eds.), TMS, Warrendale, PA, 1993, pp. 105–108.

131. K. P. Galvin, M. D. Engel, and S. K. Nicol, "The Selective Ion Flotation of Gold Cyanide from a Heap Leach Mine Feed Liquor," *Proceedings XVIII International Mineral Processing Congress*, Sydney, Australia, Austr. IMM.

132. K. P. Galvin, S. K. Nicol, and A. G. Walters, "Selective Ion Flotation of Gold," *Colloids and Surfaces*, **64**, 21–33 (1992).

133. K. Sreenivasarao, F. M. Doyle and D. W. Fuerstenau, "Removal of Toxic Metals from Dilute Effluents by Ion Flotation," *EPD Congress '93*, J. P. Hager (ed.), TMS, Warrendale, PA, 1993, pp. 45–56.

134. K. Sreenivasarao and F. M. Doyle, "The Effect of Collector/metal Ion Ratio, and Solution and Froth Height on the Removal of Toxic Metals by Ion- and Precipitate Flotation," *Extraction and Processing for the Treatment and Minimization of Wastes*, J. P. Hager, B. J. Hansen, W. P. Imrie, J. F. Pusateri and V. Ramachandran (eds.), TMS, Warrendale, PA, 1994, pp. 99–113.

135. S. Duyvesteyn and F. M. Doyle, "The Effect of Frothers and Ionic Strength on Metal Ion Removal Using Ion Flotation," *Extraction and Processing for the Treatment and Minimization of Wastes*, J. P. Hager, B. J. Hansen, W. P. Imrie, J. F. Pusateri and V. Ramachandran (eds.), TMS, Warrendale, PA., 1994, pp. 85–97.

136. K. Sreenivasarao, "Removal of Toxic Metals from Dilute Synthetic Solutions by Ion- and Precipitate-flotation," Ph.D. dissertation, University of California at Berkeley, 1996.

137. K. Sreenivasarao and F. M. Doyle, "Decomposition of Heavy Metal-dodecylsulfate Complexes," *Treatment and Minimization of Heavy Metal-Containing Wastes*, J. P. Hagar, B. Mishra, C. F. Davidson and J. L. Litz (eds.), TMS, Warrendale, PA., 1995 pp. 3–14.

138. A. W. Adamson and A. P. Gast, *Physical Chemistry of Surfaces*, 6th ed., John Wiley, New York, 1997, Chapter III.

CHAPTER 11

SOLID WASTE DISPOSAL AND RECYCLING: AN OVERVIEW OF UNIT OPERATIONS AND EQUIPMENT IN SOLID WASTE SEPARATION

Georgios N. Anastassakis
National Technical Univ. of Athens School of Mining Engineering and Metallurgy
Athens, Greece

1	**INTRODUCTION**	**308**
	1.1 Reuse and Recycling: A Necessity for Resource Conservation	308
	1.2 Scientific Basis for Solid-waste Recycling	309
2	**SAMPLING AND WASTE CHARACTERIZATION**	**310**
	2.1 Load Sampling from the Truck	310
	2.2 Incremental Sampling	311
3	**SIZE REDUCTION**	**311**
	3.1 Factors Affecting the Selection of Size-reduction Equipment	311
	3.2 Types of Equipment for Size Reduction of Solid Waste	313
4	**CLASSIFICATION**	**318**
	4.1 Vibrating Screens	320
	4.2 Rotary Screens (Trommels)	321
	4.3 Disc Screens	322
5	**HAND VERSUS ELECTRONIC SORTING**	**323**
	5.1 Design Criteria of Hand-Sorting Line	324
	5.2 Opto-electronic Sorting	325
6	**DENSITY (OR GRAVITY) SEPARATION**	**327**

	6.1 Air Concentrators	327
	6.2 Stoners	329
	6.3 Air Cyclones	330
	6.4 Heavy Media Separators	331
7	**MAGNETIC/ELECTROSTATIC SEPARATION**	**331**
	7.1 Magnetic Separation	332
	7.2 Eddy-current Separation	338
	7.3 Electrostatic Separation	340
8	**BALLISTIC SEPARATION**	**341**
9	**FROTH FLOTATION**	**342**
10	**FEEDING AND HANDLING SYSTEMS**	**342**
	10.1 Feeders	342
	10.2 Handling Systems	345
11	**PRODUCT AGGLOMERATION (CUBING AND PELLETIZING)**	**349**
12	**COMPACTION (BALING)**	**351**
	12.1 Compactors	351
	12.2 Balers	351
13	**PRODUCT STORAGE**	**352**
14	**SUMMARY**	**353**

The purpose of this chapter is to introduce the reader to the methods and equipment utilized to process recyclable streams of solid wastes in order to separate them into single constituents for recycling and recovering energy from the combustible components. The chapter describes all the unit operations that could be involved in a solid waste processing plant. This chapter is separated into three broad parts. The first part is devoted to the scientific basis of solid-waste separation techniques, as well as to the comparison between mineral processing and solid-waste separation methods. The second part describes the sampling techniques of solid waste, the working principles of the separation methods, as well as the equipment utilized to process recyclable streams of solid wastes. The third part describes all these auxiliary processes that are necessary for the smooth operation of machines and circuits (feeding, handling), as well as product agglomeration, baling, and storage.

1 INTRODUCTION

1.1 Reuse and Recycling: A Necessity for Resource Conservation

During the last 40 years, industrial and technological development has greatly increased, mainly due to the development of the world economic status and the population increase. These factors result in increased demand for more materials—and increased wastes. The production and utilization of materials follow a certain life cycle, which is determined by the economic trends prevailing in that certain period. Industrialized, developed countries mostly determine the market trends and consumption rules, as well as the economic cycle of products.

Most of the produced materials are based on primary raw materials extracted from the earth (minerals and rocks) and transformed to marketable final products upon energy consumption, which is also mainly produced from such primary sources (coal, oil, and natural gas). The primary mineral and energy resources are concentrated in defined deposits, which are nonrenewable. Consequently, resource conservation is extremely important, as it contributes to *Sustainable development* that is, development meeting the needs of the present without compromising the ability of the future generations to meet their own needs. As denoted by the definition, the meaning of *sustainability* is closely associated with resource conservation, which is achieved by maximizing the resource employment and minimizing the mass of the wastes to be disposed.

One way to minimize the mass of the wastes is by recycling and reusing the rejected end products. Although people have been conscious of the meanings of recycling and reuse for some time, it is only two decades now that recycling has drawn special attention. Although a lot of progress has been made, the level of recycling and reuse of materials from wastes is not yet widely applied. This could be attributed to some adverse factors, such as the high cost of the recycling process itself, lower cost of some primary products (e.g., plastics, paper) compared to the recycled ones, inferior properties (physical, chemical) of some products from recycled materials, variability of wastes composition, and difficulty of waste characterization and separation.

Despite these problems, recycling and reuse of materials, as well as energy recovery from wastes, is expected to increase as the population of even more countries becomes more conscious and sensitive to the arising environmental problems. Some factors expected to reinforce the global effort for continuously increasing recycling rate are related to population growth, the restriction of the available landfill space, the huge solid wastes stockpiled every year, the global demand for better life quality, and the diminishing primary raw materials resources.

1.2 Scientific Basis for Solid-waste Recycling

In principle, the existing technology meets in a high degree the recycling process of the wastes. The greater share of this technology originates and uses the same separation and processing principles as those applied to minerals. Between minerals and wastes there are similarities and differences with respect to separation processing into their useful and useless components, which are roughly summarized as follows:

Similarities

- Both deposits and wastes are resources of useful components.
- Separation and recovery of waste materials is based on the same principles as those applied to minerals.

Differences

- On a long-term basis, the composition of the feed in mineral-processing plants is more constant and homogeneous in comparison to that of waste-processing plants.
- In waste treatment, the constituents of the feed are usually liberated into single components liberation process easier.
- Size reduction is sometimes necessary, either for liberation purposes or simply for particle volume reduction. The size of the solid waste is considerably larger than the minerals or rocks.
- In contrast to the raw mineral sources, the grade of the wastes in useful and recoverable materials is higher. Consequently, the flow sheet in solid-waste treatment plants is simpler.
- In most cases, the shape of mineral particles could be considered more or less spherical, especially for small particle size, while the waste particles are of variant shape (spherical, wiry, platy etc). This shape variance in some cases hinders their effective separation, but in others it has a beneficial effect.

Although in most cases solid wastes are efficiently separated into their constituents based on well-known scientific principles and equipment from mineral processing methods, improvement and innovation of the existing equipment are sometimes necessary because of the differences in properties and characteristics between minerals and wastes.

2   SAMPLING AND WASTE CHARACTERIZATION

Solid-waste sampling and characterization is difficult because solid waste presents in varying quantities, components composition, and physicochemical characteristics. The difficulty is more intensive for municipal solid waste, because these parameters are highly dependent on the site and period of waste generation.

The information obtained from waste characterization is very important, for three reasons:

1. The data are the basis on which to determine the recoverable components and energy, process design, construction of the solid-waste processing plant, needs in equipment, and process economics in a solid waste processing plant.
2. The information will define the solid waste quantities to be landfilled.
3. The data contribute to process optimization, emissions monitoring, and determining the potential malfunctions, both in the processing and the waste-recovery facilities.

Consequently, sampling contributes highly to a precise and detailed characterization program of solid waste. The data from sampling can delineate the major constituents of the waste, as well as its content in ash, moisture, and sulfur. These parameters are of high importance for the combustion process, as they determine the heating value, gas emissions, energy recovery, and process efficiency.

Usually, a sample is obtained from a large mass of waste. There are two major concerns in waste characterization with respect to the sample. First, it must be representative and provide a confidence level. Although it is a relatively difficult task, this concern is accomplished by determining the sample weight by finding the weights of a number of smaller samples. Factors such as particle size and waste mass determine the final weight of the sample to provide a high confidence level. The human factor is very important—the sampling must be done correctly. Second, the analysis techniques of the sample must be accurate.

There are two procedures for solid waste sampling: load sampling from the truck and incremental sampling.[1]

2.1   Load Sampling from the Truck

This procedure includes the sorting of a truckload every day. The community is divided into areas. Each day the truckload from each corresponding area is separated into its constituents. The truck empties its load on a specially designed place. Subsequently, collectors separate the waste constituents into portable plastic containers. After each container is filled, it is weighed, sampled, and emptied in the corresponding bin. Solid waste is usually sorted into the following categories, which can be modified: (1) newsprint; (2) other paper; (3) diapers; (4) textiles; (5) plastic (light density); (6) plastic (high density); (7) food waste; (8) wood; (9) yard waste; (10) sweepings; (11) ferrous metals; (12) aluminum; (13) nonferrous metals; (14) glass; (15) bricks and stones; (16) oversize bulky materials. Categories 1 to 10 include combustible materials, while 11 to 16 are incombustible.

From the total sample of each waste component, collectors select a representative sample, which is sent to the laboratory for analysis only for the characteristics of fundamental interest. These characteristics usually include moisture, ash, sulfur, total chlorine, water-soluble chlorides, and heating value.

After sampling and characterization of the waste of all areas, sampling results are expressed on a daily or weekly basis. Sampling must be avoided for two to three days after extreme weather conditions.

2.2 Incremental Sampling

The second procedure includes the removal of many small-weight samples from the total mass of waste collected each day. This is accomplished by using a crane with a bucket. As the waste mass is large and the small samples must be representative, the sample must be obtained after prior mixing and homogenization of the waste with the crane. The collected small samples are deposited on a specially designed place, and the same process, as previously mentioned, follows to obtain samples for chemical analysis.

3 SIZE REDUCTION

Solid waste is usually subjected to size reduction prior to any processing stage. *Size reduction* is a mechanical process that is used to reduce the solid materials size, either for direct use and/or storage, or to make the materials amenable to further separation processes. Additionally, it makes solid waste constitution more uniform, as it imparts a blending action. Size reduction can also be applied in cases where solid waste is disposed at a landfill, with the major benefit being that of reducing waste volume and increasing the life of the disposal place.

3.1 Factors Affecting the Selection of Size-reduction Equipment

In contrast to minerals, the size-reduction process of solid wastes is relatively simple, as they are mostly composed of liberated materials. Also, the physical properties and mechanical characteristics of solid waste, which determine the material behavior in the size-reduction process, are, in most cases, strikingly different from those of rocks and minerals. These properties affect the mechanism of size reduction, as well as the equipment selection. Rocks and minerals are brittle and heterogeneous materials with structure discontinuities, while solid wastes are mostly nonbrittle and homogeneous. In general, size reduction is achieved by applying one or more of the following forces: compression, impact, attrition, and shear.

Compression is the most usual comminution force for brittle materials. Materials are compressed between two heavy-duty metal surfaces. Crushers based on compaction force are jaw, cone, gyratory, and roll crushers. These are most widely used to reduce the size of coarse rocks and minerals particles. In solid waste treatment, their use is not widespread, because compression force, in most cases, simply deforms the materials but does not change their size (e.g., metals, paper, plastic, organics).

Impact is the simplest size-reduction force. It is also applied for the comminution of brittle materials. As feed particles fall free on a high-speed solid rotor, the massive rotor impacts and breaks them into smaller particles. Secondarily, the material is struck onto a compact plate, where particle size is further reduced. Crushers based on impact force are impactors and hammer mills. They are characterized by high size reduction ratio, reaching up to 20:1.

Attrition force causes size reduction by friction contact of particles between them, with the grinding media as well as with the mill shell, and subsequent abrasion. It is effective for brittle materials of relatively small particle size. Rod and ball mills are mainly based on this force.

For brittle materials, *shear* force is restricted to hammer mills, acting along with impact and attrition, as well as to toothed crushing rolls. The prerequisite for successful size reduction of particles is the existence of macro- or microstructure weaknesses (e.g., layered structure, schistosity) or differences in the characteristics of adjacent minerals (e.g., hardness, friability). Contrary to rocks and minerals, shear force is widely applied in solid waste size reduction, as most of its constituents are soft, nonbrittle, plastic, and ductile, and, consequently, shear forces are effective. The various forms of shredders use shear force to reduce the size of solid waste constituents.

The most important factors that determine the selection of the size-reduction equipment are the mechanical characteristics (shear strength, ductility, etc.) of the feed material, as well as the size distribution of feed and comminuted product. From the aforementioned analysis, it is clear that the mechanical characteristics determine the acting force for size reduction and, consequently, the selection of the proper equipment. The size distribution of the feed stream and product determines the type of the corresponding equipment as well as the dimensions of feed and discharge openings.

Also, there are some other factors that influence the selection of the size reduction equipment. These factors are related to process economics (energy requirements, maintenance, performance, etc.), safety (e.g., explosions due to shredding of closed containers with volatile substances), environmental hazards (noise, air pollution by dust etc.), and site considerations (circuit design and installation, product storage etc.).

3.2 Types of Equipment for Size Reduction of Solid Waste

There are several types of size-reduction equipment in use with respect to solid-waste unit operations. Some of them originate from size reduction of minerals while other from solid-waste unit operations. Several and, sometimes, contradictory classification schemes exist with respect to size reduction equipment of solid waste. The equipment could be broadly classified, based on the main working principle, into the following categories: impactors, hammer mills, shredders, and grinders. The generic term *granulators* has been also used to include all these categories, as well as the many other varieties of equipment used in solid-waste size-reduction processes.

Impactors and Hammer Mills

Impactors and hammer mills accomplish material breaking and size reduction primarily by impact forces, exerted on the material by fixed or free-swinging hammers revolving about a central, high-speed rotating shaft, and, secondarily by hurling material pieces at high speed against stationary surfaces. This equipment is effective mainly for size reduction of brittle materials.

The material to be crushed enters through the feed opening, located at the top, or top side, of the equipment, and falls into the hammer circle. The first stage of size reduction is accomplished by collision of the feed particles with high-speed rotating hammers, made usually of manganese steel. The second stage of breakage occurs when the particles hit the plates or the breaker bars, which line the frame of the equipment. Hammer mills are further equipped with grid bars or a grate, placed at their bottom. Consequently, a third stage of size reduction occurs in hammer mills, relying on shearing and attrition forces exerted on oversize material between the bottom grate and hammers until its size becomes less than the grid opening.

Although there are many similarities between impactors and hammer mills, there are also several differences. The main difference is that the former rely primarily on the impact action of hammers, either fixed or free-swinging, and secondarily on particles collisions, either striking each other or on stationary steel surfaces. The hammer mills rely on both the centrifugal impact force of the hammers, as well as the attrition and shear action between hammers and grate.

Impactors.   The main components of impactors are a compact rotor with fixed hammers and one or two breaking plates. Impactors have no grates or other components at the bottom to size the end product. They are constructed to break the bulky items in the refuse stream, either for further finer-size reduction or for size preparation of the material to feed a refuse incinerator, as bulky materials are not suitable for incineration.

Hammer Mills. These are very effective for size reduction of brittle materials. Hammer mills for shredding refuse are available with fixed or swing hammers, bolted on the high-speed rotating shaft (700–1,200 RPM). Major and minor variations exist in the design of each manufacturer's hammer mill.

The type with fixed hammers is suitable for size reduction of brittle materials (e.g., glass bottles, ceramics), but, under special conditions, it can be applied for materials without brittle properties (wood, metal turnings, scrap metal, etc.), as it can be seen in Figure 1.[2] When mixed solid waste has been presorted, hammer mills provide product with the smallest particle size compared to others, under the same power consumption. The grate bar or screen, at the bottom of the equipment, prevents particle outflow until particles are reduced to the desired size. The opening size of the grates varies broadly between 5 and 50 mm. Swing hammers are commonly used for mixed and sorted solid waste. Knives or cutters can also be placed instead of hammers. In this case, the hammers extend radially from the shaft because of the centrifugal force.

In general, hammer mills are designed with horizontal or vertical shaft. Horizontal shaft hammer mills are either downrunning (Figure 2) or center-feed reversible. In the case of center-feed mills, material is fed down on the top of the hammer circle. The disadvantage of the downrunning hammer mills with horizontal shaft is the relatively increased maintenance or replacement cost of the hammers, as they always rotate to the same direction. The cost depends on the throughput and the composition of solid waste. By contrast, the reversal of the shaft rotating direction in the center-feed type results in the reduction of the replacement cost, as hammers are replaced when both cutting faces get worn. On the other side, center-feed type, along with its hopper, requires more height to avoid material ejection from feed opening of the machine.

Figure 1 Hammer mill suitable for wood fragmentation. (From Ref. 2.)

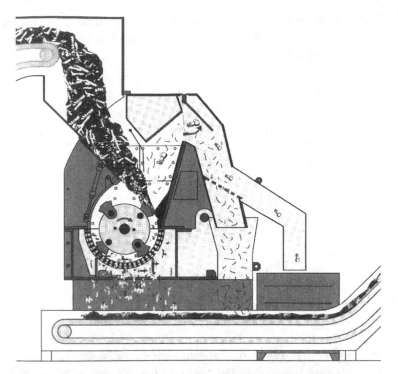

Figure 2 Horizontal shaft hammer mill downrunning. (From Ref. 3.)

Vertical shaft hammer mills have no grate at the bottom. The upper part of the equipment is conical and the lower cylindrical. The vertical shaft with the hammers is placed along the axis of the machine. Size reduction in this machine is achieved through repeated impacts of the material with the hammers and the breaking plate, as particles descend into the conical part of the machine. The final size reduction of the infeed material is accomplished at the lower part of the machine, where the distance between hammers and breaking plate is reduced. This results in the faster wear of the hammers at the lower part than at the conical part. The discharge of the material occurs through an annulus at the lower part, which also controls the maximum particle size but does so less effectively than in horizontal hammer mills with a grate at the bottom.

Shredders

Although the term *shredders* is sometimes used as a general term to include the equipment that causes size reduction of solid waste, it fits best to the equipment that uses mainly shear force to reduce the size of the materials. In such a sense, hammer mills are sometimes included in the category of shredders, as, in addition to impact force, materials are also subjected to shear force for their size reduction, in case that they are also equipped with stationary knives fixed at the wall. In this chapter, shredders are considered only the equipment that use shear as a principal force for solid waste size reduction. Under this consideration, flail mills, rotary shear shredders, and knife mills are the most popular equipment to shred solid waste.

Flail Mill. These machines are similar to hammer mills in appearance but they have some differences in construction and operation (Figure 3). Contrary to hammer mills, flail mills have no grate at the bottom and, consequently, they operate as single-pass machines. Their application is primarily for tearing bags of refuse and breaking up bundles of material in addition to providing some mixing of the waste materials. These machines should be followed by a shredder, as the product size is coarse.

Rotary Shear Shredder. Shear shredders (Figure 4) operate like scissors during materials size reduction. They are usually composed of two counter-rotating parallel shafts, mostly driven by hydraulic motors, as well as by electric motors, and a series of interlocking, toothed-cutting disks mounted perpendicularly on the shafts. Shear shredders with one, three, or four shafts are also available. They are single-pass machines, while the spacing between the cutters, or the cutter and the shell in single-shaft machines, determines the product size. In most cases, product size varies approximately between 25 and 100 mm (1 to 4 in.). The waste material is fed between the counter-rotating shafts, and its size reduction is achieved by the shearing or tearing action of the cutting discs. In case of overload, the shafts rotation is automatically reversed.

Rotational speeds are very low, in most cases ranging between 50 and 200 RPM. The low-speed hydraulic motors develop very high torques on the shafts, providing considerable shearing power to the cutting discs. This power makes shear shredders effective in difficult-to-cut materials. This equipment has been used to shred materials such as metals, truck and car tires, wood, pallets, bulky objects, and electrical and electronic devices, for example, in relatively coarse particle size. On the contrary, they are not capable of shredding materials into fine size (less than 10 mm). The main advantage of shear shredders is that

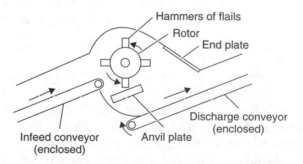

Figure 3 Typical configuration of flail mill. (With permission from Ref. 4.)

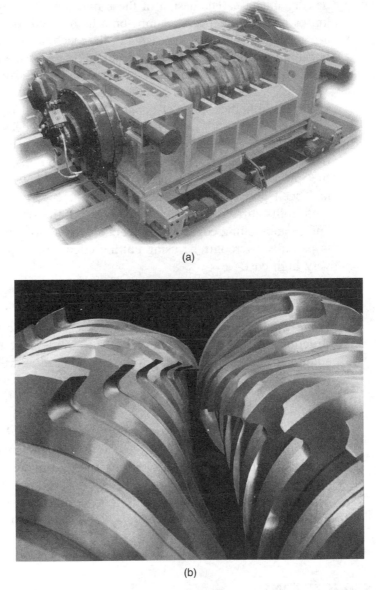

(a)

(b)

Figure 4 (a) Rotor shear shredder. (b) Details of the cutting disks. (From Ref. 5.)

the more they are fed with material, the more effective they are, possibly because the load presses the objects at the bottom layer to be easily grabbed between the shafts and the cutting discs.

Knife Shredder. Knife shredders are machines used to cut materials into smaller pieces for subsequent separation or for volume reduction, and, consequently, density increase. The design of these machines employs a series of long, sharp knifes attached in special clamps on a high-speed rotor. Cutting edges or anvils are also mounted inside the casing of the machine. The material is shredded, as it is caught between rotating knives and cutting edges. Knife shredders are mostly used for coarse size reduction, with the product size being of the order 6 to 18 mm. This equipment is suitable to effectively shred soft and plastic materials, such plastic bottles, rags, or paper. Hard materials (e.g., metals, glass) must be removed beforehand, as they destroy the cutting edges of the knives.

For very fine shredding, knife mills have been used to reduce the size of the materials down to 1.0 mm. These machines were mainly used for refuse-derived fuel production processes, under the condition that hard and abrasive materials had been previously removed. The very fine particle size obtainable rendered knife mills attractive in waste shredding, but the small distance between rotating knives and cutting edges makes them amenable to increased wear. This problem is partially overcome by using cutting edges from special hard alloys, but at a very high cost.

Wood Grinders

Wood grinders are used to shred large pieces of wood (e.g., yard wastes, construction debris, palettes) into chips. Tub grinders are the most commonly used equipment, but horizontal flail mills can be used as well.

Tub grinders are mobile machines carrying diesel engines. They consist of a large-diameter tub having a slowly rotating upper section and a stationary lower section containing a flail or hammer mill (Figure 5). The machine is usually mounted on a trailer frame so it can be easily moved. Material is fed into the top of the tub and is carried around the point where the flails or hammers intrude into the tub. The rotation of the tub ensures the continuous flow of the material to the mill. The shredded product passes a grid, which is placed below the rotor, and is removed by directly falling on a conveyor. The grid aperture determines the shredded product size. The product can be used either as a fuel or for biological treatment.

4 CLASSIFICATION

Classification or size separation is the unit operation that separates the particles according to their size. In general, it can be carried out dry or wet. The fraction of the feed having size less than the size of the screen aperture is the *undersize*, while that of greater size is the *oversize*. The fraction of the undersize that flows through the apertures is called *underflow* (or passing), while the particles that do not pass are called *overflow* (or remaining). In solid-waste treatment, classification is carried out almost always on screens and in dry, because the particles to be classified are relatively coarse and wet processes are undesirable,

Figure 5 Tub grinder. (From Ref. 6.)

as some properties of the processed materials after wet operations are inferior and an additional drying cost is required.

In solid waste-processing plants, classification has four purposes:

1. Oversized materials are removed from the processing stream.
2. The feed is separated into proper-size fractions according to the requirements of the sorting equipment.
3. Waste components are selectively separated in case that selective shredding is accomplished.
4. Oversized material is removed from combustion ash.

The equipment most commonly used for classification involves vibrating screens, rotary screens and disc screens. Air cyclones, which are mostly used as density separation equipment, are described in the corresponding chapter.

4.1 Vibrating Screens

The key element in all screens is the screening surface, as the real work of particles stratification and separation is accomplished on it. The screening surface is mounted on an inclined, or horizontal, frame and supported by bars, which are installed directly below the screening surface along the surface. This construction permits the use of light and thin screening surfaces. The screening deck is usually subjected to vibrating motion with the aid of an eccentrically rotating shaft, placed directly below the supporting bars. The rotating motion of the shaft is accomplished through an electric motor and drive mechanism, which impart motion to the shaft. The configuration of a vibrating screen is shown in Figure 6. The screening surface may be composed of one or more decks, but in most cases no more than two. Also, it may be one deck divided mostly into two parts with different aperture size. Screens are flat surfaces with a wire mesh or metal surfaces perforated with holes. The factors that affect the efficiency of the vibrating screens are solid waste composition, size and shape of particles, feed rate, stroke length, vibration frequency, screen length to width ratio, and screen tilt.

Vibrating screens are mostly used to separate fine glass, undersize material from source-separated and commingled municipal solid waste, and to process demolition and construction materials. They are not suitable to processing

Figure 6 Vibrating screen. (From Ref. 7.)

irregular-shaped, wet, and/or soft materials (such as paper or rags), as they cause clogging of the screen apertures.

In addition to vibrating, there are also shaking (or reciprocating) screens, whose screening deck is submitted to a substantially linear motion by an eccentric mechanism. For coarse particles, vibrating screens prevail over reciprocating in minerals industry and solid-waste processing because of their higher efficiency and capacity.

4.2 Rotary Screens (Trommels)

This equipment is one of the most widely used screening machines in solid waste processing. Essentially, rotary screens, or *trommels*, are large-diameter rotating drums, whose sidewall is the screening surface (Figure 7). They may separate waste materials into several size fractions, as the screening surface can be composed of cylindrical sections with different aperture size each. The screen is normally set at an inclined position of about 5 degrees. Material is fed at the front end of the inclined trommel and flows downward. The sloped position of the trommel facilitates the downward motion of the materials. As the screen rotates, material tumbles and contacts the screening surface many times as it moves downward. Small-size materials pass the openings of the trommel, with the finer particles been removed close to the feeding point, while the oversized material is removed at the end of the length. Sometimes, metal blades (lifters) protrude into the drum of the trommel. In this case, what the blades accomplish is bag breaking and better mixing of solid waste materials.

Trommel efficiency is affected both by geometrical and operational factors, such as length, diameter, angle of slope, rotational speed, and feed rate. Rotational speed is the most crucial operational parameter, as it must be maintained at an optimum value. Optimum rotational speed occurs when the waste material tumbles in the drum forming a cascade, which means that waste material is partially carried up the interior wall of the rotating drum, reaches about the maximum height, and then falls back on the bottom. The optimum rotational

Figure 7 Trommel screen. (From Ref. 8.)

speed is a percentage, usually 50 to 80 percent, of the critical rotating speed, which is the maximum speed that the material tumbles. For rotational speeds above the critical, waste material does not tumble but is centrifuged or stuck at the drum wall. Critical speed (n_c) depends on the drum diameter, according to this relationship:

$$n_c = (1/2\pi)(g/r)^{1/2} \tag{1}$$

where g = gravity acceleration
r = trommel radius

Trommel inclination, which usually varies between 2 and 5 degrees, affects the residence time of the waste material in the drum, as well as the trommel throughput. Other factors that affect the throughput are trommel diameter and filling factor. Trommel screen is a valuable item of many solid waste-processing plants, as it appears to be a clogging resistant screen.

4.3   Disc Screens

Disc screens are composed of a series of parallel, horizontal, and equally spaced bars or shafts, which are installed perpendicular to the flow of the material. Each shaft is equipped with interlocking serrated or star-shaped discs, installed at equal distances across the width of each bar. Undersized material falls in the space between the discs, while oversized material overflows the discs. The adjustment of the distance between the discs determines the materials size to be separated as underflow. After suitable adjustment, the same screen can be used to separate materials into different size.

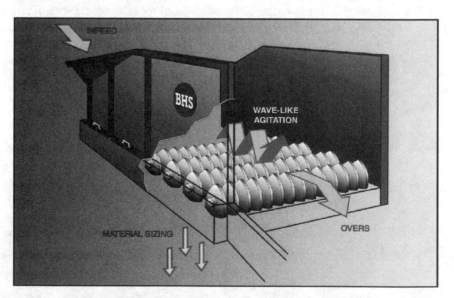

Figure 8   Configuration of patented disc screen. (With permission of Bulk Handling Systems, Inc., from Ref. 9.)

The main advantage of disc screens over vibrating ones is that they do not favor clogging of the apertures. In case that clogging occurs, due to wrapping of materials around the disc, they are usually self-cleaned by rotation. Materials such as metals, glass, and shredded plastic bottles have been separated successfully as undersize.

In a patented version of disc scree,[9] the discs are designed with a triangular shape, which results in an aggressive agitation of the material and impacts a bouncing wavelike action into it (Figure 8). Also, the discs are mounted on the shafts with an in-line configuration, which creates precise openings and results in a highly accurate material sizing. Shafts are rotating in such a way that as one disc trip is moving downward, the reciprocating disc tip on the adjacent shaft is moving upward, removing, so, the troublesome material from the potential jamming area. Additional, advantages of the disc configuration over the conventional ones with serrated or star-shaped discs are reported to be the higher sizing efficiency, increased production rates and increased life of the mechanical components.[9] A wide range of materials, that are present in residential, commercial and industrial waste streams, can be processed with the triangular shaped disc screen (e.g. municipal solid waste, construction/demolition debris, wood waste, tire, plastics, glass, paper/cardboard).

5 HAND VERSUS ELECTRONIC SORTING

Hand or manual sorting has been applied since antiquity as a mineral separation method. It has been used extensively in the past as ore deposits were of high grade and labor cost was relatively low. Hand sorting is based on solid characteristics discernible by eye, such as color, shape, transparency, and luster. The necessary requirement for the application of hand sorting is the large particle size (in general, 40–250 mm). For particle size less than 40 mm, hand sorting is inefficient, as throughput is low.

Although hand sorting has been also applied in solid-waste processing for a long time, it is widely applied in many cases nowadays as it is the simplest, most direct, and sometimes the most effective separation method. Hand sorting can be practiced both for commingled and source separated waste materials under the condition that solid waste components are of relatively large size and they possess properties discernible to the eye. At a materials recovery facility, solid waste stream is transported on a belt conveyor moving with a determined speed. Workers, placed at the side of the belt, separate the individual waste components by handpicking and dropping them into chutes that lead to storage containers. Containers are located either at the same level with the sorting line or, most commonly, below it, as it is presented in Figure 9.[10]

For efficient hand sorting, the following must be taken into consideration: (1) Prior to sorting, *liberation* of the commingled waste is necessary, which means that plastic bags with stored waste must be torn and spread out on the sorting belt; (2) the room with sorting lines must follow the rules posed by the Health

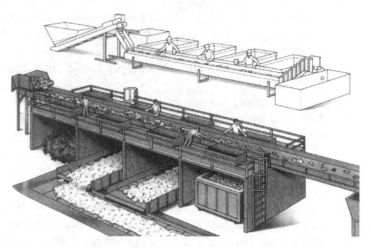

Figure 9   Setup of hand sorting lines. (With permission from Ref. 10.)

Authorities; that is, the room must be well lighted, aerated, and air-conditioned; and (3) the sorting lines must be properly designed.

5.1   Design Criteria of Hand-Sorting Line

Manual sorting of solid waste can be practiced either on static floors or on moving surfaces (e.g., belt conveyors). The use of conveyors makes the process more efficient under the condition that the sorting facilities are properly designed. Some of the factors that determine the design of hand-sorting facilities are throughput, solid-waste composition, number of components to be recovered and recycled, and the characteristics of solid waste. A sorting belt conveyor is completely determined by factors such as belt width and length, speed, and average waste thickness. In addition, the number of the workers must be determined. The critical factors are the length of the belt and the number of the workers.

To calculate the two aforementioned critical factors, a simple methodology is followed based on the composition of various practical rules, derived from experience. For the calculation, six types of data are necessary:

1. The weight of solid waste to be sorted, which covers a surface of 1 m^2 in monolayer arrangement (it is determined experimentally or from diagrams)
2. The percentage of the material to be sorted in the feed
3. The sorting capacity per worker (it is determined from field data or diagrams)
4. Belt conveyor speed
5. Distance between workers
6. Belt conveyor width

Example 1   Let's calculate the belt conveyor length for the manual sorting of plastic bottles from a commingled waste stream. The solid-waste feed rate is

10 tons per hour with a 5 percent content in plastic bottles. For calculation reasons, we assume a belt conveyor speed of 12 m/min with 1.0 m belt width and waste weight 15 kg/m of belt. The sorting capacity is considered to be 250 kg/8h&worker.

Solution:

1. Feed rate of conveyor belt: $15 \text{ kg/m} \times 12 \text{ m/min.} \times 60 \text{ min./hr.} = 10800 \text{ kg} = 10.8 \text{ tons/hr}$. So, the conveyor capacity covers the hourly rate.
2. Quantity of plastic bottles: $10 \text{ tons/hr} \times 0.05 = 0.5 \text{ tons/hr}$
3. Sorting capacity is $0.250 : 8 = 0.03125 \text{ tons/hr \& worker}$
4. Number of workers $= 0.5/0.03125 = 16$
5. If the necessary distance between two adjacent workers to work conveniently is 1 m, then the required conveyor length is 16 m. By adding 1.5 m at the feeding point and 1.5 m at the discharge point of the belt conveyor, it derives that the total length of the conveyor is 19 m.

In cases where the conveyor is long enough and the space is relatively limited, then two conveyors are established.

5.2 Opto-electronic Sorting

Although manual sorting is applied in many cases to separate waste materials, its application is not ideal due to the high labor cost and human errors. Opto-electronic sorting is an automated sorting process of waste materials in replacement of manual sorting, based on waste particle characteristics such as brightness, transparency, color, shape, and size.

Before opto-electronic sorting, the solid waste stream is subjected to screening and dedusting, so that its characteristics can be clearly distinguished. For effective waste separation, waste particles must be individualized, which is achieved by a vibrating feeder, placed below the waste bin. As the particles abandon the feeder, and during their free fall, they are scanned by a high-resolution optical system (color line camera), which classifies the waste flow by real colors. Scanning can be single or multi-sided, depending on the material to be processed. The information is rapidly transmitted and evaluated by a signal processor, which is capable of processing some tens of millions of measuring points per second. This enables the processor to identify and analyze some thousands of particles per second. Selection is performed a few centimeters beneath the detecting system by means of compressed-air pulses. This involves controlling up of several hundred separating channels (compressed-air jets) and charging them with dosed compressed-air pulses, depending on waste particle size. The process sequence is presented in Figure 10.[11]

Opto-electronic sorting systems can process waste with particle size from 5 to 250 mm. Color sorting of particles is very difficult or impossible for size less than 5 mm, under the current state of the art. Capacity ranges from 2 to 200

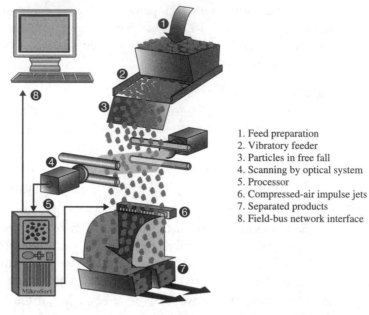

1. Feed preparation
2. Vibratory feeder
3. Particles in free fall
4. Scanning by optical system
5. Processor
6. Compressed-air impulse jets
7. Separated products
8. Field-bus network interface

Figure 10 Operating principle of electronic sorting. (With permission from Ref. 11.)

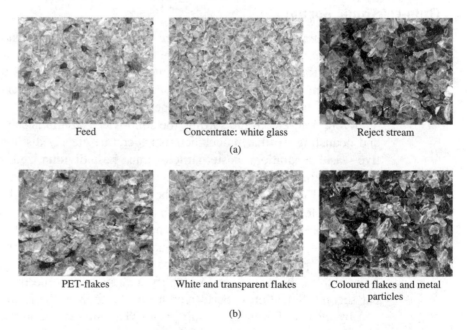

Figure 11 Products after opto-electronic separation: (a) white from colored glass, ceramics, stone and porcelain, (b) white and transparent PET-flakes from colored and metallic particles, (with permission from Ref. 11.)

tons/hour, depending on particle size and working waste stream width. For sized waste feed, the process efficiency is very high, reaching 99%, due to the high resolution of the sorting equipment. Working widths range between 0.60 and 1.0 m. In industrial scale, opto-electronic sorting has been used successfully to separate white, green, and brown glass in glass-recycling plants, plastic bottles (PET), as it is shown in Figure 11[11] while it is also possible to separate nonferrous metals (e.g., copper and brass from lead and zinc metals).

6 DENSITY (OR GRAVITY) SEPARATION

Density (or gravity) separation is a unit operation used to separate materials, among which solid waste, mainly based on the difference of their densities, as well as on hydro- or aerodynamic factors such as particle size, shape, and fluid resistance on particles motion. Gravity separation is examined almost in all cases of solid-waste treatment, as its constituents have usually efficient difference in densities and they are liberated at a coarse particle size. Density separation, along with magnetic separation, is the most widely applied process. Although the principles and equipment used in solid-waste density separation mainly originate from mineral separation, there is one striking difference. Contrary to mineral separation, dry methods prevail in solid-waste processing.

The factors that must be taken into account in the selection of the equipment and, consequently, for efficient waste separation, are waste characteristics (particle size, waste stream composition, waste components density, moisture content, tendency for aggregation, etc.), products specifications, design parameters, and space requirements. The various types of air separators (air concentrators, stoners) are the most widely used equipment. Heavy media separators are also used in some cases.

Two products are usually obtained by density separation: (1) the *light fraction*, which is mainly composed of organic materials (paper, plastics, food residue, garden waste, etc.); and (2) the heavy fraction, composed of inorganic materials (metals, construction debris, heavy plastics, etc.).

6.1 Air Concentrators

Air concentration is the unit operation used to separate light from heavy waste components, based on their density difference, with the aid of air stream. Instead of air concentration, the term *air classification* is usually used, incorrectly, because classification denotes separation according to particle size. There are mainly two designs for air concentrators: (1) with horizontal and (2) with vertical airflow.

Air concentrators with horizontal airflow are composed of a belt conveyor on which solid waste is fed. Horizontal air stream is blown behind the feeding point. The airflow lifts and removes light materials (paper, plastics, etc.) upward, while the heavy fraction remains on the belt and is removed at the end point of the

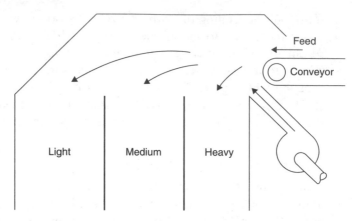

Figure 12   Horizontal air separator. (Reproduced with permission from Ref. 12.)

conveyor. In a subsequent step, the light fraction is fed into a cyclone to separate the solids from the air stream, with the solids being recovered in the underflow.

Air could be also blown at the point where solids leave the belt (Figure 12). Heavy materials are not significantly drawn by airflow and, consequently, are collected close to the point of airflow while lighter particles result in a more distant product bean. These horizontal concentrators are usually designed with more than two product beans to collect separately the various waste components. A similar separation concept is also used in air-knife separators.

In air concentration with vertical airflow, air stream is blown upward. The feeding point of solid-waste is located in the middle, or sometimes at the top, of the duct. Solid waste feeding is realized through a rotating, air-locked device, because retaining constant aerodynamic conditions (air velocity, pressure) in the separation chamber is of high importance. Light materials are drawn in the air stream and, as previously mentioned, are separated from air in a cyclone. Heavy product is recovered from the bottom of the device, usually falling on a belt conveyor. The factors that affect separation efficiency are waste feed rate, airflow

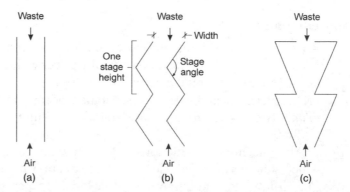

Figure 13   Duct designs of air concentrators with vertical airflow: (a) straight, (b) zigzag and (c) stacked reverse triangles. (Reproduced with permission from Ref. 4.)

rate, duct cross-section, as well as particle size and shape. There are several duct designs, including straight duct, zigzag and stacked reverse triangles (Figure 13). The design of the duct usually names the concentrator. In the first two duct designs, air velocity is constant through the entire length of the duct, as the cross-section of the duct is constant. In addition, zigzag concentrators are more effective, as their cross-section favors turbulence creation of the air stream, tumbling of the waste, and consequently, disintegration of aggregates. In ducts with stacked reverse triangles, the velocity of the air stream varies along the duct. This results in the differentiation of falling particles velocities and the collection of particles with similar terminal velocities in the same product.

6.2 Stoners

Essentially, a stoner operates as a pneumatic jig. Its operational principle is relatively simple. Solid waste is fluidized by an upward-blown air current and, in combination with the reciprocating motion of the inclined stoner deck, solid waste components are separated into a heavy and a light fraction. This principle is shown in Figure 14.

Stoners consist of an inclined deck with a porous surface, through which low-pressure air is introduced. Air is distributed uniformly below the deck while, in addition, externally regulated air-valves further regulate the locally required air-flow rate, according to the requirements of the separation. A fan, with regulating flow rate and pressure, blows air. Material is fed between the center and the upper part of the deck, closer to the discharge point of the heavy fraction. The solid waste feed is fluidized and stratified by the air current, with the light material being at the top of the bed and the heavy remaining on the deck surface. Materials separation depends on factors such as density, particle size, and shape. Above the stoner, a dust hood is installed to collect the fluidizing air along with the dust drawn from the waste.

The stoner deck is also subjected to a straight-line reciprocating motion, which affects the particles in contact with the deck. The light material flows downhill following the slope of the deck, as it is not affected by deck motion. Contrarily, the heavy fraction moves uphill, because it is in contact with the deck and is affected by the reciprocating motion, and is discharged on a takeaway conveyor.

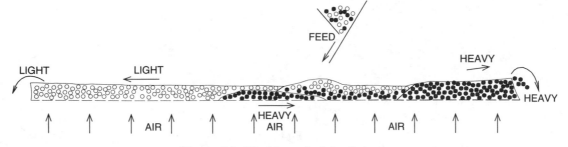

Figure 14 Working principle of stoner.

The operational variables of the stoner are the air-flow rate, frequency and stroke of the deck reciprocating motion, the slope of the deck, and air-valves regulation. For effective separation, the density difference between light and heavy fractions must be significant and the waste feed must be classified into narrow particle-size fractions. Also, the moisture content of the waste must be low, as moisture aggregates light materials and, consequently, prevents waste bed fluidization.

6.3   Air Cyclones

Air cyclones are used to separate the light fraction, recovered from air separation, into solids and air. They are also used to separate dust from air so that air to be recycled or released in the environment.

Air cyclone is a devise comprising two parts: the upper is cylindrical and the lower is conical (Figure 15). The feed stream (light fraction) is introduced tangentially into the cyclone. Solid particles are subjected to centrifugal force and axially moved to the cyclone wall. Subsequently, being in contact with the wall, they follow a spiral orbit of continuously decreasing radius until they

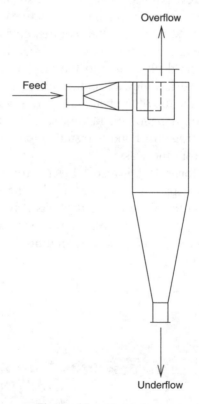

Figure 15   Air cyclone.

reach the cyclone underflow, where they exit. Air is recovered from cyclone overflow. Practically, overflow contains a small percentage of dust or very fine, light material. The material from the underflow is discharged either periodically by discharging part of the material each time or continuously through an airtight rotating system, which secures stability of aerodynamic conditions in the air cyclone chamber.

6.4 Heavy Media Separators

Heavy media separation is a wet separation method and, as such, has been applied in a limited extension in solid-waste processing. It has been mostly used to separate aluminum, which is collected to the light fraction, from other materials in car recycling.

The method is based on the creation of a heavy medium suspension of intermediate density between that of heavy and light fraction. The desired density is accomplished by mixing proper quantities of finely ground materials (e.g., magnetite, ferrosilicon, galena, etc.) and water. Shredded materials feed the heavy medium bath. The particles with a lower density than that of the suspension float (light product), while those with higher density sink (heavy product).

Factors that prevent its widespread application in solid waste treatment are the relatively complex flow sheet in comparison to dry separation methods, the high capacity requirements for economic plant operation, and, finally, its application to materials whose characteristics are not affected by water.

7 MAGNETIC/ELECTROSTATIC SEPARATION

Metal items constitute an appreciable amount of solid waste, with their percentage varying between 5 and 15 percent in most cases. As the revenue from solid-waste processing plants is derived from the sale of the separated products, among which ferrous and nonferrous metal items, the recovery and recycling of metallic objects support considerably the construction and operation of solid-waste treatment plants. The recycling of metal items is very important, as it contributes in mineral deposits conservation and in the prevention of environmental pollution from the oxidation and dissolution of various metals, being in alloyed form.

The methods used to separate and recycle metal items are mainly magnetic, electrostatic, and eddy-current separation, all three of which are based on the principles of electromagnetism. Magnetic separation is a well-known technology, as it has long been successfully applied in mineral processing. By contrast, the development of eddy-current separation is largely due to the applications in solid-waste separation. Finally, electrostatic separation is used for the separation of a variety of materials, including nonferrous metals.

7.1 Magnetic Separation

Applications

Magnetic separation is a unit operation that is applied whenever metals with magnetic properties participate in the composition of the commingled waste. Magnetic separators are encountered in almost all the recycling plants, as they are very simple in operation, have a relatively lower cost compared to the other equipment of a solid-waste processing plant, have a high efficiency (80 to 95 percent) and production rates, and short pay-back period.

Magnetic separation is used to separate and recover (1) ferrous metals from scrap and alloys, slag, mixed metals in source-separated waste (e.g., tin cans from aluminum cans), (2) ferrous items from automobile shredding, municipal solid wastes (tin cans, wires, household equipment, etc.) and combustion residues.

The position of magnets in the flow sheet of the recycling plant depends on the target to be achieved, factors such as product purity and recovery, percentage of magnetic material in the waste, as well as damage reduction of processing and recovery equipment. When magnetic separation is carried out before shredding, there is less wear on the shredders. Also, magnetic separation after shredding and before air classification may result in the entrapment of light materials (paper, plastic, etc.) into the magnetic product.

Classification of Magnets and Materials

In general, magnetic separators are classified into (1) standard- (or low-) and high-intensity, according to the strength of the applied magnetic field; and (2) dry and wet, according to the fluid medium (correspondingly, air or water). The intensity of the magnet to be used is determined by the magnetic properties of particles to be separated, while the medium is determined by the particle size of the material. Dry magnetic separation, is applied for coarse particles separation, while the wet is for fine particles.

Magnetic field is achieved either by permanent magnets or by electromagnets. Permanent magnets have very low operational and maintenance cost, as they require no power to create the magnetic field. By contrast, the magnetic field intensity in electromagnets can be regulated to the required value, according to the application. In the past, permanent magnets were considered to achieve low-intensity magnetic field (1,300–1,500 Gauss) but this is not always true nowadays, as the use of special alloys has resulted in magnetic fields of much higher field intensity. Electromagnets achieve field intensities as high as 25,000 Gauss by electric current regulation to magnetize and polarize correspondingly an iron core, which is surrounded by the electromagnetic coil.

Based on their characteristics—such as magnetic susceptibility, Curie temperature, and Neel temperature—materials are classified into diamagnetic, paramagnetic, ferromagnetic, antiferromagnetic, and ferrimagnetic. Diamagnetic materials are practically nonmagnetic, while ferrimagnetic are strongly magnetic. In practical applications, materials are classified into diamagnetic, paramagnetic, and ferromagnetic. Standard-intensity magnetic separation is applied for materials with high magnetic susceptibility (ferromagnetic), while high-intensity separation is used for weak magnetic materials (paramagnetic).

Magnet Configurations and Arrangements

Standard-intensity dry magnetic separators are commonly used in solid-waste separation, because particles are relatively coarse and those to be recovered in the magnetic product usually possess strong magnetic properties. The basic component of the separator is a rotating drum with a fixed permanent magnet or electromagnets inside of alternating polarity and equal space between them.

During standard-intensity magnetic separation, all particles of the burden are subjected to the action of a magnetic force and a mechanical or combination of forces (gravitational, centrifugal). Magnetic force affects only magnetic particles, while the mechanical forces, which act in a different direction, affect both magnetic and nonmagnetic. Magnetic particles can be separated from nonmagnetic through one of the following ways:

- *Pinning*. During this principle, magnetic particles enter the magnetic field in contact with the magnetic surface and remain attached on it until the influence of the field ceases. The nonmagnetic are not subjected to any magnetic force and fall almost vertically, under the influence of their weight or a centrifugal force. In this case, the magnet may be composed of either a compact piece (Figure 16(a)) or a set of axially oriented bars of alternating polarity.
- *Attraction*. In this case, particles enter the magnetic field through an independent carrier surface below the magnet, are attracted towards the magnet and separated, when they exit the magnetic field (Figure 16(b)). The magnetic particles move to the same direction with the carrier, which surrounds the magnet (belt or drum). For coarse particles, this configuration produces a very clean magnetic concentrate.
- *Elevation*. According to this principle, particles enter the magnetic field as previously mentioned, but move counter-current to the carrier magnetic particle surface (Figure 16(c)). This configuration is especially efficient when applied for coarse particles and with a relatively low content of magnetic particles in the feed.

The most common magnetic separators used in solid waste separation and recycling are head pulley magnets, magnetic drums, and suspended belt magnets.

Magnetic Head Pulley. This equipment uses the principle of magnetic particles pinning on the carrier, which carries the particles into the magnetic field (see Figure 17). After separation, ferrous metallic and nonmagnetic particles are carried onto conveyor belts and, subsequently, to bins or hoppers. This

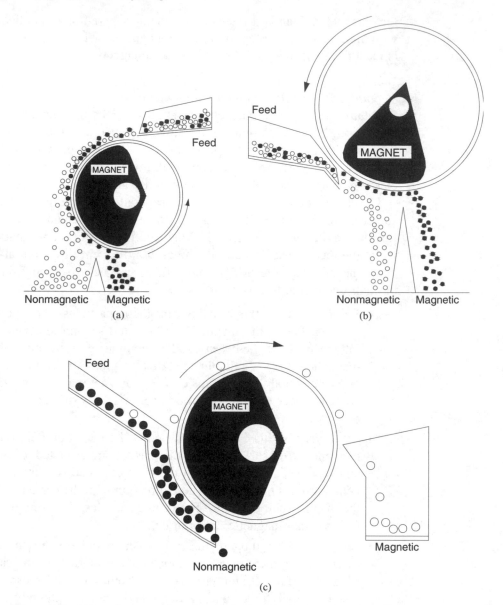

Figure 16 Magnetic separation of particles by (a) pinning, (b) attraction, and (c) elevation.

arrangement favors entrapment of materials—such as paper, plastics, textiles, and diapers—between magnetic particles and belt. Consequently, a cleaning stage is usually required before ferrous concentrate selling.

Magnetic Drum. A magnetic drum is usually composed of a permanent magnet, a compact piece, or set of axially oriented bars of alternative polarity, mounted

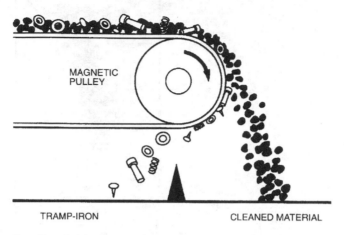

TRAMP-IRON CONTAMINATED
MATERIAL

MAGNETIC
PULLEY

TRAMP-IRON   CLEANED MATERIAL

Operating principle of a magnetic pulley

Figure 17 Magnetic head pulley. (With permission from Ref. 13.)

inside a rotating cylinder (see Figure 18). The waste material feed enters the magnetic field either on the top of the drum or below it.

In the first method, the magnetic fraction is pinned on the rotating drum and removed upon the ceasing of the field. This arrangement favors the entrapment of nonmagnetic materials (paper, plastic, etc.) and possibly a cleaning stage of the magnetic product is necessary.

In the second method, the magnetic fraction is attracted to the drum surface or elevated (Figure 16(b), (c)), depending on whether the direction of the feed is con- or counter-current to the direction of the drum rotation. The second method provides a cleaner product, especially the counter-current rotation case.

Suspended Magnet. In this case, the magnet assembly includes a permanent magnetic core surrounded by a belt conveyor. The suspended magnet configuration is either cross-belt or in-line, with respect to the waste material flow (Figure 19).

In the cross-belt configuration, the magnet is placed perpendicular to the waste material flow. This position requires a stronger magnet, and it is not recommended for excessive belt speeds or deep materials burdens. The size of the largest particles determines the distance between magnet and belt, because it must be two to three times the size of the largest particle to avoid interference of materials on the feed belt with those attached on the magnet, and subsequently, inefficient separation. In turn, the longer distance weakens the strength of the magnetic field, as the magnet is usually a permanent one and, consequently, waste materials with low magnetic susceptibility or small size may be lost to the nonmagnetic product.

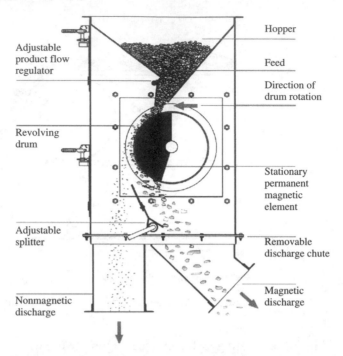

Figure 18   Magnetic drum separators top fed. (With permission from Ref. 14.)

Also, the waste material must be fed at a low thickness, so that the magnetic materials positioned at the lower part of the bed are recovered and the entrapment of laminated light materials (paper, plastic, etc.) is diminished, although, in general, entrapment is not a significant problem. In cross-belt configuration, special attention must be drawn to the problem of magnetization of the rollers, below the belt, and the rest of the metallic parts of the waste conveyor. As these components are made from ferrous metal, their magnetization exerts a counter-magnetic force to the magnetic waste particles and prevents them from collection to the magnetic product. This problem is overcome by constructing these parts from nonmagnetized materials.

In the in-line configuration, the magnet is placed parallel to the waste flow. When the magnet is in this position, it is essential that the conveyor head pulley be made of a nonmagnetic material. This arrangement succeeds a clean magnetic product, because lifting of nonmagnetic, laminated materials to the magnet is highly reduced or eliminated. It happens because the feed stream is very loose, actually suspended in the air, and, consequently, the effect of low momentum and gravity force, as well as high air resistance, prevents the aforementioned materials from lifting to the magnet. Also, the magnet assembly in the in-line configuration could be composed of a series of magnetic components with alternating polarity, leading to a much cleaner magnetic product (Figure 20).

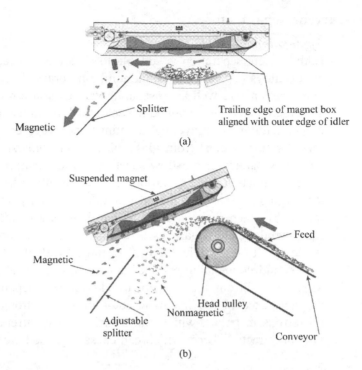

(a)

(b)

Figure 19 Suspended permanent magnet configurations: (a) cross-belt and (b) in-line configuration. (With permission from Ref. 15.)

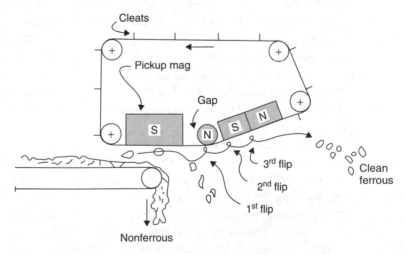

Figure 20 Typical magnets configuration with alternating polarity. (Reproduced with permission from Ref. 1.)

7.2   Eddy-current Separation

Applications

Eddy-current separation is suitable to separate and recover the nonferrous metals or alloys (e.g., copper, aluminum, zinc brass) from industrial and residential waste streams, as well as to separate various nonferrous metals from each other. This method has been applied to separate aluminum and copper from nonmetallic components in car recycling, aluminum beverage cans from either presorted or mixed refuse, metals from glass, plastic, and others. The position of the eddy-current separator in the flow-sheet of waste separation is after the removal of the light fraction (paper, cardboard, etc.) from the heavy (metals, glass, ceramics, aggregates, etc.) by air separation, and, usually, after the removal of ferrous metals by magnetic separation. When there is a requirement to separate nonferrous metals, there usually is a reciprocating need for ferrous metal separation. Although it is possible to recover three products (ferrous metals, nonferrous metals and nonmetallic residue) during eddy-current separation, it is preferable to separate ferrous metals prior to eddy-current separation. This is because, the magnets used in the magnetic rotor impart a strong, vibrating magnetic pull on ferrous materials, which can override eddy current induction and result in a violent attractive force, which can cause high belt wear.

Principles of Operation

The principle of eddy-current separation relies on Faraday's law of electromagnetic induction. When large, metal particles pass through a time-varying magnetic field, eddy currents are induced into them and they become, in effect, solenoids. According to Lenz's rule, eddy currents have such a direction that oppose to the increase of the magnetic flux, as the particles approach the magnetic field, and tend to counterbalance the cause. Consequently, metal particles that induce eddy currents into them are repelled by the magnetic field while the nonconducting particles are not. The time-varying magnetic field is generated by a fast-rotating electromagnet.

Eddy current separators consist of two drums, the head one of which surrounds an internal magnetic rotor, and a belt conveyor, around the two drums. The working principle of eddy-current separator is presented in Figure 21. The drum surrounding the magnetic rotor rotates at conventional belt conveyor speed, while the internal magnetic rotor of alternating polarity turns at much higher speeds than the external. The speed of the rotor may be as high as 3,000 to 4,000 RPM. Rotor is usually made of ferrite or rare earth. Through the induction of eddy currents and the resulting repelling forces, the alternating magnetic field selectively repels the nonferrous metals and separates them from other materials. The drum and the magnetic rotor may be either concentric or eccentric (Figure 22). The concentric design prevails, as the magnetic expulsion force over the whole area of the outer shell is maximal, in contrast to the eccentric magnetic rotors that have only one

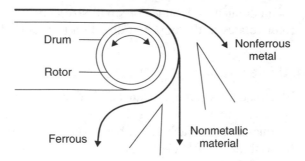

Figure 21   Working principle of eddy-current separator. (With permission from Ref. 16.)

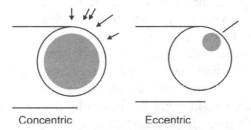

Figure 22   Concentric and eccentric design of magnetic rotor in eddy-current separators (With permission from Ref. 17.)

small point, where optimum separation can be achieved. This enables to the concentric magnetic rotors better separation.

Magnetic rotor can be powered in both directions: (1) forward, which is the same direction as the head drum; and (2) backward, which is opposite. For particle size larger than 10 mm, a forward rotation is preferable, while for smaller, a backward rotation is better. This is because smaller particles tend to rotate opposite to the direction of the head drum and the particles are caused to jump backward out of the separation zone.[16]

There are several designs of eddy-current separators, such as ramp separators and vertical drums.[18–20]

The performance of eddy-current separators depends on parameters such as belt speed, rotor speed, and splitter (or splitters) position. Separation is also affected by the characteristics of the feed such as particle size, shape and conductivity, size distribution, density, and moisture. In general, the capacity of eddy-current separators broadly ranges between 1 and 20 tons/hour per 1 m of belt width, depending on the material density and particle size. As electromechanical forces depend on particle size, for efficient separation solid waste stream must be classified into proper size fractions. In general, eddy-current separators process solid waste with particle size between 150 and 3 mm, and, sometimes as small such as 1 mm. For small size particles (less than 10 mm) the magnetic rotor rotates in the opposite direction to the head drum for efficient separation.

7.3 Electrostatic Separation

Electrostatic separation can be used to separate waste particles not contacting electricity (e.g., paper, plastic, glass, etc.) from conductors (e.g., metals) or non-conductors from each other, based on differences in their electrical conductivity. In addition to particles conductivity, electrostatic separation is also influenced by particle density, size, moisture, and purity of particles surface. Consequently, separation of waste particles is achieved under the action of electrical as well as gravity or centrifugal forces.

The commercially used electrostatic separators are those that make particles charged either by conductive induction or by charge bombardment (corona charging). During conductive induction, both good electrical conductors and dielectric particles make contact with a charged roll (Figure 23) or inclined plate. After contact with the charged surface, conductors are repelled from the roll because they are similarly charged with the surface, while the dielectric components of waste stream is not influenced by the electrostatic field.

During corona charging (or high-tension separation), the separation of the waste stream is accomplished by passing solid waste through a corona discharge (Figure 24). The separating machine is composed of a grounded rotor, on which solid waste is fed. Corona discharge is caused by a fine wire electrode or a set of needles, in parallel to the rotor, by raising their electrical potential to a determined value, which depends on the material of the wire or needles. The type

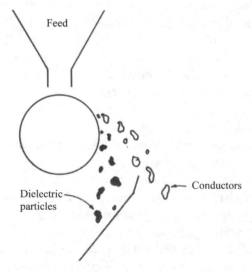

Figure 23 Electrostatic separation by conductive induction.

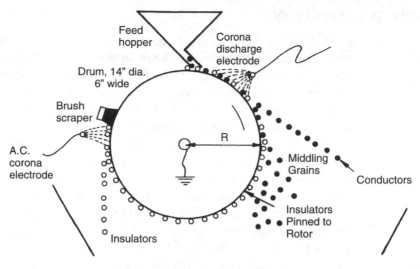

Figure 24 Electrostatic separation by corona discharge. (With permission from Ref. 21.)

of corona (positive or negative ions) depends on the polarity of the electrode. All waste particles are charged by ion bombardment from corona discharge. The conducting particles lose their charge because of their contact with the grounded rotor, while the poor- or nonconducting particles lose their charge very slowly and, consequently, are held on the rotor surface by the image force of their surface charge.

Although electrostatic separation (static or high intensity) has not been applied widely in solid waste treatment at the industrial level, it is expected to be of wider use in the future in certain sectors of applications. This separation has been successfully applied to separate plastics from paper, plastics from each other, shredded copper wires from their plastic insulation, glass from plastic, nonferrous metals from plastic, or glass, for example.

8 BALLISTIC SEPARATION

In addition to the classic waste separation methods, which are based on the difference of the corresponding physical or physicochemical property between the various constituents of the waste stream, there have been also tested methods in the past, which are based on a combination of physical properties such as density, elasticity, and friction. These methods separate organic from inorganic materials, as well as heavy and resilient from light and inelastic. *Ballistic*, *oscillating*, and *bounce and slide* separators are some of the devices based on these principles.[22] Because the necessary equipment is relatively simple, under certain materials characteristics, these separators could be used in relatively easy and low-cost waste recovery.

9 FROTH FLOTATION

Froth flotation holds a great share among the applied methods nowadays for mineral separation. It is carried out in water. Only hydrophobic minerals float, after attachment on rising air bubbles in flotation cells. Its efficiency for small particle size and its selectivity are included among the various reasons for the widespread use of froth flotation. In waste-separation plants, it is applied after the separation of coarse particles with the use of dry processing methods. Froth flotation has been successfully applied mainly for research purposes, such as to separate fine glass particles from other inorganic components as well as plastic from each other.[23–25] Its application at industrial level is rather limited because there are serious reasons that prevent its wide use in solid-waste separation plants. Among these, the following are included: waste material stream is separated, usually at a large particle size; wet processing of solid waste is undesirable in most cases; froth flotation has a relatively high operating cost; materials to be recovered are of low commercial value to compensate cost and create profit. Froth flotation has been used to separate finely sized glass particles from inorganic contaminants (ceramics, stones, and metal) in a resource recovery plant in New Orleans.[1] In this case, amine was used as a surfactant to render the fine glass particles (−0.8 mm) hydrophobic.

10 FEEDING AND HANDLING SYSTEMS

A continuous flow of the solid waste stream, from its initial feeding into the processing plant until its exit in the form of final products, is necessary so that separation is efficient and economically viable. Among other factors, the uniform and controlled feeding of process units and equipment as well as the continuous handling of the products secure the efficient operation of a solid waste separation plant.

Properly selected feeders for each case achieve continuous and smooth feeding of equipment and units, while products are handled by properly selected conveying systems. Where it is necessary, a silo is installed before separation machines or units to secure uniform feed rate and material flow.

10.1 Feeders

The most widely applied feeding devices are apron, vibrating, and screw feeder.

Apron Feeders
Basically, they are heavy-duty feeders. Apron feeders are composed of a series of steel plates, which overlap to form a continuous belt (Figure 25). The plates are linked together or mounted on a conveyor chain and driven by sprockets.

Apron feeders are mostly used to introduce solid waste in the processing plant. They can be installed either horizontally or with an upward inclination,

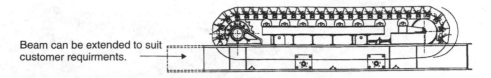

Beam can be extended to suit customer requirments.

Figure 25   Apron feeder. (With permission from Ref. 26.)

depending on the process flow sheet and the space design. Waste material is fed onto the apron feeder either from a silo, in case that waste characteristics and space design favor it, or by a mechanical loader (shovel or grab). In the last case, a horizontal loading extension of the inclined section must be added, long and wide enough to meet the loading requirements. The width of apron feeders mostly ranges between 0.8 and 2.0 m while the length is suitable to each case, being in most cases no more than 4.5 m. When the slope of the apron feeder exceeds the maximum angle of slip of solid waste material, which in most cases is of the order 25 to 27 degrees, lifters must be used. Lifters are vertical plates fitted on the top face of the steel flat plates, and their role is to prevent the down-roll of waste because of the steep slope. For most applications, the slope does not exceed 45 degrees and, under these circumstances, lifters dimensions of the order $25 \times 20 \times 10$ cm (width $\times$ height $\times$ thickness) are considered satisfactory.

The apron feeder capacity is estimated by this relationship:

$$W = S \times V \times d \times \kappa \qquad (2)$$

where W = solid waste weight (t/h)
     S = perpendicular cross-section of solids on apron feeder(m^2)
     V = apron velocity (m/h)
     d = material bulk (apparent) density (t/m^3)
     κ = material loose coefficient

Apron feeders provide several advantages over other systems. They are robust, powerful, and resistant constructions to corrosion and damages caused by impacts. They are the most suitable selection for heterogeneous, humid, heavy, irregular-shaped, and coarse materials, such as mixed waste. They are also suitable for coarse glass, metals and scrap, organics and putrescibles, textiles, wood, and timber. Their disadvantages are focused on the relatively higher cost of purchase and maintenance, in comparison to the other types of feeders, but it is the most reliable selection in heavy-duty works.

Vibrating Feeders

In general, vibrating feeders are simple and cheap constructions, usually applied to feed fluent materials. There are also heavy duty vibrating feeders, but they

Figure 26   Vibrating feeder. (With permission from Ref. 27.)

are less efficient than the apron feeders with respect to maintain a stable feeding rate, especially in case that the waste is moist and contains organics/putrescibles at a relatively high percentage.

A vibrating feeder is essentially a formed trough supported by flat springs (vibrator bars) from a stationary base (Figure 26). Vibration is imparted to the trough by an eccentric power unit.

Vibrating feeders provide long life and low maintenance cost, as the eccentric drive is the only rotating part.

Screw Feeders

A screw feeder consists of a steel plate fixed to a central shaft in such a way to form a helix (Figure 27). The shaft runs inside a U-shaped trough, with the direction of rotation determining the load travel.

Although the design of the screw feeder enables it to be also used as a screw conveyor for longer travels, up to 35 m, it is mostly recommended to infeed equipment with material than to convey materials, as there are much more efficient conveyors and there is always the possibility of blockage. Maximum travel length of screw feeders is 3 to 3.5 m in most applications.

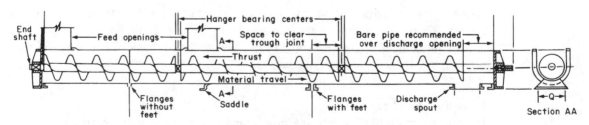

Figure 27   Screw feeder. (With permission from Ref. 28.)

They are suitable to infeed plastics, shredded paper, organics/putrescibles, and glass, but they are not suitable for scrap and mixed-waste containing textiles and plastic film, as the two last components are expected to cause immediate blockages by binding around screws.

10.2 Handling Systems

Although waste is considered as a material, in fact it is a mixture of materials, each one having its own handling characteristics, sometimes entirely different than those of the other components. There are several handling systems used to convey solid waste. The most widely used are steel plate, belt and vibrating conveyors, bucket elevators, and pneumatic systems.

There are also more types of conveyors, such as chain and moving floor conveyors, with relatively limited application in solid waste processing plants.

Steel Plate Conveyors

Steel plate conveyors are similar to apron feeders, as described in details in Section 10.1. The main difference lies in the length of the conveyor, which in any case is longer than that of the feeder but usually no longer than 35 m, as it is a heavy construction and, consequently, the demands in installed power are increased with length. They are mostly used to convey mixed-waste material to the disposal site.

Belt Conveyors

Belt conveyors are the most widely used equipment for materials handling in solid waste processing plants, applied both for materials conveying as well as for hand sorting.

A belt conveyor essentially consists of an endless belt supported and stretched by pulley assemblies at each end and conveyor idlers along the profile on both carrying and return runs. One of the pulleys is the head and drive pulley, while the other is the tail pulley. For horizontal or slightly inclined installations, the idlers are placed either horizontally or, most commonly, V- or troughlike, to increase the capacity of the belt conveyor and to reduce material spillage. Also, the idlers at the carrying run are placed in shorter distances than the return idlers, in order to support the load-carrying belt. The belt is made of rubber or, in most cases, by synthetic material. In heavy-duty works, the belt is reinforced with two to four cotton or nylon layers.

Belt conveyors are usually installed horizontally as well as with an upward or downward inclination, although there are conveyor systems capable of running both horizontally and vertically or at any angle between (Figure 28), depending on the design of the waste processing plant. The throughput of the material to be conveyed depends on the belt width, speed, and material cross-section on the belt. Power requirements depend on belt length, lift height, belt speed, throughput per hour, and the weight of the moving parts of the conveyor when it is empty.

Figure 28 Belt conveyor. (With permission from Ref. 29.)

Belt conveyors have been widely applied to convey mixed municipal solid waste, separation products, as well as for waste disposal. They can be used in almost every case unless for finely shredded, dry, and light material, as it is amenable to spill out the belt. Also, in cases where the load is moist, a cleaning mechanism, usually a brush, must be placed below the tail pulley to clean the belt before the return run.

Vibrating Conveyors

Vibrating conveyors are similar to the corresponding feeders, which are described in detail in Section 10.1. Their difference with those used as feeders lies in the size of the equipment. Feeders are usually designed with a trough length up to 3.5 m and a corresponding width, while conveyors are up to 35 m. They are not recommended to convey wet material.

Bucket Elevators

Figure 29 illustrates bucket conveyors. They are designed to elevate materials mainly vertically or for some designs with an inclination. Bucket elevators are mostly used in cases for which other conveying systems are unsuitable, either because of the ground morphology or because of the plant and process design.

The bucket elevator consists of a series of buckets mounted on an endless belt or chain between pulleys or sprockets. The buckets are loaded with material from the lowest level, which is discharged at the highest. Material discharge may be either centrifugal or continuous (Figure 29). In elevators with centrifugal discharge, the buckets load material from the bottom boot, as the buckets pass under the foot pulley, and discharge it by centrifugal force, as the buckets pass over the head pulley. In elevators with continuous discharge, the buckets are fitted to the chains or belts in such a way that the sidewall of one bucket forms a chute for the immediately following it. This way of flow is due to the greater number of buckets fitted on the chain than in the case of centrifugal discharge.

The capacity of bucket elevators is calculated from this relationship:

$$Q = S \times n \times V \times d \tag{3}$$

where Q = capacity(t/h)

S = belt or chain speed(m/h)

n = number of buckets per unit length(m^{-1})

V = volumetric capacity of one bucket(m^3)

d = material bulk (apparent) density(t/m^3)

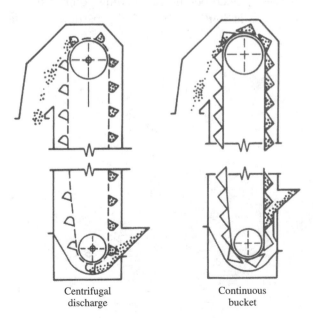

Centrifugal
discharge

Continuous
bucket

Figure 29 Configuration of bucket elevators. (With permission from Ref. 28.)

Bucket elevators are suitable to convey relatively dry, dense, granular, and free-flowing materials. Contrarily, wet or sticky materials make problems, as the material may hang up in the buckets. Also, fluffy, light, or fibrous materials make problems to machines with centrifugal discharge. Bucket elevators are suitable for dense plastics, timber/wood products, and glass. They are unsuitable for as received paper, and are rarely used for mix waste containing textiles, and wire, as these materials may create blockage. Finally, they could be used for organics/putrescibles, light plastics, and metals under the condition that spillage in the boot should be avoided or minimized.

Pneumatic Systems

Pneumatic conveying is a method of transporting materials using air as conveying medium. It is applied for dry, granulated, and low-density materials. In the case of solid waste, a properly designed and installed pneumatic conveying system is a clean, safe, and practical method of conveying.

Pneumatic conveying systems consist of a fan, feeder, piping and cyclone, used as discharge device (Figure 30). The pneumatic conveying of material is accomplished through air pressure differential, between the inlet and outlet, which can be established either by suction (negative pressure) or blowing (positive pressure). In suction systems, air is drawn through the pipe or duct from the discharge end by a fan or rotary blower in such a way that a negative pressure (below atmospheric) is created at the discharge end. In suction systems, the cyclone is installed between the fan and the end of the pipe. In blowing systems, a blower, placed behind the feeding point of the particles, blows air. So, a positive pressure is built up behind the particles, while atmospheric pressure remains

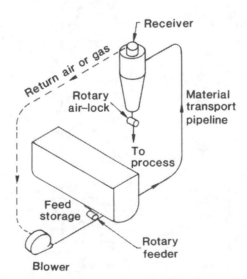

Figure 30 Configuration of pneumatic conveying system. (With permission from Ref. 28.)

before it. In this case, the cyclone is installed at the end of the pipe. For effective conveying, the velocity of the air ranges mostly between 1,440 and 1,800 m/min.

The advantages of pneumatic conveying systems are avoidance of contamination, simplicity of installation, flexibility and adaptability, ease of maintenance, and amenability to automation. The disadvantages are high capital cost, high power requirements per weight-unit of material to be transported, problems with wet or dense material, and blocking at the discharge point.

The pneumatic conveyors are suitable for refuse-derived fuel, dense and light plastics, granulated paper, milled glass, and chip wood/timber. They are not suitable for mixed municipal solid waste due to the variable characteristics of the components, organics/putrescibles due to odors, and material deposition in the rotary valves, and metal objects, except shredded aluminum.

11   PRODUCT AGGLOMERATION (CUBING AND PELLETIZING)

Fluffy products such as paper and plastic have to be further treated in order to reduce their volume and, consequently, increase their density for easier transportation, storage, and processing. This is accomplished by *agglomeration*, which occurs as briquetting (or cubing) and pelletizing. *Briquetting* and *pelletizing* have many similarities, but they are not identical. The principal difference lies in the size and shape of the agglomerates. Briquettes (or cubes) are usually of larger size and of square or rectangular cross-section, while pellets are smaller and with circular cross-section. The material from agglomeration process is used as densified refuse-derived fuel for combustion, gasification, or pyrolysis treatment.

Cubers and pelletizers consist of a cylindrical die from hard steel, perforated with holes. Inside the die there are two rollers mounted on the periphery of the die. There are two designs of the machine as regards the rotating part. In one design, the die rotates while the rollers do not, and vice versa in the second. In another design of the machine, rollers have been replaced by an eccentrically rotating cylinder.

The design and working principle of cubers and pelletizers is clear and well understood. Shredded refuse-derived fuel infeeds the machine through a short screw conveyor mounted in the casing of the main shaft. As the feed enters the agglomeration machine, it is entrapped between the die and the rollers, and, consequently, is forced to pass through the die holes. As the rollers compress the material to pass, high pressure and friction are exerted against the holes wall. These effects result in temperature increase and heat creation, which, in turn, promote fusion of the materials (e.g., plastics). Besides pressure and friction, another important parameter is the moisture content of the material to be agglomerated. Moisture content must be maintained at the level of 15 to 16 percent, as insufficient moisture results in weak bonds between the particles and, finally, in the disintegration of agglomerates after their extrusion. By contrast, high moisture content is also undesirable, as high temperature and pressure evaporate the moisture and excessive stream is generated in the agglomerate. This results in its bursting and disintegration, but for a different reason than previously. This is the reason that agglomeration machines are equipped with an automatic moisture control system as well as with water sprayers.

In another design, double screw presses are used to agglomerate shredded material (Figure 31). Also, cast iron borings, steel turnings as well as nonferrous metal chips are converted to high-density agglomerates by hydraulic pressing.

Among the advantages of product agglomeration are improved handling, reduced storage space requirements, lower transport cost, better combustion behavior, and increased heating value. The disadvantage is the additional cost, but it is counterbalanced by the benefits.

Figure 31   Double screw presses for products agglomeration. (With permission from Ref. 30.)

12 COMPACTION (BALING)

Compaction, also called baling, is a relatively simple process that does not cause any change in the physical or chemical properties of the product. It is essentially a densification process to reduce the volume of the separated products or the waste material for disposal. The advantages of baling are easy handling of the product (bale), lower transport cost, volume reduction, and lower space requirements in disposal sites. Baled material does not sustain incineration. Also, it does not favor methane generation when it is landfilled.

The equipment available for densification of products, and solid waste as well, includes compactors and balers.

12.1 Compactors

After separation and recovery of the recyclable or combustible materials, the remaining waste must be landfilled. As this material is loose, the transport cost is lower, when this material is compressed in a stationary compactor. Essentially, compactors are presses, which are divided into low- and high-pressure, if the compaction pressure is lower or higher than 6.8 atm. The compression can be carried out in the transport vehicle, containers, or specific chambers. After compression, solid waste may be transported for landfilling as it is or after baling.

12.2 Balers

Most recycling facilities employ a baler. Balers are used to densify paper and corrugated cardboard but they can also be used for aluminum, plastics (containers and bottles), ferrous and nonferrous metals, and separation residues for disposal. Baling enhances the marketability of commodities and significantly reduces transportation costs and storage space. Its disadvantage is the relatively high capital cost.

Although a baler is relatively simple equipment, the purchase of a baler is not an easy task, as many parameters have to be considered to select the type of the baler needed. These parameters are the baler's capacity, efficiency (power consumption, wire, or strap usage), design, automation, size of bales produced, materials to be processed, safety characteristics and, finally, cost. Depending on the case, more than one type of baler may be suitable.

Balers are classified into vertical and horizontal (Figure 32), according to the direction of the compressing ram. Horizontal balers are further divided into single- or double-ram, as well as into open- or closed-end. In any case, a baler includes the following basic parts: a feed bin, the compression system that includes one or more hydraulic or mechanically driven rams, the compression area, and the discharge opening. Before been discharged, bales are fastened or tied with wire or plastic straps.

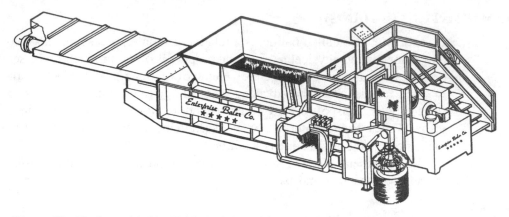

Figure 32   Horizontal baler. (From Ref. 31, with permission from the Enterprise Baler Co.)

Vertical balers have a relatively low capacity and are suitable for small-scale operations. The length of the bale ranges between 0.5 m and 2 m. In comparison to vertical, horizontal balers have lower height requirements and higher capacity, are used for large-scale operations and produce larger size bales (e.g., with length ranging between 1.0 m and 2.0 m). In most cases, the capacity of horizontal balers ranges between 10 and 40 tons/hour, depending on the material to be processed and its size.

13 PRODUCT STORAGE

After their separation and processing, materials must be stored until their purchase by a buyer or their use as refuse-derived fuel. Processed materials must be protected from adverse atmospheric conditions, so that their properties remain unaltered and marketable.

If the processed material is not affected by atmospheric conditions (e.g., plastic, aluminum), it can be stored outdoors, as baled or loose material, in containment bin walls made of precast concrete (Figure 33). Sensitive materials (e.g., paper, composting materials) must be stored indoors, in precast buildings (Figure 33) or containers. In general, containers are preferred over enclosed stores as a cheaper solution, because reloading is avoided, and they are amenable to easy transportation and shipment. Containers are filled with baled or compressed material during processing, stored at the processing plant, shipped, and transported to the buyer's facilities.

Refuse-derived fuels need special care for storage, as they maybe contain putrescibles. They are usually stored in roofed buildings and the load that comes in first also comes out first for further treatment.

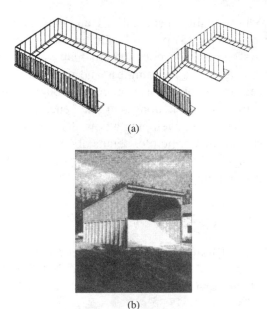

(a)

(b)

Figure 33   Products storage in containment bin walls from precast concrete: (a) outdoors, (b) indoors. (From Ref. 32, with permission from Solenberger Silos Corp.)

14   SUMMARY

This chapter describes the unit operations that could be applied in a solid-waste processing plant aiming to recover recyclable materials, refuse-derived fuel, as well as products for further chemical or biological processing. The tailings from solid-waste processing are rejected to the landfill.

The first and most important task in materials recovery facilities design is the accurate knowledge of solid waste quantity to be processed, as well as its composition. Long-term prediction concerning the future quantity and composition is a valuable tool, as it is very difficult for any waste-processing plant to be significantly modified. The knowledge of the types and amounts of materials to be processed is very important, as these parameters determine the layout and design of the processing units.

Factors that must be considered in the layout and design of each unit include: (1) estimation of solid-waste feed rate in the plant, as well as material infeed in each unit; (2) estimation of the materials types and quantities to be recovered; (3) development of an integrated flow sheet, completely balanced as regards to the weight of each material; (4) selection of proper equipment for each unit, so that a continuous flow of waste infeed and products can be established; (5) proper storage facilities of products and residues removal for landfilling; (6) control of products quality/specifications, and units performance; (7) monitoring and control of emissions; (8) public health and safety principles; (9) economics.

As solid waste composition may vary widely even in the same community or country, it is meaningless and tiresome to describe in details processing flow sheets from solid waste processing plants. The unit operations described in the previous chapters, along with the general guidelines of this chapter, aim to establish a basis for developing a flow sheet in any of solid waste-processing treatment. Based on this principle, details about equipment were considered unnecessary at this phase. The target was to provide a flexible tool to those dealing with solid waste treatment.

REFERENCES

1. E. A. Glysson, "Solid Waste", in *Standard Handbook of Environmental Engineering*, R. A. Corbitt (ed.), McGraw-Hill, New York, 1990, pp. 8.4–8.36, 8.85, 111–114.

2. Lindemann "KHA Hammer Mills", Metso Minerals, Brochure No. 1240-01-02-REC, 2000, p. 5.

3. Lindemann "ZB Turning Crushers, Model ZB 116", Metso Minerals, Brochure No. 1232-01-02-REC, 2000, p. 5.

4. G. Tchobanoglous, H. Theisen, and S. Vigil, "Integrated Solid Waste Management, *Engineering Principles and Management Issues*", McGraw Hill, New York, 1993, 257, 559.

5. Lindemann "RO Rotor Shears", Metso Minerals, Brochure No. 1243-01-02-CS, 2000, pp. 6, 8–9.

6. Vermeer, "Tub Grinder Model TG525A", Brochure TG525A, website: http://www.vermeer.com/Environmental Equipment/Tub grinders/TG525.

7. Nordberg "TS Series Screens", Metso Minerals, Brochure No. 1312-04-05-CSR, 2005, p. 2.

8. Lindemann "Trommel Screens", Metso Minerals, Brochure No. 1249-01-02-CS, 2000, p. 2.

9. Courtesy of Bulk Handling Systems (BHS) Inc., Debris Roll Screen, (DRS), cited in *Resource Recyclings*, p. 15 (Recovered Paper Supplement) (April 1998).

10. Courtesy of Hustler Conveyor Company, website: hustler-conveyor.com.

11. Mogensen GmbH&Co., KG, Mikrosort Status Report Mikrosort, February 2006, pp. 1–6.

12. W. D. Robinson (ed.), *The Solid Waste Handbook: A Practical Guide*, John Wiley, New York, Fig. 12.86, cited in: D. B. Spencer, *"Recycling"*, in *Handbook of Solid Waste Management*, F. Kreith (ed.), McGraw-Hill, New York, 1986 p. 9.53.

13. Eriez, Magnetics Europe Ltd., "Permanent Magnetic Pulleys", Brochure No SB 240 UK/4, p. 2.

14. Eriez, Magnetics Europe Ltd., "Permanent Magnetic Drum Separators, Models FA, FR, RA, RR, RAS, RASP, RRS", Brochure No SB 340 UK/5, p. 2.

15. Eriez, Magnetics Europe Ltd., "Suspended Permanent Magnet, Models CP and OP", Brochure No SB 300 UK/3, p. 2.

16. Eriez, Magnetics Europe Ltd., Eddy Current Separators, "Principle and Practice", Brochure No SB 501 UK, p. 1.

17. Eriez, Magnetics Europe Ltd., "Eddy Current Separators," Brochure No SB 500 UK, p. 3.

18. E. Schlömann, "Separation of Nonmagnetic Metals from Solid Waste by Permanent Magnets. I. Theory", *J. of Applied Physics*, **46** (11), 5012–5021 (1975).

19. E. Schlömann, "Separation of Nonmagnetic Metals from Solid Waste by Permanent Magnets. II. Experiments on circular disks", *J. of Applied Physics*, **46** (11), 5022–5029 (1975).

20. M. Lungu and Z. Schlett, "Vertical Drum Eddy-current Separator with Permanent Magnets", *Int. J. Mineral Processing*, **63**, 207–216 (2001).

21. J. E. Lawer and D. M. Hopstock, "Electrostatic and Magnetic Separation", in *SME Mineral Processing Handbook*, N. L Weiss (ed.), Society of Mining Engineers of the American Institute of Mining, Metallurgical, and Petroleum Engineers, Inc., New York, 1985, p. 6, 8.

22. R. J. Wilson, T. J. Veasey and D. M. Squires, "The Application of Mineral Processing Techniques for the Recovery of Metal from Post-consumer Wastes", *Minerals Engineering J.*, **7** (8), 975–984 (1994).

23. J. H. Heginbotham, "Recovery of Glass from Urban Refuse by Froth Flotation", *US Bureau of Mines*, IC 8826, 56–60 (1980).

24. J. Drelich, T. Payne, J. H. Kim, J. D. Miller, R. Kabler and S. Christiansen, "Selective Froth of PVC from PVC/PET Mixtures for the Plastics Recovering Industry", *Polymer Eng. Science* **38**, 1378–1386 (1998).

25. G. Dodbiba, N. Haruki, A. Shibayama, T. Miyazaki, and T. Fujita, "Combination of Sink-float Separation and Flotation Technique for Purification of Shredded PET-bottle from PE or PP Flakes", *Int. J. Mineral Processing* **65**, 11–29 (2002).

26. Metso Minerals, "World-Class Apron Feeders", Brochure No 1802–05-04-MPR, 2004, p. 7.

27. Eriez, Magnetics Europe Ltd., "Vibratory Feeders, VF Range", Brochure No VB 305 UK.

28. D. W. Scott and R. M. Hays, "Storage and Transport", in *SME Mineral Processing Handbook*, N. L Weiss (ed.), Society of Mining Engineers of the American Institute of Mining, Metallurgical, and Petroleum Engineers, Inc., New York, pp. 10.68, 10.79, 10.83 (1985).

29. Metso Minerals (Moers) GmbH, "Flexowell", Brochure No 3000-e-JWP-12/04, 2004, p.1.

30. Metso Minerals, Lindemann DSP, "Double Screw Presses", Brochure No 1238-01-02-REC, 2000, p. 1.

31. Courtesy of The Enterprise Baler Co., E2RRB Series Balers, Cited in *Resource Recyclings*, p. 32 (January 1998).

32. Courtesy of Sollenberger Silos Corp., High "Precast Concrete" Containment Bill Walls, cited in *Resource Recyclings*, p. 48 (December 1997).

CHAPTER 12

PROCESSING POSTCONSUMER RECYCLED PLASTICS

Gajanan Bhat
University of Tennessee, Knoxville, Tennessee

1	INTRODUCTION	357
2	BACKGROUND	358
3	APPROACHES TO REDUCE WASTE	362
	3.1 Reduce	363
	3.2 Reuse	363
	3.3 Recycling	364
4	ENERGY SAVINGS AND RECOVERY	366
	4.1 Pellet Fuels	367
	4.2 Challenges to Recycling	367
5	POSTCONSUMER RECYCLING OF PLASTICS	368
	5.1 Collection, Cleaning, and Separation	368
	5.2 Material Recovery Facilities	369
	5.3 Processing Separated Plastics	370
	5.4 Mixed Plastics	375
	5.5 Chemical Recycling	377
	5.6 Pyrolysis	377
	5.7 Degradable Plastics	378
6	PRODUCTS FROM POSTCONSUMER RECYCLED POLYMERS	378
7	LIFE-CYCLE ANALYSIS	380
8	DESIGN FOR SUSTAINABILITY	381
9	ECONOMICAL ASPECTS AND FUTURE PROSPECTS	382

1 INTRODUCTION

The continuing growth of the use of plastics has created problems in municipal solid-waste (MSW) disposal. Plastics continue to replace traditional materials in many applications and are being used in several new products. Many advantages of plastics in their performance as well as in the ease of their fabrication have contributed to their growth. Since large shares of the products are one-time or short-time usage products, they end up in the waste stream. In the beginning, most of the plastic waste was discarded in landfills. Because of their low density, they occupy high volumes and are highly visible in the waste stream, attracting more attention from the critics. Thus, it is important to find a suitable way of dealing with plastic waste. Over the years, the share of plastics recovered and recycled has increased. Alternatively, there are also approaches to enhance the degradability of some plastics. A combination of these efforts should help sustainability. The problems, issues, and opportunities in the post-consumer recycling of plastics are discussed in this chapter.

2   BACKGROUND

The consumption of plastics continues to increase all around the world, especially in the developed and developing countries. According to the U.S. Plastics Council, the total sales and use of plastics in North America was close to 114 billion pounds in 2004, an increase of 7 percent over that of 2003. The details of different plastic resins produced/used are shown in Table 1. The data in the table include thermoplastics, thermosets, and engineering resins. These plastics are used for a variety of applications and have different ranges of use life. Thermoplastics have a major share of the total, and are also likely to be used in a higher percent of short life products. The breakdown of these different markets is shown in Figure 1 for two major resins polyethylene terephthalate (PET) and high density polyethylene (HDPE). The total of all these was 86 billion pounds for 2004. The change in market size for these different products from 2000 to 2004 is shown in Table 2. It is obvious from the chart that all major applications have increased in use. The total consumption of plastics increased at a compounded rate of 2.9 percent, but packaging products showed maximum increase of 5.5 percent a year. Of all the products, packaging materials have the shortest life and more than likely end up in the waste stream soon. The average usage period for packaging materials is less than a year, whereas for other products, it is 5 to 10 years, and for building materials about 25 years (Table 3).

The plastic content of the MSW is 10 to 11 percent, amounting to about 40 billion pounds per year (Table 4), which is less than 50 percent of the fiber/resin produced. However, this is a significant amount of plastics going into the landfill. This problem of increasing consumption of plastics and its share in MSW for Europe, Japan, and Australia is comparable to that of the United States. There are limited data on developing countries, but the growth rate for plastic consumption in China is very high, in the order of 20 percent per year.

Advantages of plastic waste reduction and recycling are multifold. In addition to savings in valuable landfill space, there is a significant reduction in the energy devoted to the manufacture of plastic parts and products, since plastic products consume roughly 3.5 percent of total U.S. energy consumption. For a long time, recovery and recycling were not attractive because of a lack of suitable technology and cost considerations. In many instances, the price of recycled plastics is higher than that of virgin polymers. Also, the lack of a consistent supply of recycled plastics has been a constraint on the growth of plastics recycling. The current rate of recycling of the resins that go into solid waste is much below the potential. Most of the recycling is from PET and HDPE bottles (30–50%), but the recycling of other resins is negligible, with a total recovery of less than 6 percent of the plastics in the MSW.

Table 1   Production and Sales of Resins by Type

APC Year-End Statistics For 2004

PRODUCTION, SALES & CAPTIVE USE
2004 vs. 2003
(millions of pounds, dry weight basis)(1)

Resin	Production			Total Sales & Captive Use		
	2004	2003	% Chg 04/03	2004	2003	% Chg 04/03
Epoxy (2)	645	578	11.6	658	587	12.1
Urea and Melamine (3)	3,316	3,174	4.5	3,294	3,152	4.5
Phenolic (3)	4,633	4,442	4.3	4,200	4,015	4.6
Total Thermosets	**8,594**	**8,194**	**4.9**	**8,152**	**7,754**	**5.1**
LDPE (2)(3)	8,295	7,804	6.3	8,208	7,814	5.0
LLDPE (2)(3)	12,434	11,137	11.6	12,182	11,101	9.7
HDPE (2)(3)	17,548	15,709	11.7	17,519	15,906	10.1
PP (2)(3)	18,552	17,665	5.0	18,523	17,497	5.9
ABS (2)(4)	1,364	1,262	8.1	1,370	1,324	3.5
SAN (2)(4)	134	121	10.7	136	121	12.4
Other Styrenics (2)(4)	1,734	1,596	8.6	1,742	1,636	6.5
PS (2)(3)	6,750	6,393	5.6	6,765	6,478	4.4
Nylon (2)(4)	1,340	1,279	4.8	1,357	1,306	3.9
PVC (3)	15,985	14,702	8.7	15,883	14,938	6.3
Thermoplastic Polyester (2)(4)	8,263	7,587	8.9	8,632	7,950	8.6
Total Thermoplastics	**92,399**	**85,255**	**8.4**	**92,317**	**86,071**	**7.3**
Subtotal	**100,993**	**93,449**	**8.1**	**100,469**	**93,825**	**7.1**
Engineering Resins (3)(5)	2,708	2,068	30.9	2,013	1,767	13,9
All Other (6)	11,506	10,913	5.4	11,458	10,874	5.4
Total Engineering & Other	**14,214**	**12,981**	**9.5**	**13,471**	**12,641**	**6.6**
GRAND TOTAL	**115,207**	**106,430**	**8.2**	**113,940**	**106,466**	**7.0**

(1) Expect Phenolic resins, which are reported on a gross weight basis.
(2) Sales & Captive Use data include imports.
(3) Canadian production and Sales data included.
(4) Canadian and Mexican production and sales data included.
(5) Includes: acetal, granular fluoropolymers, polyamide-imide, polycarbonate, thermoplastic polyester, polyimide,
    modified polyphenylene oxide, polyphenylene sulfide, polysulfone, polyetherimide and liquid crystal polymers.
(6) Includes: polyurethanes (TDI, MDI and polyols), unsaturated (thermoset) polyester, and other resins.
Sources: APC Plastics Industry Producers' Statistics Group, as compiled by VERIS Consulting, LLC; APC

Note: From Ref. 1.

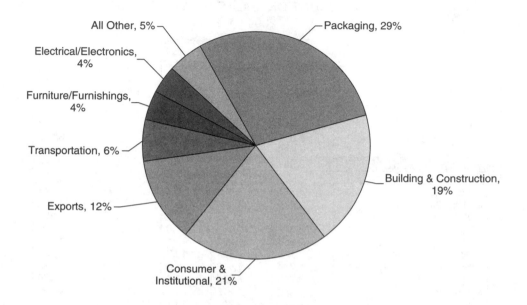

2005 Percentage Distribution of Resin Sales & Captive Use by Major Market

- All Other, 5%
- Electrical/Electronics, 4%
- Furniture/Furnishings, 4%
- Transportation, 6%
- Exports, 12%
- Consumer & Institutional, 21%
- Packaging, 29%
- Building & Construction, 19%

Resins Comprising Market Distribution:

- Low Density Polyethylene (LDPE)
- Linear Low Density Polyethylene (LLDPE)
- High Density Polyethylene (HDPE)
- Polypropylene (PP)
- Acrylonitrile-Butadiene-Styrene (ABS)
- Styrene-Acrylonitrile (SAN)
- Other Styrenics
- Polystyrene

- Styrene Butadiene Latexes (SBL)
- Thermoplastic Polyester
- Nylon
- Polyvinyl Chloride (PVC)
- Engineering Resins
- Polyurethanes
- Epoxy

SOURCE: Major market volumes are derived from plastic resin sales and captive use data as compiled by Veris Consulting, LLC, and reported by APC's Plastic Industry Producers' Statistics Group, includes APC estimates.

Figure 1 Resin sales distribution by major markets. (From Ref. 1.)

Most of the household packaging is contaminated by mixtures of paper labels, metal caps, food residues, inks, and adhesives, which makes mechanical recycling very difficult. Agricultural packaging is contaminated with soil, organic matter, and transition metal ions, making it impossible to recycle. Although many of these contaminants can be washed, an additional problem for disposing of them is created and the cost increases. Uncontaminated and single-family plastics can

Table 2 Growth Trend of Plastics in Important Markets

Total Sales & Captive Use of Selected Thermoplastic Resins* by Major Market, 2000–2004 (millions of pounds dry weight basis)

Major Market	2000	2001	2002	2003	2004	Compound Growth Rate 2000–2004
Transportation	4,389	4,207	4,736	4,732	4,899	2.8%
Packaging	20,941	22,847	24,170	24,087	25,952	5.5%
Building & Construction	14,439	13,988	14,729	14,485	15,876	2.1%
Electrical/ Electronic	2,787	2,801	3,037	2,862	3,096	2.7%
Furniture & Furnishings	3,572	3,226	3,507	3,361	3,458	−0.8%
Consumer & Institutional	16,487	16,510	17,549	17,571	18,714	3.2%
Industrial/ Machinery	1,084	968	996	962	1,042	−1.0%
Adhesives/Inks/ Coatings	1,167	1,143	1,165	1,170	1,196	0.5%
All Other	3,003	2,705	2,283	2,021	2,108	−7.8%
Exports	9,771	9,295	10,048	9,009	9,900	0.3%
TOTAL SELECTED PLASTICS:	77,640	77,390	82,324	80,270	96,101	2.0%

*Selected thermoplastics are:

Low Density Polyethylene	Nylon	Acrylonitrile-Butadiene-Styrene(ABS)
Linear Low Density Polyethylene	Polyvinyl Chloride	Styrene-Acrylontrile(SAN)
High Density Polyethylene	Engineering Resins	Other Styrene-Based Polymers
Polypropylene	Polystyrene	Styrene Butadiene Lataxes

be collected and separated easily and can then be recycled or even blended with virgin polymers.

Other than packaging, collectible items are automobile, light truck and large appliance shredded residue; carpets; wire and cable coverings; and tailings—plastic containing streams recoverable from MSWs.

Over the years, the recycling of plastics has increased. Some of the highly recycled materials are beverage bottles and other similar packaging materials, car bumpers, and battery cases. What is needed for recycling is good infrastructure for collecting, sorting, cleaning, reprocessing, and manufacturing new products for marketing to end users.

Table 3 Average Useful Life of
Plastics in Different Applications

Product Category	Years
Packaging	<1
Adhesives and others	4
Consumer and institutional	5
Furniture and fixtures	10
Transportation	11
Electrical and electronics	15
Industrial machinery	15
Building and construction	25

Note: From Ref. 4.

Table 4 Estimates of Plastics in Municipal Solid Waste[a]

	1990	1995	2000	2010
	(Billions of pounds)			
MSW (Total)	33	38	45	60
Fraction (1)[c]	0.7[b]	2	5	12
Fraction (2)[d]	19	22	27	33
Fraction (3)[e]	13.5	13	13	15
"Tailings" from reprocessing (1) and (2)[f]	0.2[b]	5–10	6–12	8–16

[a]All numbers have been rounded off; the complete methodology used to develop these estimates is given in Volume 2, Appendix 1 of the DOE Report.
[b]The quantity of plastics actually returned to market as products made from recycled plastics was approximately 0.5 billion lb; 0.2 billion lb were "tailings".
[c]Plastics recoverable by source separation and hand-sorting as practiced today.
[d]Additional plastic potentially recoverable by a combination of known technology, mostly source separation and hand/mechanical sorting of MSW. This represents a significant but attainable advance over today's practices. Therefore, the 1990 number represents a lost opportunity.
[e]The remainder of the plastics in MSW.
[f]Included in the numbers for MSW Fractions (1) and (2).

Note: From Ref. 4.

3 APPROACHES TO REDUCE WASTE

Plastic waste can be either landfilled, reused, converted to energy by combustion, reclaimed and reprocessed into new products, or converted into monomers or simple chemicals. The U.S. Environmental Protection Agency (EPA) advocates three Rs: reduce, reuse, and recycle. The approaches to handle the plastic waste problem are: reduce the amount of waste to be discarded; reuse a significant amount of the waste discarded; and recycle as much of the waste as possible.

3.1 Reduce

According to the U.S. EPA, *source reduction* is the process by which a package or product is made using fewer resources, creating less pollution and utilizing fewer potentially toxic ingredients. The EPA also defines source reduction as increasing the functionality and durability of a product. Plastics help product manufacturers do more with less material—which is known as *resource efficiency* or source reduction. In addition to helping the manufacturers' bottom line and keeping consumer prices down, source reduction also yields significant resource conservation benefits. For example, advances in plastics have helped manufacturers make products using less material by lightweight and thin-walled consumer product packaging. Just two pounds of plastic can deliver 1,000 ounces—roughly 8 gallons—of a beverage such as juice, soft drinks, milk, or water, compared to 3 pounds of aluminum, 8 pounds of steel, or 27 pounds of glass to bring home the same amount.

Plastics help make packaging more efficient, thereby conserving resources. As a result, the consumer can buy larger, economy-size products (e.g., laundry detergents) that are convenient to use. It also means that it takes fewer trucks, and therefore less fuel, to get products to the consumer from its source. For example, using plastic bags instead of paper bags to transport products can reduce the truckloads by more than half. Because of their unique characteristics, such as light weight, formability, and durability, there is a reduction of materials used, energy consumed, and waste generated for many products, ranging from disposables to durables. Thus, plastic products help resource conservation in a big way.

3.2 Reuse

Reuse is another approach to conserve resources. The durability of plastics allows for many products and packaging materials to be reused again and again. In a 1997 survey, Wirthlin Worldwide found that more than 80 percent of Americans reuse plastic products and packaging in their homes. The durability of plastics makes them the material of choice for commonly reused items, such as food storage containers and refillable sports bottles. Reuse of plastics reduces trash disposal costs and extends landfill capacity. As much as 40 percent of certain plastic parts from damaged or discarded cars are repaired and reused, reducing the amount of automotive components sent to landfills. Laundry products are also being packaged in reusable plastic bottles and small, refill packages of concentrated products, helping to reduce packaging waste.[1]

Many U.S. businesses have made the decision to receive their supplies and ship their products in reusable plastic shipping containers (RPSCs) rather than single-use corrugated boxes. Over a period of two years, the Ford Motor Company has eliminated more than 150 million pounds of wood and cardboard packaging that would have gone to a landfill by asking its suppliers to use returnable plastic shipping containers and plastic, rather than wood pallets. Returnable containers are also making major inroads in the produce and meat packaging industries.

3.3   Recycling

Plastic recycling has been in practice for a long time. Whereas recycling of waste in an industrial setting is easy, as they are well identified and not contaminated, postconsumer recycling is more difficult. Just as glass and metal have been recycled from household waste, there has been a continuing effort to encourage the recycling of plastics, especially since share of plastics in the MSW is continuing to grow. There are several approaches to managing plastic waste, as summarized in Figure 2.[2] The term *recycling* refers to the process of recovering plastics from the waste, and the term *recycled* refers to any material containing some quantity of recovered plastics, normally 25 percent or higher by weight. To help this effort, a coding system was developed to identify the different plastics in the products. The numbered coding system is shown in Figure 3. Products worldwide have these labels, making it easier for consumers to recycle as well as for agencies to collect and sort the plastic.[3] The numbering is also based on the ease of recyclability. For example, PETE, number 1, is the most recycled resin, and number 7 is the least recycled.

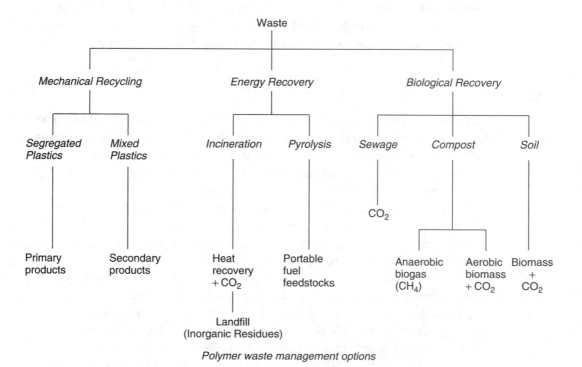

Polymer waste management options

Figure 2   Options for polymer waste management. (From Ref. 2.)

PETE   HDPE   V   LOPE   PP   PS   OTHER

SPI resin identification code. The seven symbols in the SPI code can be found at the
website http://grn.com/grn/library /symbols.htm (and elsewhere). However, the acronyms
below the images are not part of the website images and have been added separately here.

Figure 3   Plastic coding system to help identify recyclable polymers. *Note:* From Ref. 3.

According to the American Plastics Council (APC), more than 80 percent of all U.S. households have access to a plastics recycling program, and about 17,000 communities now collect plastics for recycling, half of which offer residents curbside collection. However, all types of plastics are not collected for recycling in all communities. It varies from place to place. Of course, the items most commonly recycled are plastic bottles.

In 2004, the amount of plastic bottles recycled reached an all-time high of 1.9 billion pounds, and that was 247 million pounds more than the previous year.[1] Both HDPE and PET bottle recycling rose to record levels.

The number of plastics recycling businesses has quadrupled since 1990, with more than 1,700 businesses handling and reclaiming postconsumer plastics. The data in Figure 4 clearly show the continuing growth in plastic bottle recycling. APC's Recycled Plastic Products Source Book lists approximately 1,300 commercially available products made with or packaged in postconsumer recycled plastics, including office supplies, park benches, sweaters, jeans, videocassettes, detergent bottles, and toys.

Based on the generation of each stream, polymer recovery program trends, and expansion of recycling programs, it is estimated that close to 15 billion pounds of polymer streams will be available for recycling or recovery (Table 5).[4] These

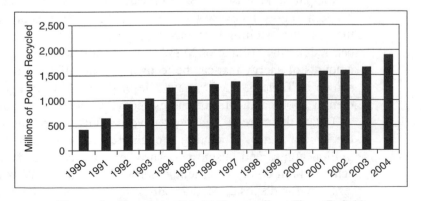

Figure 4   Growth trend in bottle recycling. (From Ref. 1.)

Table 5   Polymer Streams Available for Recovery/Recycling[a]

	1990	1995	2000	2010
	(Billions of pounds)			
ASRs	1.8	2.1	2.3	3.0
Carpets	2.1	2.3	2.5	3.0
Wire and cable covers	0.4	0.5	0.6	0.8
Tailings from MSW[b]	0.2	5.0	6.0	8.0
Total	4.5	9.9	11.4	14.8

[a]Does not include plastics recovered/recycled in current (1993) Recycle Programs; by the year 2010, these plastics are projected to be an additional 11 billion lb.
[b]Conservative estimate from data in Table 2-2. This stream is the largest and most dispersed; its use has the most demanding technical and economic challenges.
Note: From Ref. 4.

estimates are based on polymer waste generation, the current recycling rate, and trends in recycling.

4 ENERGY SAVINGS AND RECOVERY

Another advantage of recycling is the potential energy savings. By recycling instead of using virgin polymers, energy is not required to produce new polymers. As it is, plastics consume less energy than other traditional materials to produce. For example, a plastic beverage container requires only 0.11 kWh per container, compared to 3.00 kWh for an aluminum can and 2.00 kWh for a glass bottle.[2] However, the plastics are an excellent source of energy. The calorific value of plastics is higher than that of coal and similar to that of fuel oil. Also, because it simply replaces fossil fuel, there is no CO_2 burden on the environment.

Thus, an important way to conserve resources is to recover the energy value of products after their useful life has ended. One such method involves combusting municipal solid waste (MSW) or garbage in waste-to-energy (WTE) facilities. Modern energy recovery facilities burn MSW in special combustion chambers, then use the resulting heat energy to generate steam and electricity. This process can reduce the volume of MSW to be landfilled by as much as 90 percent. The amounts of energy that can be recovered from different components in MSW are shown in Table 6. Plastics have one of the highest energy content, thus making them a good candidate for energy recovery when they cannot be recycled.

Today, there are 103 energy recovery plants, operating in 32 states throughout the United States, generating enough electricity to meet the power needs of 1.2 million homes and businesses. These energy recovery facilities are designed to achieve high combustion temperatures that help MSW burn cleaner and create less ash for disposal. Modern air-pollution control devices are used to control and reduce potentially harmful particulates and gases from incinerator emissions.

Table 6   Energy that Can Be
Recovered from Different
Components in the MSW

Material:	btu/lb
Mixed Food Waste:	2,370
Mixed MSW:	4,800
Mixed Paper:	6,800
Newsprint:	7,950
MIXED PLASTIC:	14,100
POLYSTYRENE:	16,419
POLYETHYLENE:	18,687

Note: From Ref. 1

The EPA estimates that energy recovery plants will dispose of 15 percent of the nation's MSW.

One of the drawbacks in converting plastics into energy is that the energy-recovery facilities must operate within existing state and federal regulatory standards for human health and safety and environmental protection. The Clean Air Act Amendments of 1990 tightened existing emissions standards and provided an additional margin of security. Then in 1995, additional air pollution control standards were promulgated by the EPA administrator for municipal waste combustors, including WTE facilities. These standards, combined with states' and EPA's strict enforcement policies, make the new rules among the toughest in the world for these power plants. Plastics are a valuable feedstock for these facilities, and most plastics, when properly combusted, produce energy, water, and carbon dioxide as the principal products of combustion, similar to other petroleum-based fuels such as home heating oil.

4.1   Pellet Fuels

Processed-engineered fuels (PEF), also known as *pellet fuels*, are produced from a mixture of plastic with other recycled materials. The other waste can be unrecyclable paper, tires, or other scrap materials. The amount of plastic is varied to yield a pellet fuel possessing the desired combustion characteristics. The idea is to design PEFs to provide highly predictable and uniform combustion characteristics. Essentially, the plastic that can be recovered or recycled by other means is targeted for this approach.

4.2   Challenges to Recycling

Just as there seem to be good opportunities for plastic recycling, there are many technical, marketing, and economical challenges as well. One of the major issues is the diversity in the composition, as the discarded products are made from a wide variety of resins and additives. The polymers listed in the Table 1 give an idea of the different classes of plastics, and there are many variations within those groups. The waste stream of plastics is further contaminated by labels, glue, printing, and product residue. Essentially, most of the waste streams include several immiscible polymers, including six major thermoplastics in the waste, which are not miscible, either. Unless these are sorted and cleaned, conventional melt processing of mixed plastics yields poor-quality products. Also, with differences in melting and processing temperatures of these different resins, it is a challenge to process them without causing degradation of some components. Also, multiple melting cycles can further complicate the contamination. Thus, separation and cleaning is the key to having a sufficient supply of clean feedstock for a viable recycling industry. Though recycling is good for the environment, one has to remember that unless it is economically viable, the recycling industry does not survive and thrive. The lack of a continuous supply of clean feedstock, and the high cost of transportation and reprocessing, are some of the economic challenges for the recycling industry. The cost of collecting and separating plastics is still high compared to that for aluminum and paper. The recovery/recycling processes must also be environmentally safe and nonpolluting.

5   POSTCONSUMER RECYCLING OF PLASTICS

Recycling of postconsumer plastics involves the process of collection, cleaning, separation, and then processing them into various products. These different steps will be elaborated in the following sections.

5.1   Collection, Cleaning, and Separation

For successful recycling, there must be a good way of collecting recyclable waste. Collection and separation can be quite complex because of many issues involved. In the early stages, the collection of recyclable materials was done at drop-off or buy-back centers. Here people could take their materials for recycling, either for their inherent value as in case of aluminum cans, or purely because of environmental concerns and awareness. Many times, these were the results of increased involvement and education from voluntary organizations such as Boy Scouts, and less than 5 percent of the population participated in this program. Plastics were not a part of such programs in the initial stages, and there were limited collection centers without easy access to majority of the population. In the late 1980s, several states passed a *bottle bill*, and after that there were several buy-back centers in those states with a redemption value of about five cents per beverage container (glass, aluminum, steel, or plastic). This led to an increase in PET bottle recycling.

In the mid-1980s, residual curbside recycling program started in many states, especially in areas where landfilling costs were skyrocketing due to the lack of available landfill spaces nearby. In fact, by 1995, about 15 percent of the U.S. population had access to this program, and this led to a significant increase in recycling. Even in curbside recycling programs, plastics were not part of the collectibles in the initial stages, with paper, glass and metals being the main ones. However, by mid-1990s, plastics became a part of this service. For the successful recycling of plastics, a key factor was education and promotion. Communicating the program details, including the day of pick up and type of materials accepted, was important. Once a routing is established, people start separating plastics for recycling and then storing them in special plastic containers provided by waste-collecting agencies. Well-established collection programs help increase in bottle recycling. It was observed that in the United Kingdom, with an increase in collection facilities, bottle recycling increased by 100 percent in two years from 2003 to 2005.[5]

Although initial collection centers accepted only PET bottles, curbside collection programs were extended to other plastics, including HDPE milk jugs and PVC containers. When recycling was extended from PET and HDPE bottles to all plastic bottles, the bottle collection increased by 44 percent in *bottle bill* states and by about 13 percent in *non–bottle bill* states. Because of the large volume-to-weight ratio of plastics, it is important to have proper collection and transportation vehicles with compacting systems. In some places, plastic and glass bottles are collected together, and plastics reduce the breakage of glass in the mixed stream. However, this practice requires separation systems as well. More than 95 percent of the current production of plastic bottles is from PET and HDPE, and the recycling of plastic bottles is also the maximum for these two resins (Figure 5).

Special collection systems are generally provided for large-scale or commercial postconsumer applications such as for agricultural films, chemical containers, automotive parts, carpets, and polystyrene foam packaging. Since most of the plastic is collected as multimaterial or in commingled forms, the collected plastic waste has to be sorted, separated, and cleaned, and most of this is done at material recovery facilities (MRFs).

5.2 Material Recovery Facilities

Sorting is a critical step, as successful recycling depends on the efficiency of separating plastics in the waste stream by resin type, color, and so on. Sorting is done at a home site, in a commercial location, at pick-up during collection, at recycling centers, or at material recovery facilities. MRFs are a critical part of large-scale recycling programs. In the 1990s, when cities were putting together such facilities, it was conceived that a community with a population of about 100,000 should have some form of an MRF. MRFs function as the place to separate various materials, and in some cases, clean and prepare the separated materials for the market. This involves separating plastics from glass, metal,

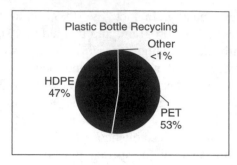 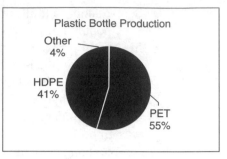

Figure 5   Plastic bottle production and recycling by resins. (From Ref. 1.)

labels, and other impurities, and then separating different types of plastics as much as possible. One example of such a separation schematic is shown in Figure 6. Here, both hand sorting and automatic sorting techniques are used. To help automated separation, coding systems that help in identifying and separating individual plastic bottles/containers using the barcode or similar marking systems for different plastics have been developed, and they are being tried. Most of the MRFs still use hand sorting to separate plastics by type and color. However, some systems separate plastics by their density difference. For hand sorting, bottles from bales are used. For mechanical sortation, many times the bottles are shredded and then the shredded plastics can be separated by density differentiation techniques. The purpose of sorting is to produce segregated plastic streams with sufficient purity for reclaiming the resins.

5.3   Processing Separated Plastics

PET Recycling

Processing of recycled PET into various products is very common, and is practiced in many countries around the world. Well-separated, cleaned, and recycled PET has good value and is the most profitable of all the recycled polymers. An important issue in processing PET is the removal of contaminants that can hydrolyze it. Processors must avoid the use of cleaning agents, such as caustic soda, that are commonly employed to remove labels and adhesives. Even after washing, some of these chemicals may be entrapped, further leading to hydrolysis. PET has to be dried to a moisture content of 0.005 percent or less. In spite of this, there will be a drop in intrinsic viscosity (IV) of 0.02 to 0.03 units during melt processing. Melt processing is the most common technique used to convert recycled PET into products.

The majority of the recycled PET is consumed to produce fibers that are used in all types of clothing, carpets, fiberfill, and technical textiles (Figure 7). In addition to being economically viable, the recycled PET bottles have a higher viscosity than the typical fiber-grade virgin polymer. This results in stronger fibers, and also provides a wider margin for possible degradation during processing.

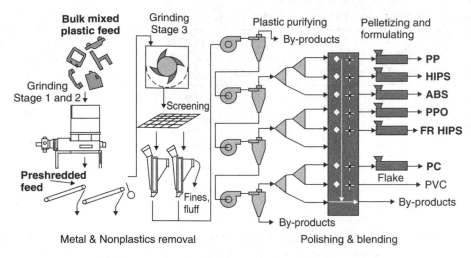

Figure 6   Plastic recovery scheme in an MRF. (From Ref. 4.)

Slightly higher processing and melt temperatures are used to get the right viscosity for fiber formation. Also, better melt filtration systems and more frequent filter changes are required because of the presence of contaminants, however thorough the cleaning process might have been. As far as such issues are addressed, these can be processed in normal PET fiber spinning lines. As most of the applications have been using staple fibers, short spinning systems have been considered more suitable for such a system (IFJ articles). Many companies have been successfully adopting this system to produce fibers from recycled PET, with Wellman being the world leader. Mohawk, formerly Image Carpets, has been the pioneer in using postconsumer recycled PET for making carpets.[6,7] In the case of fiberfill and many traditional textile applications, clean and clear recycled PET is preferred. For some of the nonwoven and geotextile applications, green PET can also be used, as it can be colored black easily. These days, recycled PET is also used as a continuous filament and as spunbond nonwovens for some applications. Recycled PET has been successfully processed to produce good-quality melt-blown nonwovens that can be used as filters or part of the protective garments. Although fine fibers with diameters less than 5μ can be successfully produced, because of the lack of stress-induced crystallization in the process, the webs have a high boiling shrinkage, which may limit their application potential. It was also shown that for the melt-blown process, undried PET can be successfully used since the hydrolytic degradation helps in reduction in viscosity and thus fine-fiber production (Table 7). Studies have been done to find ways to enhance the shrinkage resistance of PET nonwovens using nucleating agents.[8]

Another market for recycled PET is plastic strapping, which is replacing the traditional metal strapping in packaging industry. Since extrusion of PET into straps requires high IV PET, recycled-bottle-grade PET is ideal. Because most

Table 7   Comparison of Dried and Undried recycled PET in Fiber Processing [Bhat]

Samples	Basis Weight (oz/yd^{12})	Thickness (um)	Shrinkage (%) MD	CD	Filtration Efficiency (%)	Tenacity (mN/Tcx) MD	CD	Break Elong. (%) MD	CD	Bursting Strength (kPa)	Stiffness (mg.cm)
1A	0.67	285	26.3	3.8	39	6.1	3.4	44	96	21.1	83
2A	0.74	259	26.5	17.2	45	8.6	6.0	50	78	25.8	73
1B	1.15	312	35.3	11.3	51	11.5	4.8	25	140	33.6	332
2B	1.28	309	32.0	11.3	61	9.8	6.9	50	95	34.5	300

IA—Dried Recycled Day-PET; 2A—Undried Recycled Day-PET; 1B—Dried Recycled FE-PET; 2B—Undried Recycled FE-PET

Note: From Ref. 11.

of the straps are black, this can use green PET, as well. In some instances, where viscosity is not high enough, solid-state polymerization, where the recycled PET flakes/chips are heated in a controlled condition to increase their molecular weight, can be used to increase the IV. Another growing application is that of sheeting, where recycled PET can replace traditional PVC and high-impact polystyrene sheets in blister packaging, food trays, and cups. In some of these extrusion processes, postconsumer recycled PET may be blended with virgin or industrial recycled PET. Recycled PET is also finding a lot of applications as part of alloys and compounded products used in engineering applications that have very high values. Some examples are blends of PET and polycarbonate that are easily injection moldable, and fiberglass-filled PET.

Polyolefins

After PET, polyethylenes, especially HDPE, are the most recycled resins. The recovery process is quite similar to that of PET, and similar operation combinations are used to clean and separate polyolefins. Remember that there are many varieties of polyolefins, HDPE, LDPE, PP, LLDPE, copolymers, and co-extruded products that make the separation and reclamation more complex and challenging. The cleaned resins that are granulated or further palletized can be fabricated into products using one of the melt processing techniques such as molding. Since many of the recycled polyolefins are pigmented, attempts are made to separate them by color as well, which gives some value to the recycled material, compared to the ones with mixed colors. There is increasing activity in recycling polyolefin films. Recycled polyolefin films are hard to sort and separate by polymer type, and are mostly reprocessed as mixed or commingled plastics.

During melt processing, the molecular structure of the polyolefin resins changes, due to high temperatures, shear, and oxidizing atmospheres. This can lead to a reduction in the molecular weight by degradations or increase in molecular weight due to cross linking. There is also change in molecular weight distribution. The changes in structure due to melt histories have an impact on

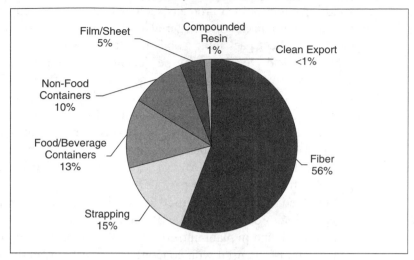

Domestic Recycled PET Bottle End Use

Figure 7 Products made from recycled PET bottles. (From Ref. 1.)

rheological properties, which, in turn, affect processing and further the mechanical properties of the formed products. In many cases, recycled resin is blended with virgin resin to make up for the loss in performance. Packaging materials produced from recycled polyolefins are used by many companies, as there is an effort to use at least 25 percent recycled content in much of the non-food packaging. Figure 7 shows the breakdown of product categories that use recycled HDPE from bottles. Some examples are bottles used for packaging household liquid chemicals, detergents, cleaning chemicals, fabric softeners, and motor oil. The challenge here is to produce containers that have properties that are indistinguishable from those of virgin polymers, but can still be economically competitive.

Among the polypropylene products, the most recycled are the battery cases. Almost all automotive and truck battery cases are made of PP, and there is a well-established industry to recycle batteries in the United States, due to its lead contamination. Because of the combination of land disposal ban, and mandated delivery/take-back provisions, automotive batteries are recycled at a very high level. About 40 percent of the recovered PP is used again in battery production and the rest is used for making other injection-molded products. Some of the cleaner recycled PP can be used in fibers, such as spunbond or meltblown nonwovens for geotextiles or filter applications.[9]

Other Resins

Polystyrene (PS) is another resin widely used in food packaging and other cushion packaging applications. These include extruded foam sheets such as meat trays, poultry trays, egg cartons, and produce trays; molded foam articles such as cups; solid injection-molded products such as knives, forks, cup lids, salad boxes, etc.; and loose fill packaging materials—peanuts and custom-molded shapes. It is easier to collect packaging goods than the service materials. Recycled PS, after separation and cleaning, may be densified or used as foam. Depending on the waste type, the molecular weight will also be different. There is some drop in molecular weight on recovery and grinding, but change in mechanical properties is small. The recycled resins can be blended with virgin resins to fabricate quality products. Because of the FDA regulations, the recycled PS is used to fabricate nonfood contact items such as desk trays, card file boxes, waste baskets, coat hangers, key chains, and toys. These resins can also be used to make foamed products for insulation and packaging.

Shredded foam may be used as void fill material or thermal insulation or construction products. Ground PS foam is used in floral vase stuffing, lawn furniture fill, and in plant nurseries as soil lightener to improve aeration. PS foam beads can also be used with concrete to make lightweight insulating material for use in highway road beds, railroad beds, and airport runways. PS foam can also be an oil spill clean-up material.

Polyvinyl Chloride (PVC) recycling has been much slower to catch up compared to other polymers. PVC is one of the harder polymers for collection and sortation. As PVC has density close to that of PET, it is hard to separate them, and the two polymers are incompatible as mixtures. However, once separated and cleaned, PVC can be ground and processed, just like virgin PVC.

Engineered plastics (EP) are not commonly found in MSW, and collecting and sorting them is more difficult. Automobiles use a large share of EPs, such as polycarbonates, nylons, and Polyesters. Recycling of shredded automobile parts is continuing to increase. Lack of continuous and reliable supply of recycled EPs is a problem, however. The European Union mandate on automobile recycling might help increase the recovery and recycling of EPs. Polycarbonate (PC) five-gallon water bottles are also being recycled. Recycling of electronic goods and appliance parts is still developing.

Acrylic polymers have unique property advantages because of clarity, toughness, chemical resistance, and weatherability. These are used in automobiles, houses, appliances, lighting, and fixtures.

Automotive scraps are a good source for postconsumer recycling of acrylics and EPs. However, automobile scrap will contain both metallic and nonmetallic contaminants. The metallic component is removed magnetically, leaving automotive shredder residue (ASR) that consists of mixed plastics and other impurities. ASR will have glass, fabric, rubber, and small amounts of metal. The acrylic content in ASR may be as much as 30 percent. The most common approach for recycling of PMMA is depolymerization and recovery of monomers. Generally, the monomer yield is about 95.5 percent. ASR can be milled and then compression molded into panels. Compression-molded panels have good rigidity but moderate strength and impact resistance. Several additives such as virgin material, HDPE, and PP can be used to improve the impact strength of these products, so that they are comparable to that of virgin materials.

Residential carpets are continuing to be more visible in MSW. Carpets are quite complex with polyamide/PP/PET face fibers, PP backing, latex adhesive, and calcium carbonate as filler. In addition, the recycled carpet will have a higher amount of dust particles. There have been efforts in the United States to collect residential carpet. As it is very difficult to separate the different components of the carpet, the reasonable way is to shred the carpets and pelletize it so that the pellets can be used in products such as plastic lumber. Only a small percentage of the face materials (nylon6 and nylon 66) are recovered and recycled.

5.4 Mixed Plastics

Although PET, HDPE, and, to small extent, other polymers are recovered from the waste stream, that is only a small share of the total plastics in MSW. The rest of the plastics in the waste stream can be processed into useful products. This mixture of different plastics is called the *commingled plastics*, which may include films containers, packaging materials, toys, and others. Many of the plastic containers are not easy to identify, and it becomes more difficult and expensive to separate them by individual resins. Even when they are identifiable, they may have some pigments or additives that make then unsuitable for individual resin recovery. These plastics left out after recovering PET and HDPE are also called *tailings*. Instead of sending this lower-value mixed plastics to landfill, it may be possible to process then as commingled. Commingled plastic processes can essentially use many feedstocks, including plant scraps, postconsumer plastic waste, and various additives and fillers. When mixed plastics are collected as in curbside recycling, it is easy to separate clear and green PETs, and clear HDPE, which leave remaining mixed plastics, the tailings. The composition of commingled plastics varies from one facility to another and year to year, depending on collection and sortation practices. Most of these have a high percentage of HDPE, with the rest being a mixture of several resins. Sometimes this mixture can be further modified by blending with plant plastic scrap.

Because of the heterogeneous nature of the mixture, and presence of contaminants, the articles produced should have large cross-sections, so that small imperfections do not hinder the mechanical performance of the products. Products can be manufactured by continuous extrusion, compression molding, or by Klobbie-based intrusion processes. Intrusion process is a cross between conventional injection molding and extrusion. In this, the extruder first works and softens the thermoplastic mixture, which is poured into one of the molds without using screen packs or nozzles. Today's recycling equipments are capable of processing most types of mixed thermoplastic material, even with up to 30 to 40 percent contamination as unmelted polymers or nonpolymer materials such as glass, paper, metal, or dirt. Since the major component (60% or higher) is polyolefins, they soften or melt during processing and encapsulate the contaminants. The most common value-added product made from commingled plastics is plastic lumber. Plastic lumber has some unique properties that make it more appealing than wood for several applications. Plastic wood is manufactured by pressing the commingled plastics at a temperature sufficient to melt the majority of its components. Color additives can be added to mask the multicolored appearance of commingled plastics.

In another (Reverez) process, commingled waste plastics are softened in a hopper and then mixed in a screw. The fluidized plastic can be processed by blow molding, extrusion, or compression molding. When commingled plastics are processed as is, the properties of the products are generally poor, with more brittleness. To improve the properties of the products, both mechanical and chemical means are used. The poor performance is the result of incompatibility of the components in the mixture, leading to phase separation and aggregation. The remedy is to improve blending of components to achieve compatibility. The mechanical means consist of shearing of the molten polymer using counter-rotating twin screw extruder. Of course, one of the challenges is the differences in melting temperatures and viscosities of the different components. Also, if temperature is increased to melt all the major components, low-melting-temperature polymers start degrading. Chemical compatibilizers are added quite often to improve the mixing of different components. These may be maleic anhydride–grafted polymers or rubbery (block and graft) copolymers, including elastomers such as Kraton. Some of the copolymers may be produced in situ by blending suitably functionalized polymers. Also, specific low-moleculars weight compounds that promote copolymerization or cross-linking may be added. Selection and use of appropriate compatibilizer system requires a good understanding of the composition of the commingled plastics. The addition of compatibilizers will help in achieving better mixing, with reduced phase separation. This will result in improvement in properties even on a smaller scale, and will definitely result in improved toughness of the product. The aggregate sizes will be much smaller than the unmodified plastic mixture. The size and morphology of the dispersed phase is determined by blend composition, viscoelastic properties of the components, and interfacial adhesion.

Electrical/Electronic Products

Electrical and electronic products such as computers, cell phones, TVs, and stereos are becoming a more visible part of the MSW. Some of the resins used in electronic products are PS, HIPS, ABS, PC, PP, PU, PV, PVC, polyamides, phenol formaldehyde, and blends of some of these polymers. Several technologies are being developed for the separation of different plastic types. Since the electronic parts are made from many engineering plastics, and with many different additives, it is more difficult to identify and separate the individual resins.

After separation and removal of nonplastic contaminants, the plastic can be granulated and then used for fabrication of similar parts. Recycling technologies are available today for handling plastic waste coming from the electronic industry. The challenge is to have an infrastructure to effectively manage collection and transportation so that high enough volume of recycled material is available to make the industry economically viable.[10]

Plastic Film

Recycling films is more complicated than recycling other commodity plastics. There is a need for efficient removal of contaminants and consolidation of materials. Because of very high volume-to-weight ratio, it is less economical to transport films. Films are more difficult to sort because they do not have a good marking system. Also, different additives such as plasticizers, antislip agents, and lubricants are used during film manufacture, and it is hard to sort them out. Films exposed to the outdoors for a long time will degrade making recycling and melt extrusion more difficult. Film recycling seems to work better in situations where large volumes of similar films with very low level of contamination are collected. Stretch film recycling in warehouses and distribution centers, grocery store chains, and bulk-mail facilities are quite effective. In these places, there is enough material to justify densification and transportation, and these are not contaminated with food or dirt. Once recycled, films can be used to make products such as nonfood application films, trash bags, liners, construction film, grocery sacks, and retail bags, and to produce household items such as waste baskets and recycling bins. The mixed plastic material can be used in the production of plastic lumber.

5.5   Chemical Recycling

Plastics with a carbonyl group can be converted to monomers by hydrolysis or glycolysis. Condensation polymers such as polyesters and nylons can be depolymerized to form monomers. For Polyurethanes (PURs), what is obtained is not the initial monomer, but a reaction product of the monomer diamine, which can be converted to diisocyanate. For PURs, hydrolysis is attractive as they can be easily broken down to polyols and diamines. The only issue is to separate them later. Steam-assisted hydrolysis has been shown to yield 60 to 80 percent recovery of polyols from PUR foam products. A twin screw extruder can be used as a reactor for hydrolysis. Glycolysis of PURS, yields mixture of polyols that can be reused directly.

5.6   Pyrolysis

In addition to chemical conversion, thermal energy can be used to breakdown polymers. *Pyrolysis* of plastics to monomers requires that the polymers be sorted and reclaimed to a reasonable purity, product gases be sufficiently purified, and the monomers be polymerized to resins. Mostly vinyl polymers are likely to depolymerize, yielding monomers. As discussed earlier, PMMA can be pyrolized to a MMA yield of 90 percent. PS yields some monomers. PEs produce a mixture of unsaturated hydrocarbons. One of the problems in chemical and thermal means of recovering monomers is the possible environmental pollution. Also, they need additional thermal and chemical energy to break them apart, and then additional energy has to be spent to put them back together. If PVC is present, the presence of chlorine is a problem because of the production of HCl during pyrolysis, which is highly corrosive. Pyrolysis may also be used to produce liquid and gaseous fuels instead of monomers. Unlike in combustion, during pyrolysis there is a huge reduction in volume of gases produced, which reduces the complexity of exhaust gas purification.

5.7   Degradable Plastics

When recovery and recycling is not feasible due to certain applications, the better approach is to allow the plastic to degrade in the landfill or in a composting atmosphere. Both photo- and biodegradability is possible. Biodegradable polymers, both natural and synthetic, are becoming more important these days. Some of the degradable polymers include polylactic acid (PLA), polyvinyl alcohol, polyvinyl ethanote, many aliphatic polyesters and polyurethanes, copolyesters, and polyhydroxy buterate. Some of the issues with degradable materials is that some degrade too slowly, others produce hazardous substances during degradation, and some materials require higher shelf life and cannot be degradable. Since there is plenty of literature on that subject, it will not be discussed in more detail here.

6   PRODUCTS FROM POSTCONSUMER RECYCLED POLYMERS

Many products can be produced from recycled polymers. When they are clean, almost every product that is produced from virgin polymers can be manufactured using recycled plastics. The only place where these products cannot be used is in food packaging. However, if the plastic is converted to monomers, and then repolymerized, it can be part of the food and beverage packaging market. Some of the possible products are summarized in Table 8, which again indicates a large opportunity for recycled plastics.

The best use for recycled plastic is the same as it was used originally, or some other application with high value. For many of the sorted plastics, reuse is possible. Some examples are LDPE in films, PP in auto parts, PS in insulation/packaging, and PVS in detergent bottles and pipes. Fiber applications

Table 8 Some of the Markets and Products Made from Processing of Postconsumer Recycled Plastics

Apparel & Accessories	*Industrial, Shipping & Warehouse*
Aprons, Backpacks, Boxer Shorts...	Baskets, Caster Wheels, Foam Packing...
Automotive Accessories	*Janitorial Supplies*
Funnels, Ice Scrapers, Mud Flaps...	Buckets, Wringers, Pails...
Bags & Liners	*Landscape & Garden Design*
Dispensers, Can Liners, Trash Bags...	Decking, Fencing, Rakes...
Bins & Containers	*Marine*
Bus Boxes, Carts, Dumpsters...	Bollards, Decking, Sea Walls...
Building & Construction	*Office Supplies*
Bulkheading, Insulation, Shingles...	Badges, Clipboards, Paper Clips...
Carpet, Fabric & Fiber	*Packaging*
Carpet, Geotextile Fiber	Bottles, Carryout Containers, Trays...
Custom Products	*Planting Materials & Accessories*
Molded, Structural Foam...	Planters, Window Boxes...
Farming & Agriculture	*Premium & Promotional Items*
Egg Filler Flats, Feed Carts, Sheds...	Bookmarks, Can Openers, Key Chains...
Film & Sheet	*Recreational Products & Toys*
Pallet Covers, Sheet, Slip Sheets...	Bicycle Racks, Cameras, Sandboxes...
Furniture & Accessories (Indoor)	*Road, Highway & Parking*
Desks, Lamps, Love Seats...	Barricades, Parking Stops, Speed Bumps...
Furniture & Accessories (Outdoor)	*Signage*
Benches, Hammocks, Tables...	Sign Posts, Signs
Housewares	
Blankets, Clocks, Pillows...	

Note: From Ref. 1.

continue to be the number one market for recycled PET bottles. Although some of the recycled PET goes to food packaging as well, nonfood containers have been increasing their share of the recycled PET market. This is partly due to the regulations such as California's Rigid Plastics Packaging Container Law, which mandates the use of recycled plastics in rigid nonfood products packaging. The primary market for recovered clear HDPE bottles is for nonfood applications such as detergents, motor oils, and household cleaners. Pigmented HDPE is used for the production of PE pipes and many lawn and garden products. Plastic lumber consumes many plastics, and this market continues to grow, consuming a significant share of the recycled plastics. Plastic lumber has its highest value in applications where the natural wood is likely to degrade or rot. Use of plastic lumber will eliminate maintenance cost as well as use of toxic antifungal agents in natural wood, thus increasing the value of plastic wood. With product development, it is anticipated that plastic lumber can be used in structural applications such as railroad ties, marine pilings, and small bridges. Plastic lumber is about 20 percent costlier than natural wood and is only economical in applications where natural wood requires additional maintenance.

7 LIFE-CYCLE ANALYSIS

Life-cycle analysis (LCA) covers every phase from raw materials, processing, and manufacture of resin, transportation, processing, recycling, and disposal, and has a lot of environmental implications with respect to plastic waste. LCA involves several steps, as shown in Figure 9[3] Typically, it involves the conversion of crude oil into petrochemicals, followed by synthesis of resins that are processed into various products. After their useful life, the products are either recycled or landfilled. By this analysis, it is possible to maintain a high performance value for the product. Plastic recycling has a mixed response so far. Whereas many plastics cannot be recycled economically and are landfilled, emerging technologies and new attitudes are providing a new optimism. Continuing commitment for a greener environment worldwide has resulted in increased community involvement and government regulations, which are likely to increase plastic recycling in the future.

Research by Franklin Associates, Ltd. compared the life-cycle energy impacts of plastics and alternative materials.[1] This study compared the energy required to manufacture, use, and dispose of common packaging items—including bottles, tubs, wraps, and foam clamshells—with the most likely nonplastic alternatives. Franklin found that by using plastic instead of alternative packaging, product manufacturers save enough energy each year to power a city of 1 million homes for roughly three and a half years.

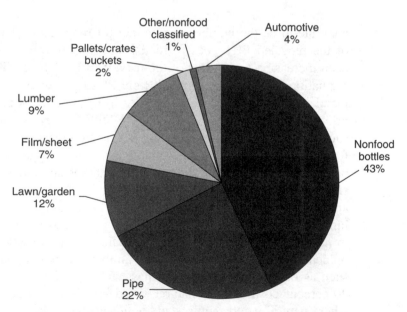

Figure 8   Products from recycled HDPE bottles. (From Ref. 1.)

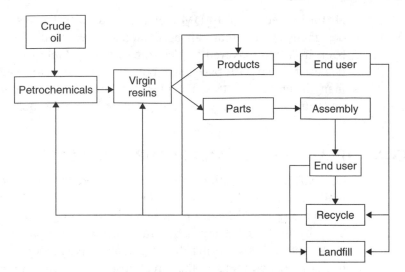

Figure 9   Life-cycle analysis of plastics. (From Ref. 3.)

Plastics also help conserve energy in our homes. Because of the superior insulation properties of plastics, vinyl siding and windows help reduce energy consumption and lower your utility bills. Vinyl windows, for instance, can save the average homeowner between $150 and $450 each year on heating and cooling costs compared to other types of windows. House wrap is shown to be very effective in home heating and cooling, and it has revolutionized the whole building industry.

The same principles apply in modern appliances such as refrigerators and air conditioners. Fifty-three billion kilowatt hours of electricity are saved every year by improvements in major appliance energy efficiency made possible, to a great extent, by plastics. Without the benefits provided by polyurethane foam insulation, these appliances could use as much as 30 percent more energy, driving up our electricity bills and increasing spoilage.

8   DESIGN FOR SUSTAINABILITY

As the issues relating to plastic waste problem and difficulties and challenges in recovery and recycling become important, one of the approaches is to design plastic parts so as to reduce the amount of plastic waste generated. This involves several approaches. Some of the approaches are to extend the part of a life through improved performance; reuse the plastic for a secondary application; or use closed-loop recycling, where the recycled plastic is put back in the same product. Designers should use compatible plastics so that the difficult part of disassembly can be eliminated. Automakers are adapting a single-material approach for large plastic parts such as bumpers and for fixtures and fittings—for example, all PP instead of PP/EPDM blend, as it used to be. BASF has been working on designing carpets that can be completely recycled. The ketchup industry went back from multilayered bottles to easily recyclable PET bottles. Soft-drink bottles used to have PE base cups, but newer bottles are made of all PET. There are many instances like this. The plastic industry is continuing to design and produce parts that can be easily recycled, thus reducing the burden on landfill.

9 ECONOMICAL ASPECTS AND FUTURE PROSPECTS

Of course, for recycling to be successful, economics should be favorable. The main reason for the success in PET bottle recycling is that the companies can make a profit. Although PE clear bottle recycling is profitable, colored PE recycling by itself is not profitable. In fact, many companies that got into recycling in the early 1990s, during the heightened environmental awareness, went out of business. If one looks at the direct cost and benefits, recycling business is often not profitable. That is where the help from legislation, tipping fees, and so on come into play. A good example is the bottle bill. The extra five cents charged for the bottles helped in setting up facilities to reclaim and recycle bottles. The EU legislation on automotive recycling and on packaging materials has tremendously helped the progress and growth in recycling.

The current recycling technology is very well advanced, and can handle many types of plastic wastes. Depending on the waste stream, the available technology can be used to produce high-value recycled plastics suitable for making a variety of products. Also, machinery manufacturers are meeting the demands of industry with equipment that can handle different types of recycled plastics and turn-key processing facilities can be set up. Innovations in recycling machinery are continuing in all aspects of the process, and the cost of reprocessing can be reduced with the ability to add value to recycled products.

REFERENCES

1. American Plastics Council, http://www.americanplasticscouncil.org http://www.plastics-resource.com http://plasticrecycling.org EPA website
2. G. Scott, Polymers and the Environment. Royal Society of Chemistry, Cambridge, UK 1999
3. N. Khait, "Recycling Plastics," in *Encyclopedia of Polymer Science & Eng.*, Vol. 7, pp. 657–678.
4. A. L. Bisio and M. Xanthos, *How to Manage Plastic Waste, Technology and Market opportunities*. Hanser Publishers, New York, 1994.
5. APC, "Understanding Plastic Film: Its uses, benefits and waste management options," 1997, http://www.americanplasticscouncil.org.
6. D. Keser, "Processing Lines for Production of Polyester Fiber from Recycled Bottles," *International Fiber Journal* (June 1995), pp. 4–12.

7. B. Baker, "Staple Fiber Spinning with Reclaimed Polymers," *IFJ* (June 1995), pp. 34–50.

8. G. S. Bhat, V. Narayanan and L. C. Wadsworth, "Dimensionally Stable Polyester Films, Fibers, Yarns and Melt Blown Nonwovens," U.S. Patent 5,753,736 (issued May 19, 1998).

9. G. S. Bhat, V. Narayanan, L. C. Wadsworth, and M. Dever, "Conversion of Recycled Polymers/Fibers into Meltblown Nonwovens," *Journal of Polymer-Plastics Technology and Engineering*, **38** (3) 499–511 (1999).

10. J. Biancaniello, L. Headley, M. M. Fisher, and T. Kingsbury, *Setting the Record Straight: Busting Common Myths about Plastics from Recovered Consumer Electronics,* APC-GPEC 2003.

11. G. S. Bhat, Y. Zhang, L. C. Wadsworth, and M. Dever, "Processing of Post-Consumer Recycled PET into Melt Blown Nonwoven Webs," *International Nonwovens Journal*, 3,6 54–61 (1994).

CHAPTER 13

SUPERCRITICAL WATER OXIDATION

Philip A. Marrone
Science Applications International Corp. (SAIC), Newton, Massachusetts

Glenn T. Hong
General Atomics, San Diego, California

1	INTRODUCTION	385
2	PROPERTIES OF SUPERCRITICAL WATER	386
3	THE SUPERCRITICAL WATER OXIDATION (SCWO) PROCESS	392
4	HISTORY AND CURRENT STATUS OF SCWO	398
5	SCWO RESEARCH EFFORTS	408
	5.1 Physical Property Determination	408
	5.2 Global Kinetics and Elementary Reaction Network Modeling	409
	5.3 Catalysis	414
	5.4 Salt Phase Equilibrium	414
	5.5 Salt Nucleation and Growth	415
	5.6 Corrosion	415
6	PRACTICAL ASPECTS OF SCWO PROCESSING	417
	6.1 Feed Materials	417
	6.2 Reactor Type	421
	6.3 Reactor Conditions	424
	6.4 Solubility and Phase Behavior	425
	6.5 Corrosion Handling	430
	6.6 Salts and Inert Solids Handling	434
	6.7 Effluent Quality	436
7	SCWO PERMITTING	438
8	SCWO ECONOMICS	439
9	SUMMARY	443

1 INTRODUCTION

Our world faces an ever-worsening dual problem of increasing waste quantities and decreasing options for disposal. Traditional methods of waste disposal, such as landfill and air incineration, face greater restrictions and scrutiny from an increasingly skeptical public and litigious society, driving up costs. Supercritical water oxidation (SCWO) is an innovative and rapidly maturing treatment technology that has shown promise in providing safe, rapid, and efficient destruction of a wide variety of wastes in an environmentally responsible way. It also has a growing record of public acceptance. Thus, SCWO technology can be a viable alternative to standard waste-treatment options. However, while SCWO has been shown to be effective at treating many types of waste, it is important to understand its capabilities, strengths, and limitations when considering whether it is the best technology to use in a particular application.

The purpose of this chapter is twofold: (1) to provide a description of what SCWO is and can do, including its scientific basis and current commercial worldwide status; and (2) to allow one to determine if SCWO can be effective for a particular waste stream of interest, and if so, to present what needs to be considered in determining the optimum system configuration. Toward this end, information is provided on how and why the SCWO process works, typical system configurations and variations, development history, current full-scale and major pilot-scale operating plants, key results from several decades of research, feed characteristics, salt behavior and control techniques, corrosion control options, typical effluent characteristics, permitting record and history, and available cost data.

For additional background on SCWO technology, the reader is referred to earlier review articles by Kritzer and Dinjus, Shaw and Dahmen, Schmieder and Abeln, Gloyna and Li, and Jayaweera.[1–5,31]

2 PROPERTIES OF SUPERCRITICAL WATER

In general, any substance that is above the temperature and pressure of its thermodynamic critical point is called a *supercritical fluid*. A critical point represents a limit of both equilibrium and stability conditions, and is formally defined as a point where the first, second, and third derivatives of the energy basis function for a system equal zero (or, more precisely, where $\partial P/\partial V|_T = \partial^2 P/\partial V^2|_T = 0$ for a pure compound). In practical terms, a critical point is identified as a point where two or more coexisting fluid phases become indistinguishable. For a pure compound, the critical point occurs at the limit of vapor-liquid equilibrium where the densities of the two phases approach each other (Figures 1a and 1b). Above this critical point, no phase transformation is possible and the substance is considered neither a liquid nor a gas, but a homogeneous, supercritical fluid. The particular conditions (such as pressure and temperature) at which the critical point of a substance is achieved are unique for every substance and are referred to as its critical constants (Table 1).

Because a supercritical fluid does not undergo any phase change, its physical properties vary continuously with temperature and pressure, achieving values intermediate to those typical of liquids and gases in the usual region of interest. Most supercritical fluids have a density (or solvating power) closer to that of a liquid but mass transport properties (such as viscosity and diffusivity) closer to that of a gas (Table 2). It is often possible to obtain a dramatic change in the value of these properties with a relatively small change in pressure, particularly when passing through or near the critical point. This combination of a liquid-like solvent ability and gaslike transport properties, and the ability to "tune"

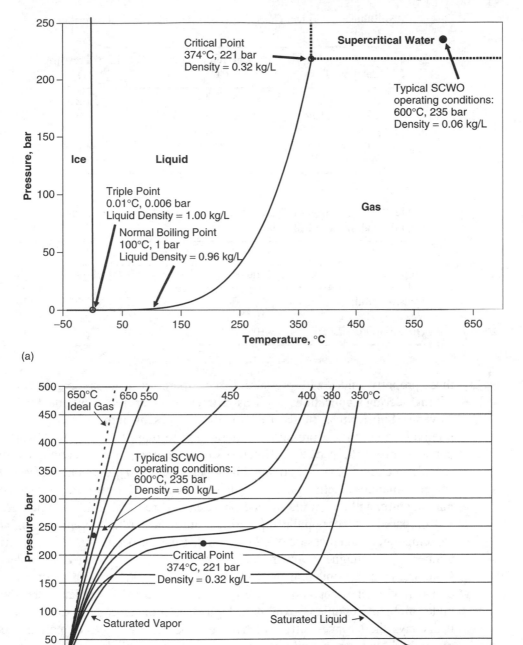

Figure 1   (a) Pressure-Temperature (P-T) and (b) Pressure-Volume (Density) (P-V) phase diagrams for pure water.

Table 1 Critical Constants for Commonly Used Supercritical Fluids

Compound	Critical Temperature (°C)	Critical Pressure (bar)
Ethylene	9.9	51.2
Carbon dioxide	31.1	73.9
Ethane	32.2	48.8
Propane	96.8	42.6
Water	374.1	221.2

Table 2 Comparison of Typical Values for Some Physical Properties of Gases, Liquids, and Supercritical Fluids

Property	Phase		
	Liquid	Supercritical Fluid	Gas
Density (g/ml)	1	10^{-1}	10^{-3}
Viscosity (Pa sec)	10^{-3}	10^{-4}	10^{-5}
Diffusivity (cm^2/sec)	10^{-5}	10^{-3}	10^{-1}

these property values with pressure, make supercritical fluids ideal for a number of engineering applications such as extraction or as replacements for organic solvents. Supercritical fluids used for these types of applications often have a relatively mild critical pressure and temperature (e.g., CO_2, C_2H_6, C_3H_8) so as not to damage the species of interest during extraction or reaction. Examples include decaffeination of coffee using supercritical CO_2, extraction of oils from spices, isolation of active plant ingredients (e.g., taxol) for use in pharmaceutical manufacture, and use as the mobile phase in supercritical fluid chromatography.

Compared to other popular supercritical fluids, water has a relatively high critical temperature and pressure of 374°C and 221 bar, respectively. This high critical temperature, combined with the unique characteristics of the water molecule's structure, leads to some unusual additional effects on the solvent nature and general behavior of water above its critical point. Water is a polar molecule that under ambient conditions exhibits a high degree of hydrogen bonding. In fact, it displays the strongest effect of hydrogen bonding of any compound because of the high electronegativity of the oxygen atom, coupled with the fact that the number of hydrogen atoms available per molecule are exactly matched by the number of oxygen lone electron pairs with which to bond (i.e., 2). This hydrogen bonding is the main reason for the high normal boiling point of water compared to analogous compounds such as H_2S (Figure 2). It is responsible for water existing in liquid form under ambient conditions and the structure of DNA, both of which literally all life on Earth depends. It also plays a significant role in

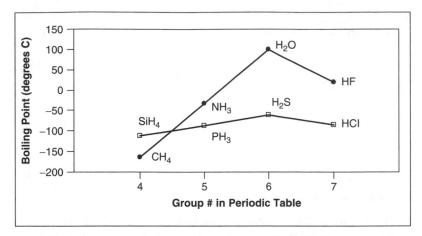

Figure 2   Boiling point trends of nonmetal hydrogen compounds. Boiling points increase with increasing molecular size (greater van der Waals forces) and with increasing electronegativity of the atom paired with hydrogen. Note that compounds with the most electronegative elements (N, O, F) have notably higher boiling points than their analogs in other periods, with water having the highest because of maximum pairing of available H atoms and oxygen lone pair electrons

the readily observable and commonly known phenomena that room-temperature water will dissolve other polar substances such as inorganic salts easily, while nonpolar substances such as hydrocarbons are insoluble.

As temperature and pressure are increased from ambient conditions up to the critical point, the density of water decreases by just over an order of magnitude. While still remaining liquid-like, the decrease in density is enough to cause a significant loss of water's ability to hydrogen bond to other water molecules, since hydrogen bonding forces are strong only when molecules are in close proximity. The number of hydrogen bonds per water molecule, normally 3.5 under ambient conditions (and 4 in theory), is only 1.8 in dense supercritical water,[6] and even less under conditions of industrial interest (Figure 3). Also under these conditions, the water molecules are sufficiently separated from each other (i.e., large enough void volume) that their dipole-dipole interactions are greatly diminished. With the loss of hydrogen bonding and weakened interactions between molecules, water in its supercritical state starts to behave more like a nonpolar rather than a polar solvent. In this case, the energy cost associated with separating the water molecules to create a cavity and add a nonpolar solute molecule is diminished or eliminated. Conversely, the energy recovery from charge stabilization by the water molecule that normally occurs with insertion of a polar solute molecule also disappears. As a result, nonpolar compounds such as most hydrocarbons and gases (e.g., O_2, N_2) are highly soluble in supercritical water and polar compounds such as inorganic salts are insoluble in supercritical water—a complete reversal of water's behavior under ambient conditions. As shown in Figure 1(b), at typical

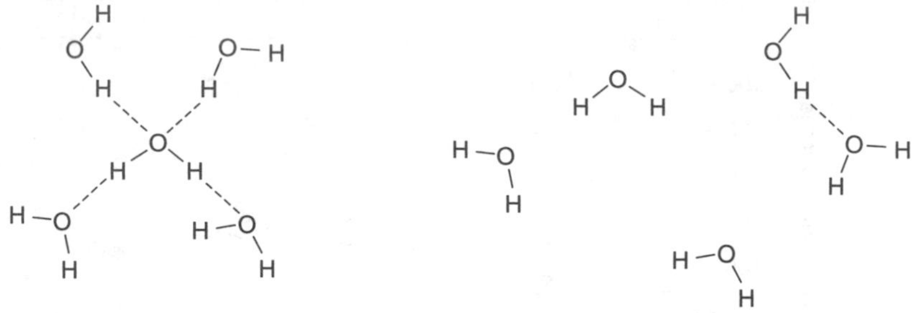

Representation of water at ambient conditions, showing full tetrahedral network of hydrogen bonds (indicated by dashed lines)

Representation of supercritical water ($\rho \sim 0.05$ g/ml), consisting of single molecules and/or binary clusters

Figure 3   Hydrogen bonding in sub- and supercritical water. These models show hydrogen bonding among water molecules at ambient and supercritical conditions. Numerous studies have been performed in recent years to determine the structure of water under both sub- and supercritical conditions. Both experimental spectroscopic data (Nuclear magnetic resonance (NMR), Infra-red (IR) absorption, and X-ray diffraction) and computational models (molecular dynamics (MD) and *ab initio*) have been generated by these investigations. While there has been debate in the literature over whether hydrogen bonding truly disappears in the supercritical state, as some experimentalists have claimed,[159] or whether hydrogen bonding persists to some degree as some modelers have claimed,[160] the conclusions tend to depend on what density range was explored or modeled. It is unlikely that hydrogen bonding behaves as a step function (i.e., being either fully "on" or "off" relative to a threshold condition), but rather that the change is gradual, with more evidence of hydrogen bonding seen as density increases and less as density decreases. All researchers agree, however, that the amount of hydrogen bonding is considerably diminished in supercritical water compared to that in ambient water. At conditions of industrial interest ($T = 400–700°C$, $P = 230–250$ bar, $\rho = 0.05–0.2$ g/ml), water is likely to exist as monomer units (i.e., no hydrogen bonding exhibited) or as dimers (clusters of two hydrogen bonded molecules),[6] far different from the three-dimensional tetrahedral network of hydrogen bonds exhibited by ambient water.

SCWO reactor conditions of 250 bar and 600°C, the density deviation of SCW from an ideal gas is fairly small. The compressibility factor Z ($= 1$ for an ideal gas) at these conditions is 0.88.

The dramatic change in the nature of water as it is heated and pressurized up to and beyond the critical point is best expressed by noting the trends of several key physical properties over this range (Figure 4). The dielectric constant of water (ε) has a value of 80 at ambient temperature, characteristic of a highly polar bulk material with a high degree of charge alignment. With increasing temperature at 250 bar, the value of the dielectric constant decreases steadily and then sharply in the vicinity of the critical point. At the critical temperature, it has a value of about 10, which is comparable to that of a weakly polar solvent such as acetone. It then further decreases to a value of about 2 at 450°C and above, which is comparable to that of a nonpolar solvent such as cyclohexane (Table 3). The ion product or dissociation constant of water (K_w) starts at its familiar value of 10^{-14} $(mol/kg)^2$ at ambient conditions and initially increases to a maximum of about 10^{-11} $(mol/kg)^2$ in the range of 200 to 300°C before eventually dropping down to

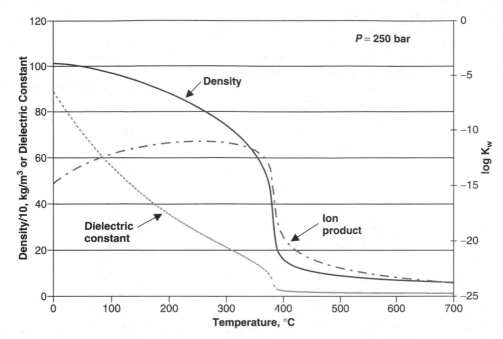

Figure 4 Solvation properties of water as a function of temperature at 250 bar. Note that as water is heated from ambient, the dielectric constant drops off more rapidly than the density. This reflects the fact that the water molecules are still relatively close together, while the thermal motion of the molecules disrupts the alignment of the dipoles

$<10^{-23}$ (mol/kg)^2 above 550°C. This change of over 12 orders of magnitude over a relatively small temperature range has a significant effect on the role water plays in such phenomena as acid/base catalyzed reactions, corrosion, and the ability to solubilize salts. As an example of typical salt behavior, the solubility of NaCl drops from a value as high as 37 wt% even up to 300°C down to only about 120 ppm by 550°C and 250 bar. Conversely, the solubility of a typical nonpolar organic species such as benzene increases from 0.07 wt% (i.e., insoluble) at room temperature, up to 7 to 8 wt% at 260°C, to completely miscible above 300°C at 250 bar. A more extensive discussion of phase behavior and solubility in the context of SCWO systems is given in Section 6.4.

Although supercritical water is an excellent solvent for nonpolar organic compounds, the high critical temperature of water can be a problem when used in extractive applications, since many of the organic compounds of interest break down under these conditions. There has been some recent interest in the use of water as a replacement solvent for reactions involving organic species, but this has been primarily limited to subcritical or near-critical temperatures, particularly where the ion product is at its maximum. For supercritical water, the combination of high temperature and its ability to bring organic species and oxygen together in a single, homogeneous, dense phase provides all of the conditions necessary

Table 3 Comparison of Dielectric Constant Values for Water at Various
Temperatures and Organic Solvents at Ambient Conditions

Dielectric Constant (ε)	Temperature at Which Water Achieves ε Value (°C)[a]	Comparable Compound (i.e., same ε value) under Ambient Conditions[b]
78.5	25	Water
58	94	Formic acid[c]
42.5	162	Glycerol
37	192	Glycol
32.6	219	Methanol
24.3	278	Ethanol
20.7	306	Acetone
15.0	349	Cyclohexanol
12.3	366	Pyridine
10.3	375	1-Octanol
9.1	379	Methylene chloride
6.2	384	Acetic acid
4.3	386	Ethyl ether
3.0	392	Furan
2.5	400	o-Xylene
2.3	407	Benzene, Carbon tetrachloride
2.0	424	Cyclopentane, Cyclohexane
1.8	447	n-Pentane

[a] At 250 bar
[b] Ambient conditions = 1 bar, 20 or 25°C
[c] At 16°C

for a combustion-like reaction to take place. As a result, the most popular application of supercritical water has been its use as a medium for the destruction of organic compounds rather than for extraction or solvation.

3 THE SUPERCRITICAL WATER OXIDATION (SCWO) PROCESS

In the SCWO process, organic compounds and an oxidant are brought together and react in a water-rich medium above the critical temperature and pressure of water. The organic feed can be either pure or in solution, but the process is particularly attractive and most competitive for dilute aqueous solutions of organics. The organic species can also be in any phase as long as it can be dissolved or slurried into a form that can be pumped to the reactor under pressure. For solids in particular, this often involves some form of size reduction as a pretreatment step. The oxidant used is typically oxygen or air, but others such as hydrogen peroxide and nitrate salts have been used in the laboratory and may be practical commercially under special circumstances. The choice of oxidant usually depends on the economics associated with the particular application. For example, the choice of pure oxygen is the most straightforward, but typically requires delivery and cryogenic storage in liquid form, along with added safety precautions. By contrast, the use of air requires large compressors and a larger-size reactor to accommodate the significant volume of nitrogen that must pass through the system.

At the temperatures and conditions characteristic of supercritical water, organic compounds are rapidly oxidized to CO_2 and water (Figure 5). The process in many ways is similar to a combustion reaction carried out in water, and in fact, visible hydrothermal flames can be generated at organic concentrations as low as 10 wt% methanol, with oxygen as the oxidant.[7,8] However, there are a number of significant differences between SCWO and traditional combustion reactions. The high pressure (although only moderate by chemical industry standards), complete miscibility of nonpolar feed species, and single phase all enable good contact among reacting species and allow a much smaller reactor volume than is typical for air incineration. The single homogeneous phase also eliminates interphase mass transfer limitations for nonsolid waste materials.

Under typical SCWO process operating conditions (i.e., 550–650°C, ~250 bar), virtually all organic compounds can be completely destroyed with

Reaction:

$$\text{Organic} + O_2 \longrightarrow CO_2 + H_2O + N_2 + \text{mineral acids} \xrightarrow{\text{base}} \text{carbonate} \downarrow + H_2O + N_2 + \text{salts} \downarrow$$

Typical operating conditions: 550–650°C
230–250 bar
< 10–120 sec residence time

Advantages / Benefits	Disadvantages / Drawbacks
Single, dense, homogenous phase with tunable properties	Corrosion
Gas-like transport properties	Salt precipitation/accumulation
Complete organic, oxidant miscibility	Potential erosion from solids
Rapid oxidation kinetics (residence time ~ seconds)	Operates at moderate/high pressure
Complete oxidation; minimal or no products of incomplete combustion (e.g., CO)	Potentially expensive materials of construction for high-temperature components
Usually > 99.99% destruction efficiency for wide range of organic compounds	Feeds must be in form that can be pumped (i.e., liquid, slurry, solution)
No interphase mass transfer limitations	
Relatively small reactor volume	
No NO_x or SO_2 formation	
Lower operating temperature than incineration	
Products inherently scrubbed by water medium	
Potential energy and water effluent recovery	

Figure 5 SCWO characteristics.

residence times of only one minute, and typically much less (often on the order of seconds). Despite the short residence time, little or no CO or other gaseous hydrocarbons characteristic of incomplete or inefficient air combustion are produced. Also unlike typical combustion processes, no NO_x compounds or SO_2 are produced. While SCWO temperatures are high enough to ensure fast kinetics, they are well below the temperatures common to a conventional incinerator or car engine (e.g., 900–1300°C), which are needed to favor formation of undesirable NO_x species, as shown through equilibrium calculations.[9] Instead, any nitrogen contained in organic feed compounds is converted primarily to N_2, with possibly smaller amounts of N_2O. Similarly, other compounds characteristic of incomplete combustion that plague air incineration, such as dioxins and furans, appear to occur at much lower levels in SCWO. Other heteroatoms (e.g., Cl, S, P) present in organic feed compounds are converted to their corresponding mineral acids (e.g., HCl, H_2SO_4, H_3PO_4) in the SCWO process.

The use of water as the reaction medium in the SCWO process adds further advantages over that of air incineration processes. The water simultaneously acts as an inherent scrubber for acid and other product gases, resulting in a clean effluent gas phase after process conditions are returned to ambient. Because water is the medium, the considerable energy expense for incineration of vaporizing water in high-water-content feeds is minimized in SCWO. Hence, SCWO is particularly ideal for processing dilute aqueous organic feeds.

Although the increased solubility of nonpolar organic species in supercritical water is a major advantage of the SCWO process, the corresponding decreased solubility of polar species such as inorganic salts can be a problem. In fact, the two main difficulties encountered by the SCWO process in general are corrosion and salts/solids handling. Both issues often arise when handling organic species containing heteroatoms. The acids produced during breakdown of these feeds can be particularly aggressive under SCWO process conditions. Ironically, the greatest risk of corrosion is frequently in the heat-up and cool-down sections of the system, where temperatures are hot but still subcritical, and particularly where the dissociation constant of water is at its maximum. At supercritical temperatures in the center of the reactor, water does not support dissociation of acidic species into the corresponding ions usually responsible for corrosion. However, corrosion can occur in the microenvironments that can form underneath a salt layer near the reactor walls, where dissociation is possible due to formation of dense brines or salt melts.

Most SCWO designs utilize high nickel alloys such as Inconel or Hastelloy as the materials of construction for components and tubing exposed to supercritical conditions. This is due to the improved corrosion resistance and higher temperature strength of these materials over stainless or carbon steels under typical operating conditions. Other materials can and have been used in SCWO for specific corrosion resistance for the particular mixture composition being processed, such as the use of titanium when processing feeds with high chloride content.

A way to further minimize corrosion is by adding base to the feed or reactor, so that acids formed during the oxidation reaction are immediately neutralized. However, one must then deal with the resulting salts. Whether formed during reaction or already contained in the feed, salts will quickly precipitate in supercritical water. As these salts tend to adhere to and accumulate on the reactor walls and other surfaces within the reactor, they can inhibit and ultimately block process flow unless they are removed or their accumulation is controlled. Non-salt solids (e.g., metal oxides, grit), by contrast, have little tendency to stick to process surfaces but can be a problem with respect to erosion and system pressure control. Methods that have been developed to manage and/or minimize the impact of corrosion, salt precipitation/accumulation, and solids handling are discussed in Sections 6.5 and 6.6.

The most common application of SCWO is for destruction of organic hazardous wastes. Figure 6 shows a general process flow diagram of a SCWO process. Although there are many specific variations that can be made, depending on the particular application, design philosophy, and economics involved, all SCWO processes consist of these basic steps. Most feeds consist of an aqueous waste stream containing 1 to 20 wt% organic. This targeted range is where SCWO is most competitive because it represents a concentration too dilute for incineration to handle efficiently but too concentrated for practical application of an adsorption-based process such as activated carbon treatment. Solid organic waste materials (e.g., plastic, rubber, wood) need to be size-reduced and mixed

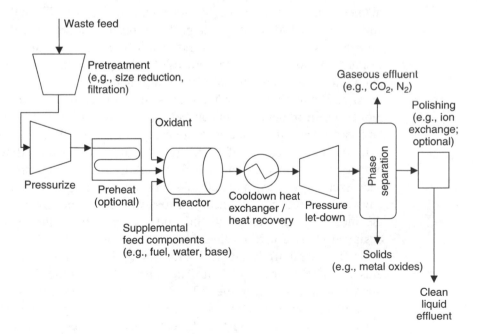

Figure 6   General process flow SCWO diagram.

into a homogeneous slurry or suspension that can be pumped up to operating pressure prior to processing. The pressurized feed may be preheated or may be fed unheated (i.e., cold) and utilize the heat of reaction to reach operating temperature upon entering the reactor.

In some designs, reactor effluent is used to preheat the incoming feed in a regenerative heat exchanger. Preheating the feed is the most direct approach to ensuring that the feed reaches the desired temperature, but labile feeds may begin to hydrolyze or decompose prematurely if preheated, forming char or corrosive species in the reactor inlet line. Feeding without preheat avoids this problem and the risk of corrosion while passing through hot subcritical temperatures, but requires a more complex feed mechanism and good mixing conditions to allow for rapid heating and reaction initiation once inside the reactor.

The compressed oxidant can be added to the feed prior to reaching the reactor or added directly to the reactor through a separate feed line. Similarly, several other supplemental feed streams may be added to the organic feed at any point or fed separately to the reactor. These other feeds are typically added to allow for better control of conditions in the reactor. Examples include auxiliary fuel such as kerosene (to maintain reactor temperature for a low-heating-value waste), diluent water (to maintain reactor temperature for a high-heating-value waste), and base (to neutralize acids formed during oxidation).

The most common types of SCWO reactors are cylindrical vessels or pipes. The vessel-type reactors (i.e., length/diameter ratio <20) typically have a vertical orientation with process flow directed downward. Pipe reactors (i.e., length/diameter ratio >100) typically are mounted horizontally and are often coiled, since they need to be longer than vessel reactors to achieve the same residence time. SCWO reactors may have external heating (e.g., heating coils, steam jacket) or may rely solely on heated process fluid and the heat of reaction to reach and maintain the desired internal temperature under steady-state conditions.

After exiting the reactor, the process fluid undergoes temperature and pressure reduction. In principle, this can be done in any order, but usually the temperature is reduced first since it is preferable to handle liquid water effluent rather than steam, particularly if there are salts or solids in the effluent. Temperature reduction is usually accomplished by passing the effluent through some form of heat exchanger, such as shell and tube or double pipe. Many SCWO designs have utilized regenerative heat transfer, where the hot effluent is used to preheat the incoming feed, although this adds complexity to the system. For systems designed to handle feeds that contain or produce salts, there may be an added flow of water introduced just before or after the end of the reactor that also aids in temperature reduction. The purpose of this flow (often referred to as quench water) is to drop the effluent temperature from super- to subcritical prior to reaching the main heat exchanger, so that any precipitated salts exiting the reactor will redissolve in the effluent. Such a strategy increases the volume of effluent produced, however.

Pressure reduction can be achieved through a variety of means, such as back pressure control valves or capillary tubing. This can be done in one or more stages and on either the full effluent flow or on separate gas and liquid phases (if phase separation is performed before full pressure reduction). Effluent containing a large quantity of solids can lead to significant erosion of pressure reduction components such as valve stems, unless the pressure let-down subsystem is properly designed to handle these types of flows. Energy recovery of the pressurized effluent (e.g., through a turbine) has been proposed in the past and is certainly possible in principle, but it has, to our knowledge, never been effectively or economically implemented in a working SCWO design.

Separation of phases is typically the last step in the SCWO process. Gaseous products, the major ones being CO_2, N_2, and excess O_2, begin to come out of solution once conditions become subcritical. They can be separated from the liquid flow through use of a simple separator or flash vessel. There may be more than one gas/liquid separation step, depending on the number of pressure reduction steps. The total product gas flow is combined and usually monitored by several on-line analyzers such as those measuring CO, total hydrocarbons (THC), and O_2 concentrations. These species (or lack thereof for the first two) are good indicators as to conditions in the reactor and whether the process is running properly. The clean and inherently scrubbed vapor phase can be released directly to the atmosphere or to a facility ventilation system. For low solids feeds, the remaining liquid phase often has a consistency similar to mineral water, due to a residual amount of CO_2 and the presence of dissolved solids. Depending on the feed, it can also be more like a brine solution (i.e., high dissolved salt concentration) or slurry (i.e., high suspended solids concentration).

Solid/liquid phase separation may or may not be necessary, depending on the amount and type of solids contained in the effluent and the ultimate means of disposal. For example, it may be advantageous to separate out certain regulated heavy metals in the effluent (which will exist as oxides after passing through SCWO) in order to minimize costs. One would also want to separate out the solids if certain compounds (oxides or salts) were desirable or valuable. Solids can be separated by any standard devices such as filters or cyclones. Salts can be concentrated via techniques such as reverse osmosis or evaporation/crystallization. Depending on the discharge location and requirements, the liquid effluent may undergo a final polishing step such as passage through an ion exchange column. In most cases, the clean liquid effluent can be disposed directly to a public sewer line or be reused as process-grade water without further treatment.

There are a few technologies related to the SCWO process that are worth mentioning. Wet air oxidation (most commonly known commercially as the Zimpro Process) is a process similar to SCWO that operates at sub- or supercritical pressures but always subcritical temperatures (e.g., 150–300°C).[10] An advantage to operating at lower temperatures is the use of less expensive materials of construction and retention of many salts in solution. The trade-off, however, is longer residence times (on the order of tens of minutes or hours rather than seconds) and lower destruction efficiency of organic feeds. Another process related to SCWO is Assisted Hydrothermal Oxidation (AHO), which was developed by personnel at SRI International.[11] AHO operates at near supercritical temperatures (e.g., 380–420°C) and either sub- or supercritical pressures. The process utilizes a reactor consisting of a bed of fluidized solids (typically sodium carbonate, which is insoluble under these conditions) that functions as both a reactant and adsorptive surface for salt control.

4   HISTORY AND CURRENT STATUS OF SCWO

Although the unique properties of supercritical fluids have been known for over a century, the development of SCWO has occurred only over the past few decades. There have been earlier applications of supercritical water, most notably its use as a heat transfer medium in nuclear reactors (pressurized water reactors), which has been utilized in commercial power plants since the 1950s. In the area of waste destruction, the subcritical water-based process known as wet air oxidation predates SCWO, with initial development going back to the early 1900s and the first commercialization of this technology beginning in the 1950s. It was not until the mid-1970s that the unique and superior properties of supercritical water for organic destruction were discovered and documented.

While performing reforming experiments with glucose in aqueous solution in 1975, Modell and co-workers at MIT observed unusual results. As in all reforming reactions, the purpose was to break down a complex organic compound into smaller, more useful species such as H_2 or CH_4, although here in the presence of pressurized liquid water rather than the more traditional steam or vapor phase. At subcritical temperatures, they noticed a considerable amount of char formation (undesirable partially oxidized organic solids), which increased with increasing temperature up to the critical point. For those experiments conducted above the critical temperature of water, however, little to no char formation was observed. The reason for this dramatic change in results has been attributed to the increased solvency of supercritical water for the organic reaction intermediates, preventing char and tar precursors from agglomerating to form these undesired byproducts. A patent for this discovery was issued in 1978.[12] Incidentally, it should be noted that reforming or gasification reactions in supercritical water is another application that is under development and has seen increased interest in recent times as the need to find alternative, cleaner energy sources increases.[13,14]

Building on the results of these initial studies, Modell began adding oxygen in order to explore the effect of the supercritical water environment on oxidation reactions of organic species. In 1980, he started MODAR as the first company established to develop and commercialize the SCWO process. Many of the characteristic attributes that are commonly associated with SCWO were first discovered and described by personnel at MODAR, and they established an extensive database of organic destruction efficiencies and material compatibility. Later in the 1980s, MODAR was joined by MODEC (a second company started by Modell) and Oxidyne (Table 4). At the same time, basic research programs were begun in several universities and national labs to more fully develop the underlying science behind supercritical water and SCWO in hopes of being able to better optimize SCWO system designs. Of particular note were those programs established at the University of Karlsruhe in Germany under Dr. E. U. Franck, at MIT under Dr. J. W. Tester, at the University of Texas at Austin under Dr. E. F. Gloyna, at the University of Michigan at Ann Arbor under Dr. P. E. Savage, and at the Forschungszentrum Karlsruhe (FZK) in Germany and Los Alamos and Sandia National Laboratories in the United States. Support for early research and commercial development of SCWO in the United States came from venture capital and government agencies such as the Army, the Environmental Protection Agency (EPA), NASA (for waste treatment during extended space missions[15]), and the Department of Energy (for destruction of food processing and industrial wastes).[16]

Research efforts intensified in the 1990s as the potential of SCWO as an effective waste destruction technology became more widely recognized. The need to understand and develop approaches to mitigating the effects of corrosion and salt plugging also drove much of this continued research. Some of the targeted applications of SCWO initially inspiring this research were volume reduction of mixed radioactive/organic wastes from nuclear clean-up programs,[17–18] pulp and paper mill waste,[19] and human waste/sewage sludge treatment.[20–21] Several branches of the U.S. Department of Defense became interested in the potential of SCWO during the 1990s and initiated a number of programs to explore and develop its use for a variety of military wastes, both at the university and commercial level. In particular, the U.S. Defense Advanced Research Projects Agency (DARPA) became interested in SCWO for destruction of chemical and obsolete conventional munitions. The Army became interested in SCWO for destruction of chemical weapons after a favorable review on its potential as a replacement for incineration by the National Research Council (NRC) in the early 1990s.[22] The U.S. Navy also became interested in the use of SCWO on its ships for destruction of shipboard wastes,[23] and the U.S. Air Force tested the use of SCWO for destruction of rocket motor propellant during this time period.[24] Most of the work being done at this time was still at the bench or pilot-scale.

By 1994, Eco Waste Technologies (established in 1990) had designed and built the first commercial SCWO plant for Huntsman Chemical Co. in Austin, Texas.[25,26] This plant was designed to treat laboratory wastewater containing alcohols, glycols, and amines. Although this feed is relatively benign from a SCWO standpoint (i.e., no corrosive species or salt formation), it was nevertheless an important milestone in the progress of SCWO commercialization. The plant operated successfully for several years before being shut down by Huntsman in 1999. During the 1990s, several other established companies became

Table 4 Past and Present Companies Involved in Commercializing SCWO

Company	Dates of Involvement in SCWO Commercialization	Unique Reactor Design or Process Feature	Licensees or Partners
MODAR, Inc.	1980–1996[a]	Vertical reverse flow vessel reactor	Organo Corp.
MODEC (Modell Environmental Corp.)	1986–1995	Horizontal tubular/pipe reactor; high throughput velocity and periodic mechanical brushing	Organo Corp., Hitachi Plant Engineering & Construction, Ltd., NGK Insulators, Ltd., NORAM Engineering and Constructors, Ltd.
Oxidyne Corp.	1986–1991	Cased subsurface deep well for reactor	
Eco Waste Technologies, Inc.	1990–1999[b]	Tubular/pipe reactor	Chematur Engineering AB, Shinko Pantec (Kobelco)
GeneSyst International, Inc.	1990–2003	Below ground gravity pressure vessel	
Abitibi-Price, Inc.	1991–1997	Tubular/pipe reactor	General Atomics
General Atomics	1991–present	Vertical vessel reactor with replaceable liner	Komatsu Ltd., Kurita Water Industries Ltd.
Summit Research/ Turbosystems Engineering, Inc.	1992–present	Transpiring wall reactor (porous liner)	
Foster Wheeler Development Corp.	1993–2003	Transpiring wall reactor (platelet liner)	Aerojet Gencorp Corp., Sandia National Laboratory
SRI International	1993–present	Fluidized bed reactor (AHO process)	Mitsubishi Heavy Industries Ltd.
KemShredder, Ltd	1993–1996[c]	Use of H_2O_2 as oxidant	
Chematur Engineering AB	1995–present	Tubular/pipe reactor; multiple oxidant and quench injection points	Johnson Matthey, Feralco AB, Stora Enso
Hanwha Chemical Corp.	1994–present	Have used both vessel and tubular reactors	
HydroProcessing, L.L.C.	1996–2002	Tubular/pipe reactor; patented pressure let-down design	
Hydrothermale Oxydation Option	2000–present	Multistage tubular/pipe reactor	

[a] Acquired by General Atomics
[b] Acquired by Chematur Engineering AB
[c] Acquired by HydroProcessing, L.L.C.

involved in SCWO commercialization work, including many still active today, such as General Atomics, SRI International, and Chematur Engineering AB. Each of these companies has developed a unique variation of the basic SCWO process, which differs from the others primarily by how corrosion and salt precipitation/accumulation are controlled, and/or what type of feed is targeted (e.g., liquid vs. slurry, high salt vs. no salt). Many of these SCWO vendors have also licensed their core technology and process design to other companies to develop for a particular geographic region such as Asia. Others have partnered with firms that provide either a key design component or a unique application in order to develop a customized version of the SCWO process.

By the late 1990s/early 2000s, nearly two decades of research and development had resulted in a significant maturation of SCWO technology, with a better understanding of the core principles and improving techniques to mitigate the corrosion and salt-plugging problems. As is often the case with maturing technologies, this in turn resulted in a shift in emphasis of SCWO activity from one of fundamental research to full-scale application. Between roughly 1998 and 2002, seven full-scale SCWO plants were built in locations around the world for a wide variety of feeds, ranging from industrial wastewater and sewage sludge to explosives (Table 5). Since then, three of these plants have been shut down permanently, with mechanical or operational problems being a factor in at least two of these shutdowns. Two of the associated SCWO vendors (Foster Wheeler and HydroProcessing) have also since left this field. However, two additional SCWO plants became operational in 2004 and 2005 in France and Japan, respectively, and one is expected to begin operation in South Korea in 2006. Thus, while the transition of SCWO technology to large-scale application has not been without difficulty, it nevertheless has continued to garner more interest from an increasing number of potential clients, with a small but steady number of new plants being commissioned.

In addition to Foster Wheeler and HydroProcessing, several companies are no longer active in SCWO, including the original SCWO companies (MODAR, MODEC, Oxidyne, and EcoWaste Technologies), as well as Abitibi-Price and KemShredder. The legacy of some have survived in some sense through acquisition by other companies or through licensees. For example, General Atomics acquired MODAR in 1996 and the Swedish company Chematur Engineering acquired the assets of EcoWaste Technologies in 1999. Both companies have since become very active in furthering SCWO development and commercialization. Also, the Japanese firm Organo, a licensee of MODAR, has built two full-scale SCWO plants utilizing the reverse flow vessel reactor design that was the heart of the MODAR process. A number of additional firms and their licensees are also still active in SCWO, including SRI International, Hanwha Chemical, and Hydrothermale Oxydation Option.

Table 5 Commercially Designed Full-scale SCWO Plants

Company (Process)	Client, Location	Feed	Reactor Type	Capacity (tons/day)	Notes	Reference
Active						
General Atomics	U.S. Army McAlester Army Ammunition Plant, McAlester, OK	Pink water	Vessel	6.5	Operating since 2001	General Atomics (unpublished)
Chematur Engineering	Johnson Matthey, Brimsdown, UK	Spent catalyst (recovery of precious metals)	Pipe	80	Operating since 2002; taken over by JM in October 2004	Grumett, 2003 (Ref. 27)
Hanwha Chemical	1. Namhae Chemical Corp., Korea 2. Samnam Petrochemical Co., Korea	1. Waste water from DNT production 2. Waste and wastewater from TPA production	1. Vessel 2. Pipe	1. 53 2. 145	1. Commissioned in 2002 2. Initial start-up of plant to begin in mid-2006	Han, 2006 (Ref. 33)
Organo (MODAR)	A national university, Japan	Laboratory wastewater	Vessel	Not available	Operating since 2002	Urai, 2005 (Ref. 32)
Hydrothermale Oxydation Option	SYMPESA, southern France	Food industry wastewater	Pipe	2.7	Began operating in April 2004	Bonnaudin, 2005 (Ref. 34)
Mitsubishi Heavy Industries (SRI International)	Japan Environmental Safety Corp. (JESCO), Tokyo Bay	PCBs	Vessel (fluidized bed) followed by pipe	306 (3 trains)	Began operating in late 2005	Jayaweera, 2003 (Ref. 31)

(continued overleaf)

Table 5 (*continued*)

Planned

General Atomics	1. U.S. Army Blue Grass Chemical Activity, Richmond, KY	1. Chemical weapons (agent and energetic hydrolysates)	1. Vessel	1. 36 (3 trains)	1. Currently in design phase; operation to begin in 2011	1. Marrone et al., 2005 (Ref. 35)
	2. Tooele Army Depot, Tooele, UT	2. Hydrolysate of conventional explosive devices	2. Vessel	2. 18	2. Currently in design phase; operation to begin in 2007	2. General Atomics (unpublished)

Inactive

EcoWaste Technologies	Huntsman Chemical, TX	Waste alcohols, glycols, and amines	Pipe	29	In operation from 1994–1999	Svensson, 1995 (Ref. 25); McBrayer et al., 1996 (Ref. 26)
Foster Wheeler Development Corp.	US Army Pinebluff Arsenal, AR	Obsolete conventional weapons (smokes, dyes)	Vessel	3.8	Built in 1998; shut down in 2001 due to liner mechanical problems and limited funding	Crooker et al., 2000 (Ref. 23)
Organo (MODAR)	Nittetsu Semiconductor, Tateyama, Japan	Semiconductor manufacturing wastes	Vessel	2	Built in 1998; stopped operation due to business decision by client	Oe et al., 1998 (Ref. 161); Urai, 2005 (Ref. 32)
Hydro-Processing	Harlingen Wastewater Treatment Plant, TX	Municipal/industrial wastewater sludge	Pipe	150 (9.8 dry tons/day of sludge) (2 trains)	Built in 2001; shut-down in 2002 due to corrosion in cooldown heat exchanger	Griffith and Raymond, 2001 (Ref. 132)

As of late 2005, there were six full-scale SCWO plants in operation around the world, one under construction, and at least two in the design phase:

1. Chematur Engineering AB has partnered with the chemical company Johnson Matthey in developing and building a full-scale SCWO plant for processing spent catalyst, replacing an incineration-based process.[27] This plant, which became operational in the fall of 2002, is located in Brimsdown, United Kingdom, and treats a 5 percent slurry feed of spent catalyst. The process, referred to as AquaCat®, destroys the organic contaminants and carbon backing, while allowing for recovery of the precious metal components as precipitated solid oxide particles for reprocessing and reuse. Chematur has also been working with client/partner Feralco AB on a version of SCWO (Aqua Reci®), which will enable recovery of phosphate from processing of sewage sludge[28] and recovery of iron or aluminum coagulants from processing of drinking water treatment sludge.[29] They have been working with Stora Enso on a version of SCWO that will enable recovery of paper filler materials from processing of paper mill deinking sludge.[30]

2. Mitsubishi Heavy Industries Ltd., a licensee of the SRI International AHO process, has built the most recent full-scale SCWO plant. The plant was built for the Japan Environmental Safety Corp. (a company owned by the Japanese government) to destroy that nation's large stockpile of polychlorinated biphenyl (PCB) waste.[31] It consists of three reactor systems, each having a capacity of 2 tons PCB/day and 100 tons water/day, and a treatment requirement of <3 ppb PCB in the liquid effluent (representing six 9s destruction efficiency). The plant began operation in 2005.

3. A second full-scale SCWO plant in operation in Japan was built by Organo Corp. to treat laboratory wastewater at a national university. This plant is based on the MODAR design and has been operating since 2002.[32]

4. Hanwha Chemical Corp. built a full-scale SCWO plant for Namhae Chemical Corp. in Korea. The plant was commissioned in 2002 for treatment of wastewater from production of dinitrotoluene (DNT). Hanwha has constructed another SCWO plant for treatment of waste and wastewater from terephthalic acid (TPA) production. This plant was preparing for initial start-up in mid-2006.[33]

5. The newest company to venture into SCWO commercialization, Hydrothermale Oxydation Option (HOO), designed and built a plant in France that began operating in April 2004.[34] HOO is affiliated with the Institute of Condensed Matter Chemistry Bordeaux (ICMCB) at the Centre National de la Recherche Scientifique (CNRS) in France, and was established to commercialize SCWO research and patents developed by ICMCB. The plant was built for a consortium referred to as SYMPESA to treat food and pharmaceutical industry wastewater. Although it is referred to as the first

Table 6 Notable SCWO Pilot-scale Systems (≥ 10 kg/hr capacity)*

Organization	Location	Description	Reference
Forschungszentrum Karlsruhe (FZK)—Institute of Technical Chemistry (ITC-CPV)	Karlsruhe, Germany	System equipped with both a pipe reactor (10 kg/hr capacity) and transpiring wall reactor (20 kg/hr capacity)	Abeln et al., 2001 (Ref. 133)
Swiss Federal Institute of Technology (ETH)	Zurich, Switzerland	Wall-cooled hydrothermal burner operated with hydrothermal flame; 10 kg/hr capacity	Wellig et al., 2005 (Ref. 134)
University of British Columbia	Vancouver, BC, Canada	Pipe/tubular reactor system built in partnership with NORAM Engineering and Contractors; outfitted for heat transfer and fouling research; 120 kg/hr feed rate	Teshima et al., 1997 (Ref. 135)
University of Valladolid	Valladolid, Spain (CETRANSA Treatment Center for Waste Management)	Cooled wall, packed bed reactor; 30 kg/hr capacity; recently scaled up to 200 kg/hr capacity	Cocero et al., 2000 (Ref. 136)

*Does not include pilot-scale systems owned and operated by commercial SCWO companies or their licensees or partners listed in Table 4.

industrial SCWO plant in France, the relatively low feed rate of 100 kg/hr implies that the plant is more similar to a large pilot-scale unit.

6. General Atomics constructed a small commercial plant for the McAlester Army Ammunition Plant in McAlester, Oklahoma, which has been operating since 2001. The plant processes pink water from explosives processing using General Atomics' design.

7. General Atomics is currently contracted to supply a SCWO system to the Tooele Army Depot in Tooele, Utah. The plant will process hydrolyzed conventional explosive devices using General Atomics' design. The program is currently in the design stage, with the expected completion date for the plant projected for 2007.

8. General Atomics is part of a team contracted by the United States Army to design, build, and operate a facility to destroy the stockpile of chemical weapons located in Richmond, Kentucky.[35] In the proposed process, all chemical agent and explosive materials will first be hydrolyzed in hot caustic or water, with the resulting hydrolysate then destroyed by SCWO. The SCWO portion will be based on General Atomics' design.

Table 7 Examples of Substances Successfully Destroyed by SCWO

Specific Organic Compounds

Acetamide	Cyclotrimethylene nitramine (RDX)	Hexanol	Nitrotoluene
Acetic acid	DDT (1,1,1,-trichloro-2,2-bis(p-chlorophenyl) ethane	Hydrazine	Octahydro-1,3,5,7-tetranitro-1,3,5,7-tetrazocine (HMX)
Acetone	Decachlorobiphenyl	Hydrogen	Pentachlorophenol
Acetonitrile	Dextrose	Hydroquinone	Pentaerythritol tetranitrate (PETN)
p-Aminophenol	Diethyl ether	o-Hydroxyaceto-phenone	Phenol
Ammonia	Diethylene glycol	m-Hydroxyaceto-phenone	Polychlorotrifluoro-ethylene
Ammonium chloride	1,2-Dichlorobenzene	p-Hydroxyaceto-phenone	Polyvinyl chloride (PVC)
Ammonium hydroxide	1,4-Dichlorobenzene	Isopropanol	n-Propanol
Ammonium perchlorate	2,4-Dichlorobenzene	Isopropyl amine	Propylene Glycol
Ammonium sulfate	1,2-Dichloroethylene	Lactic acid	Pyrene
Ammonium sulfide	2,4-Dichlorophenol	Malathion	Pyridine
Aniline	2,4-Dichlorophenoxy-acetic acid	Methane	Resorcinol
Benzene	Diethanolamine	Methanol	Sodium isethionate
Biphenyl	Dimethyl methyl phosphonate (DMMP)	m-Methoxyphenol	Stearic acid
2-Butanone	2,4-Dimethyl phenol	p-Methoxyphenol	Tetrabromobisphenol A
Carbon	2,4-Dinitrotoluene	Methylamine	Tetrachloroethylene
Carbon monoxide	Dodecane	Methylene chloride	Tetramethyl ammonium hydroxide
Carbon tetrachloride	EDTA (ethylene diamine tetraacetic acid)	Methyl ethyl ketone	Thiodiglycol
Chemical agent hydrolysate (HD, GB, VX)	Ethane	Methylphosphonic acid (MPA)	Toluene
2-Chlorobiphenyl	Ethanol	Naphthalene	Tri-n-butyl amine
3-Chlorobiphenyl	Ethylene Glycol	Naphthol	Trichloroacetic acid
4-Chlorobiphenyl	o-Ethylphenol	p-Nitroaniline	1,1,1-Trichloroethane
Chloroform	m-Ethylphenol	Nitrobenzene	1,1,2-Trichloroethane
4-Chloro-3-methylphenol	p-Ethylphenol	Nitroethane	Trichloroethylene
2-Chlorophenol	Fluorescein		2,4,6-Trichlorophenol
4-Chlorophenol	Formic acid	Nitromethane	Trinitrotoluene (TNT)

(*continued overleaf*)

Table 7   (*continued*)

o-Chlorotoluene	Hexachlorobenzene	2-Nitrophenol	VX (o-ethyl-S[2-(diisopropylamino)-ethyl]-methylphos-phonothiolate)
o-Cresol	Hexachlorocyclohexane	4-Nitrophenol	o-Xylene
Cyclohexane	Hexachlorocyclopenta-diene	1-Nitropropane	

Complex and/or Mixed Organic Wastes

Adumbran	Crude Oil	Ion Exchange Resins (Styrene; Divinyl benzene)	Polyaromatic hydrocarbons (PAHs)
Alcohol distillery wastewater	Diesel Fuel	Lignite	Polymeric materials
Arochlors (PCBs)	Dog food	Lubricating Oils & Greases	Protein
Asphalt	Drinking water treatment sludge	Malaria Antigen	Pulp & Paper Mill Sludges
Bacillus Stearothermophilus (Heat-Resistant Spores)	Dyes	Medical Wastes	Radioactive Mixed Wastes
Bran Cereal	Dyestuff wastewater	Metal Cutting Fluids	Rocket Propellants
Brewery wastes	E. coli	Microalgae	Sewage Sludge
Carbohydrates	Electronic component manufacture waste	Military Smokes & Dyes	Solvents
Casein	Electronic scrap	Natural & Synthetic Rubbers	Soybean Plants
Cellulosics	Emulsified Wastes and Oils	Natural Gas	Spent catalyst
Chemical & Biological Warfare Agents	Endotoxin (Pyrogen)	Navy shipboard wastes	Sulfolobus Acidocaldarius
Chlorinated dibenzo-p-dioxins	Flame retardants	Paper	Surfactants
Chlorofluorocarbons (CFCs)	Food Processing & Agricultural Wastes	Peat	Textiles
Coal & Coal Slurries	High Explosives	Pesticides	TNT Waste (Red and Pink Water)
Coal Waste	Human Waste (black/gray water)	Petroleum Refining & Petrochemical Wastes	Transformer Oil
Corn Starch	Hydraulic Fluid	Pharmaceutical Wastes	Yeast

Note: Adapted from Ref. 36.

Of importance to note was the considerable level of public involvement in the extensive testing and evaluation of potential technologies as alternatives to incineration for this application, and the strong public support demonstrated for SCWO. The program is currently in the design stage, with the expected completion date for the plant projected for 2011.

There are many pilot-scale SCWO systems currently in existence at both research and commercial entities. All of the major SCWO companies listed in Table 4 have or had at least one pilot-scale system used to conduct scoping studies for interested clients and for internal research investigations. Other notable pilot-scale SCWO systems are listed in Table 6.

Over the past several decades of SCWO work ranging from research to commercial applications and from lab to full-scale in size, a long list of feeds has been shown to be effectively destroyed by this technology (Table 7). The variety of compounds treated is considerable, ranging from simple hydrocarbons to complex multi-substituted mixed wastes. Under normal SCWO operating conditions, nearly all of these compounds have been destroyed to at least 99 percent destruction efficiency (and typically much higher for the vast majority of cases) without the noxious byproducts typically produced by other waste destruction technologies such as air incineration.

5 SCWO RESEARCH EFFORTS

Basic research efforts with respect to supercritical water and SCWO have been directed within a number of areas that are critical for design and optimization of a SCWO reactor and ancillary equipment. These areas include physical property measurement and correlations, kinetics and reaction mechanisms, salt equilibrium and transport behavior, and corrosion.

5.1 Physical Property Determination

Many physical properties undergo dramatic changes in value as water is heated and pressurized from sub- to supercritical conditions, particularly in the region of the critical point where some properties such as heat capacity reach a singularity. This change in behavior means that more familiar correlations of properties measured at subcritical conditions are likely to be inaccurate when applied at supercritical conditions. There have been some experimental studies performed to measure, tabulate, and in some cases correlate values of key properties of supercritical water, such as the self-diffusion coefficient,[37-38] viscosity,[39-40] thermal conductivity,[41] heat capacity at constant volume,[42] dielectric constant,[43] and self-dissociation constant.[44] Far more work has been devoted to calculation of property values from models fitted empirically to data or developed more rigorously through molecular simulation. For PVT data and its derivatives, several attempts

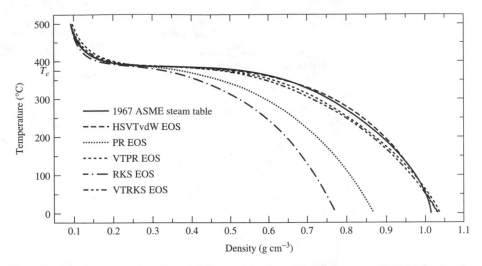

Figure 7   Temperature-density plot for water at constant pressure (253.31 bar) calculated from various equations of state. HSVTvdW is the hard-sphere, volume-translated van der Waals equation of state developed by Kutney et al. (Reprinted from Ref. 48). Other equations of state shown: PR = Peng-Robinson, VTPR = volume translated PR, RKS = Redlich-Kwong-Soave, VTRKS = volume-translated RKS

have been made to modify standard equations of state to accommodate the properties of supercritical water[45–46] or develop new ones.[47] In particular, the hard sphere, volume-translated van der Waals equation of state of Kutney et al. has been specifically developed for use under SCWO conditions (Figure 7).[48] The latest formulation of thermodynamic properties of water issued by the International Association for the Properties of Water and Steam (IAPWS-95)[49] allows reasonably accurate calculation of a number of important properties over a wide range of conditions encompassing the subcritical, near critical, and supercritical state (up to 1000°C and 1000 MPa), including energy (e.g., internal, enthalpy, Gibbs, Helmholtz), thermodynamic derivative quantities (e.g., heat capacity, isothermal compressibility, density, dielectric constant), and transport properties (e.g., viscosity, thermal conductivity). Empirical heat transfer correlations that account for properties and phenomena associated with supercritical water have also been developed for various reactor configurations and geometries.[50]

5.2   Global Kinetics and Elementary Reaction Network Modeling

Research into reaction kinetics under well-defined conditions has been one of the most active areas of investigation, as this knowledge is directly applicable to optimized reactor sizing and design. Much of this work has been performed at the lab scale in universities and national laboratories around the world. The focus in these kinetic studies has typically been on model compounds rather than actual wastes. These compounds have been chosen because they are easier and

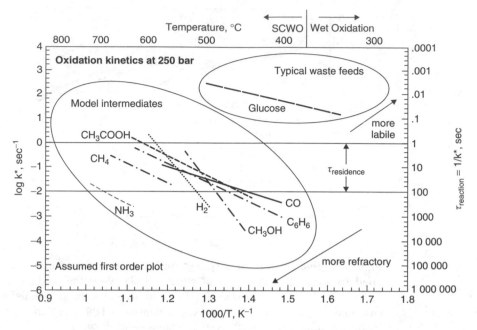

Figure 8 **Assumed first-order Arrhenius Plot for oxidation of several model organic compounds in supercritical water, showing relative ease or difficulty of destruction.** The band denoted as $\tau_{residence}$ represents a typical time scale that can be investigated in a lab-based SCWO system. Note that the ordinate is the base ten logarithm and K* represents the assumed first order rate constant. (From Refs. 51–57.)

less hazardous to work with yet still contain key functional groups of importance and/or represent the rate limiting step in the breakdown of a wider range of more complex waste species. Examples of many of these compounds are listed in the top half of Table 7 and in Figure 8, with some of the most studied being methanol, acetic acid, and phenol. Much of the kinetics work at the macroscopic level has centered on developing global rate expressions of the generic power law rate form:

$$d[\text{Organic}]/dt = -A\exp(-E_a/RT)[\text{Organic}]^a[\text{Oxidant}]^b[\text{H}_2\text{O}]^c \qquad (1)$$

where A and E_a are the preexponential factor and activation energy, respectively, in the terms constituting the Arrhenius-form rate constant, and a, b, and c are the order with respect to the organic species, oxidant (typically O_2), and water, respectively. These terms are typically regressed from the experimental data. Modifications or correction factors to this standard rate form have been applied in certain cases—for instance, where catalytic effects[51] or solvation effects[58] have been observed. Examples of rate expression parameters from past studies are provided in Table 8. More recently, there have been attempts to generate global rate expressions based on feed mixtures more applicable to real-world wastes using total organic carbon to track overall conversion instead of a single specific organic species concentration in the rate expression.[59]

Table 8 Global Rate Expressions for Oxidation of Select Model Organic Compounds in Supercritical Water*

Compound	Log A $[(mol/L)^{1-a-b-c}, s^{-1}]$	E_a (kJ/mol)	a	b	c	No. of data points	Conditions	Reference
Carbon monoxide	8.5 ± 3.3	134 ± 32	0.96 ± 0.30	0.34 ± 0.24	—	82	420–570°C 246 bar 5–14 s	Holgate et al., 1992 (Ref. 53)
Methane	11.4 ± 1.1	179 ± 18	0.99 ± 0.08	0.66 ± 0.14	—	14	560–630°C 245 bar 6.2–11.6 s	Webley and Tester, 1991 (Ref. 52)
Hydrogen	24.4 ± 4.9	390 ± 60	1.1 ± 0.25	0.02 ± 0.29	—	55	495–600°C 246 bar 3.4–11.0 s	Holgate and Tester, 1993 (Ref. 54)
Methanol**	13.2	199	1	0	—	—	440–589°C	Vogel et al., 2005 (Ref. 137)
Ethanol	17.23 ± 1.65	214 ± 18	1.34 ± 0.11	0.55 ± 0.19	—	33	433–494°C 246 bar 2.0–11.8 s	Schanzenbacher et al., 2002 (Ref. 138)
Acetic acid	9.9 ± 1.7	168 ± 21	0.72 ± 0.15	0.27 ± 0.15	—	53	425–600°C 246 bar 4.4–9.8 s	Meyer et al., 1995 (Ref. 56)
Acetamide	3.6 ± 2.5	68.8 ± 3.8	1.15 ± 0.06	0.05 ± 0.06	—	65	393–528°C 243–339 bar 9.1–59.7 s	Lee and Gloyna,1992 (Ref. 139)
Lactic acid	18.7 ± 4.2	226 ± 46.6	0.88 ± 0.11	0.16 ± 0.19	—	39	300–400°C 276 bar 3.4–44.1 s	Li et al., 1999 (Ref. 140)
Benzene	13.7 ± 1.0	270 ± 10	0.40 ± 0.07	0.17 ± 0.05	1.4 ± 0.1	107	479–587°C 138–278 bar 3–7 s	DiNaro et al., 2000 (Ref. 141)

(continued overleaf)

411

Table 8 (*continued*)

Compound	Log A [(mol/L)$^{1-a-b-c}$, s^{-1}]	E_a (kJ/mol)	a	b	c	No. of data points	Conditions	Reference
Phenol	2.34 ± 0.28	52 ± 4	0.85	0.5	0.42 ± 0.05	96	380–480°C 253–282 bar 3–15 s	Gopalan and Savage, 1995 (Ref. 150)
2-Chloro-phenol	2.0 ± 1.2	46 ± 16	0.88 ± 0.06	0.41 ± 0.12	0.34 ± 0.17	62	300–420°C 187–298 bar 3.6–69.8 s	Li et al., 1993 (Ref. 142)
2,4-Dichloro-phenol	1.86	56.9	0.97	1.16	—	Not available	400–550°C 250 bar 5–10 min	Lin et al., 1998 (Ref. 143)
o-Cresol	5.7 ± 2.2	124 ± 35	0.57 ± 0.15	0.22 ± 0.22	1.44 ± 0.60	51	350–500°C 203–304 bar 0.5–46.3 s	Martino et al., 1995 (Ref. 144)
Pyridine	14.1 ± 0.8	229 ± 9	0.66 ± 0.88	0.28 ± 0.69	—	54	426–527°C 276 bar 2.1–10.7 s	Crain et al., 1993; 1995 (Refs. 145, 146)
Biphenyl	24.4 ± 0.3	294 ± 5	2.02 ± 0.05	0	—	10	400–550°C 250 bar 3.8–30 s	Anitescu and Tavlarides, 2005 (Ref. 147)
Arochlor 1248***	17.0 ± 0.1	186 ± 2	2.09 ± 0.02	0	—	20	450–550°C 253 bar 5.1–18.1 s	Anitescu and Tavlarides, 2000 (Ref. 148)
Methylphos-phonic acid	14.0 ± 1.6	228 ± 22	1	0.30 ± 0.18	1.17 ± 0.30	67	478–572°C 138–277 bar 3.0–9.5 s	Sullivan and Tester, 2004 (Ref. 149)

*The kinetic terms A, Ea, a, b, and c are as defined in Equation 1, and were determined from regression of experimental data under the conditions indicated. Uncertainty values represent 95% confidence intervals.
**Methanol kinetics have been studied by a number of different researchers. The kinetic parameters cited here were determined by Vogel and co-workers. from regression of several data sets judged to represent the best behavior of methanol SCWO in a plug flow reactor.
***Mixture of 76 PCB congeners

412

While useful in correlating data for engineering design, global rate expressions provide little information about the reaction mechanism. As a result, computational studies have been performed in parallel with experimental work to elucidate the underlying reaction mechanism for the oxidation of several model compounds in supercritical water. Because of the low dielectric constant characteristic of supercritical water, most SCWO reactions proceed via free radical rather than polar ionic reaction pathways. Therefore, the typical starting point for mechanistic development has been to use the extensive body of elementary free radical reaction networks developed by the combustion field and adjust these to their high pressure limits. The most significant difference between the supercritical water environment and the lower-density/higher-temperature conditions typical of air combustion is the important effect of water as a third body with respect to molecular collisions. This can be seen, for example, by the increased prominence of peroxy radical species such as HO_2 in many elementary reaction networks developed under SCWO conditions. This, in turn, requires identification of new intermediates and pathways and their associated rate data to evaluate under supercritical water conditions in the model development. Ploeger et al. provide a good review of past and present efforts in SCWO elementary reaction network modeling.[60]

Some general trends have been observed from the many kinetics studies over the years:

- To a good approximation, most global kinetic rate expressions tend to be first order with respect to the organic and zero order with respect to oxidant and water, due to the fact that these latter two species are often in significant excess.

- Large, complex organic species that are typical of real world wastes typically are fairly labile under SCWO conditions and tend to break down rapidly to a series of simpler but more refractory compounds (see Figure 8). Examples of these refractory compounds are acetic acid and CO for hydrocarbons, ammonia for nitrogen-containing wastes, and methyl phosphonic acid (MPA) for phosphorus-containing wastes. The more refractory the compound, the more extreme the conditions needed to destroy it, although even the most refractory compounds are destroyed by 700°C (at the high end of normal SCWO operating conditions).

- Different breakdown mechanisms can be more or less favored, depending on the specific operating conditions. Reaction mechanisms active under slightly polar conditions in the near-critical temperature region may be deactivated in favor of a different free radical mechanism at 600°C, depending on how key intermediates (i.e., transition state species) interact with the water solvent environment.

- Catalytic effects from metal reactor walls (typically consisting of a high-nickel alloy) have been observed during SCWO of some compounds.

- Results to date with elementary reaction network modeling have been mixed, with early attempts in particular having difficulty in matching experimental data over the full range of interest. Nevertheless, these efforts have still been successful at providing valuable insight into the understanding of SCWO reaction mechanisms and interpretation of experimental data and observed phenomena from a semi-quantitative and qualitative perspective. With increasing computational power, the predictive accuracy of elementary reaction network modeling has improved and been extended from simple C1 compounds and H_2 to larger species such as phenol, methylamine, and MPA.

5.3   Catalysis

Many researchers have also been exploring the addition of heterogeneous catalysts for further facilitating SCWO reactions. In principle, the use of a heterogeneous catalyst eliminates the advantage of the single-phase reaction medium and the lack of interphase mass transport limitations characteristic of supercritical water. However, its use can be beneficial if the overall reaction rate with the catalyst is faster than the uncatalyzed reaction. Such a possibility would mainly be associated with more refractory organic species that normally would require temperatures higher than 500°C for complete destruction. Typical catalyst materials that have been investigated are metal oxides. A particular focus of catalyst research has been finding materials that have both good activity toward the organic species of interest, as well as good stability under SCWO conditions. Ding et al.[61] and, most recently, Savage et al.[62] provide thorough reviews of catalyst use in SCWO for both full and partial oxidation.

5.4   Salt Phase Equilibrium

Because of the impact of salt nucleation, precipitation, and accumulation on SCWO reactor operation, there has been obvious interest in understanding phenomena associated with salt thermodynamics and transport. Phase diagrams for salt systems are an important tool in determining optimal operating conditions and effective strategies for salt management. There are relatively little phase data available, however, for multicomponent water, gas, organic, and salt systems that would replicate the composition expected during oxidation of a typical waste. Nevertheless, studies have been performed to map out phase boundaries of several key binary and ternary systems of common salts under temperatures and pressures typical for SCWO that still provide valuable information in analyzing real applications. Examples of systems that have been studied include $NaCl$-H_2O and Na_2SO_4-H_2O,[63] and Na_3PO_4-H_2O.[64] Knowing salt solubility as a function of temperature and composition as is provided by these phase diagrams can be of use in adjusting feed composition or reactor conditions for improved salt behavior and transport during SCWO of real wastes. Further discussion of salt phase behavior as applied to SCWO operation is provided in Section 6.4. Experimental data on phase equilibria can be used to determine mixing parameters for use within equations of state developed specifically for supercritical salt-water systems[65] and to validate Molecular Dynamics simulations,[66] as has been done in both cases with the model system of NaCl-H_2O. More recently, the NaCl-H_2O equation of state has been extended to include other salts and nonelectrolytes in supercritical water, which is more applicable to the reactor contents of real-world SCWO processes.[67]

5.5 Salt Nucleation and Growth

Salt nucleation and growth studies have also been performed to better understand kinetics and mechanisms associated with these phenomena. Most of these types of studies involve sudden contact of a dissolved salt stream with another fluid or solid body whose temperature is sufficient to cause precipitation of the salt. This approach allows one to observe the morphology of salt particles as they form and grow. The particle morphology can have a significant impact on the particular macroscopic transport characteristics exhibited by the salt as it moves or accumulates within a SCWO reactor. Notable salt nucleation and growth studies in supercritical water have been performed within an optically accessible cell,[68] within a heated tubular reactor,[69] and on a heated solid cylinder.[70]

5.6 Corrosion

Corrosion studies in SCWO environments have been carried out by a number of organizations, most notably at FZK[71,72] and MIT.[73] Research on corrosion with respect to SCWO has been driven by the need to better understand regions of corrosive susceptibility, so as to identify operating conditions and materials best suited for safe use at a duration that is economically practical. Such information is typically contained in Pourbaix diagrams (plots of electrostatic potential vs. pH), where the thermodynamically stable phase of a material (e.g., dissolved ion, neutral metal, or oxide) at any given condition can be determined (Figure 9). Unfortunately, not much data are available at temperatures applicable to SCWO for construction of these Pourbaix diagrams. The situation is complicated by the fact that (1) the nature of water, and thus its ability to support ionic species that are most responsible for corrosion, changes with density, and (2) the ability to measure critical parameters such as pH and potential under typical SCWO conditions is difficult because of the harshness of the environment on traditional measuring instruments or probes.

Nevertheless, numerous materials tests under SCWO conditions (typically in the form of coupons or a flow tube) and analyses of failures performed over the years have contributed to improved strategies for corrosion mitigation.[73–75] This experimental data, coupled with existing Pourbaix diagram data at lower temperatures (e.g., 300°C), can provide rough guidance to operating conditions that

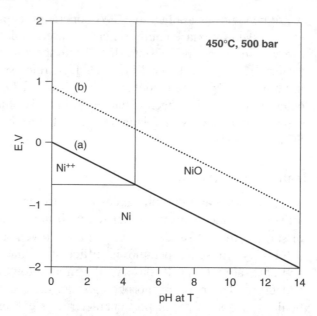

Figure 9 Potential-pH (Pourbaix) diagram for nickel at 450°C and 500 bar. Ni species indicated represent stable form in that region. Lines (a) and (b) represent equilibrium conditions for reduction and oxidation of water, respectively. (Reprinted from Ref. 77.)

bracket where the material in question is likely to exist in a stable form. In some cases, existing data have been extended to higher temperatures and pressures using equations of state, as Kriksunov and Macdonald have done for iron, for example.[76] Also, progress has been made on development of more robust electrochemical probes that allow more accurate measures of key corrosion-relevant data in the SCWO environment.[77,78] Kritzer has provided an extensive review of the SCWO corrosion literature, including a substantial range of metals and ceramics exposed to a variety of SCWO environments.[79]

The following are some general conclusions based on corrosion research as applied to SCWO:

- Corrosion is most severe at the highest temperatures where the density is high enough to support dissociated ionic species and reaction mechanisms.[80] This is not necessarily in the supercritical regime, but typically occurs in the region near but below the critical point (e.g., 250–350°C) and is typically encountered in heat-up (e.g., preheater) or cool-down (e.g., heat exchanger) components in SCWO systems. In the reactor, ionic fluids can occur in areas containing charged species that are shielded from the bulk supercritical environment, such as beneath a moving molten salt layer.
- Salts and strong acids provide the greatest risk of corrosion in the SCWO reactor regardless of the material of construction. Corrosion due to alkaline species such as sodium carbonate has also been observed.

- While certain materials perform particularly well under certain conditions (e.g., titanium in the presence of chloride-containing feed), there appears to be no one material that is sufficiently corrosion resistant to all feeds of interest to SCWO.[81] Material choices must therefore be made based on the most corrosive and/or most common feed component anticipated, and may vary for different parts of the SCWO system (including the reactor). Periodic replacement of components may be necessary as well.

- Metals, particularly high nickel-chromium alloys, provide the best overall combination of corrosion protection and mechanical strength for use in SCWO systems. Ceramics (including coatings) have generally proven to have acceptable corrosion resistance but insufficient mechanical integrity or durability.

- If data are available, feed chemistry should be adjusted to achieve a desired pH and electrochemical potential favoring a stable material state (e.g., oxide). This can be an effective method of minimizing corrosion, but may be difficult to achieve if the feed composition varies over time or is not well enough known.

6   PRACTICAL ASPECTS OF SCWO PROCESSING

Prior sections have reviewed the scientific basis of SCWO, provided a general description of the process, described historical context and current status, and summarized the research efforts that have been carried out over the years. The intent of this section is to relate this broad background of material to practical applications of SCWO technology. This section should in large part enable the reader to determine if a particular waste stream is well-suited to the SCWO process and what particular challenges might be encountered.

6.1   Feed Materials

Waste
Suitability of waste feed to a SCWO system is determined by several factors, including heating value/concentration, chemical composition, and material form. Considerations in regard to these factors are provided as follows.

Heating Value/Concentration.   The SCWO process is applicable to a very broad range of waste feed types and concentrations. However, process economics are most likely to be favorable for waste feeds with heating values in the range of 1000 to 4500 kJ/kg, which, depending on the particular species involved, generally corresponds to organic content in the range of 1 to 20 wt%. Use of SCWO outside this range involves increased costs for fuel and/or oxidant. For wastes more dilute than this range, technologies such as biotreatment or activated carbon adsorption may be more competitive, while for wastes more concentrated than this range, technologies such as incineration may be more competitive. While a useful guideline, exceptions to these limits are not uncommon. For example, certain wastes may be toxic or refractory to biotreatment microorganisms, and environmental regulations may make incineration difficult or impossible to permit in many locations.

In some cases, auxiliary processing options may be used to bring the primary waste of interest into the optimum range for SCWO treatment. Some examples of such processing options include the following:

- Dilution of high concentration wastes
- Addition of fuel to low concentration wastes
- Coprocessing of dilute and concentrated wastes[26]
- Partial sludge dewatering[82]
- Preconcentration of dilute waste using reverse osmosis or evaporation[83,84]

A waste feed with a heating value that will raise the temperature from ambient to the reactor operating temperature of the SCWO reactor is defined as *autogenic*. The autogenic heating value thus varies with reactor operating temperature. In addition, the autogenic heating value varies with oxidant type, since for air oxidant a significant amount of nitrogen must also be heated to the reactor temperature. For the common reactor operating temperature of 600°C, autogenic feed has a heating value of ~4200 kJ/kg when air is the oxidant and ~3500 kJ/kg when pure oxygen is the oxidant. For waste materials composed primarily of carbon, hydrogen, and oxygen, about 12,900 kJ of heat are released for every kg of oxygen consumed. Thus, for a reactor operating temperature of 600°C, autogenic feed has a chemical oxygen demand of about 325,000 ppm when air is the oxidant and about 270,000 ppm when pure oxygen is the oxidant.

These heating values are approximate and do not distinguish between higher heating value (HHV, assumes reaction product water is present as liquid) and lower heating value (LHV, assumes reaction product water is present as vapor). The enthalpy of water in the SCWO reactor is closer to that of water vapor than that of liquid water, so that lower heating value gives a better representation of the heat available to reach the peak reactor temperature. Wastes with less than autogenic heating value are frequently most economically treated by using regenerative heat exchange (i.e., by using hot reactor effluent to preheat incoming feed). Although the reactor effluent is at a temperature of about 600°C, regenerative preheat of incoming feed is limited to a temperature of about 400°C due to temperature pinching in the heat exchanger caused by the high heat capacity of water near its critical point.

Chemical Composition. With regard to chemical composition, feed materials containing only carbon, hydrogen, oxygen, and nitrogen (CHON) are readily treated without difficulty from corrosion or solids separation. The presence of heteroatoms in the SCWO feed, defined as non-CHON elements for the purposes of this discussion, is generally of minor importance at levels up to tens of parts per million (ppm). In the range of 100 ppm and higher, however, most heteroatoms are accompanied by process effects that can influence system design and/or operation. Commonly encountered effects include corrosion, solids plugging or scaling, and erosion.

SCWO system corrosion is caused by a variety of compounds, including acidic corrosion by mineral acids (e.g., HCl, H_2SO_4), corrosion by neutral salts (e.g., $NaCl$, Na_2SO_4), and corrosion by alkaline salts (e.g., Na_2CO_3, K_2CO_3). Acidic corrosion is generally the most aggressive mechanism, but corrosion by neutral and alkaline compounds can also be very significant. As previously mentioned, the zone of highest corrosion is frequently in the high density, intermediate temperature region found in feed preheaters or cooldown heat exchangers. However, ionic molten phases or under-deposit microenvironments can also result in high corrosion rates in the reactor itself. No single material resistant to the entire range of SCWO conditions exists, although titanium and high nickel alloys have fairly wide applicability and are commonly specified. A more detailed discussion of corrosion and SCWO materials of construction is provided in Section 6.5.

SCWO system solids plugging or scaling can be caused by all common salts, with sodium and calcium salts being most frequently encountered. These materials are referred to as 'sticky salts' in recognition of their tendency to adhere to process surfaces. Organic, nitro, or ammonium salts in the waste feed are not themselves a problem, as the organic or nitrogen moiety is converted to water and/or gas. However, residual hetero-cations such as sodium will form sticky carbonate salts in the reactor while residual hetero-anions will form corrosive acids. A major fraction of the salts precipitating in a SCWO reactor or heat exchanger can adhere to the wall and lead to rapid plugging or degraded performance. A more detailed discussion of methods for dealing with salt precipitation is provided in Section 6.6.

Erosion in SCWO systems can be caused by hard particles inherent in the waste feed, such as the silica in sewage sludge, or newly formed hard particles such as titanium dioxide corrosion product. Metal oxides are the most commonly encountered abrasive particles. Erosion is of particular concern where the process stream changes direction and at sudden contractions or expansions. The liquid pressure letdown system can be particularly vulnerable to erosion due to small openings and the corresponding high velocities. In addition, the pressure letdown system can experience cavitation due to entrained or dissolved gas. A control valve intended for slurry service is recommended for applications with significant erosive solids. One such valve that has given excellent service utilizes a freely rotating ball as the control element, rather than the more common stem valve. Other approaches to dealing with abrasive solids include the use of capillaries to obtain a more gradual pressure drop, and use of an auxiliary water stream as a pressure control means.[85]

Some waste feeds may contain a combination of abrasive particles and sticky salts. This can sometimes provide a benefit as the abrasive particles can scour the process surfaces and keep them sufficiently clear of solids buildup. U.S. Patent No. 5,620,606 teaches the addition of inert particles to SCWO feed to mitigate solids build-up.[86]

Material Form. The need to compress the waste feed to high pressures makes liquids or slurries the preferred feed form for the SCWO process. Liquid wastes composed of two phases (e.g, an oily layer and an aqueous layer) may be supplied from an agitated tank that keeps the material in homogeneous form. In some cases, pump-around recirculation loops, possibly with a static mixer, may suffice. For rapidly separating liquids, it may be preferable to pump the phases separately, either through separate high-pressure pumps or through separate low-pressure pumps supplying a single high-pressure pump.

Solids-containing feeds that have particle sizes in the range of 1 mm and less are readily fed to a SCWO system using a high-pressure slurry pump, although the acceptable particle size varies with feed nozzle design and system size. Generally, larger systems with larger clearances can tolerate larger particle sizes, but if the particles are organic, longer residence time will be needed to attain high destruction efficiencies.

A preferred pump type for slurry feeding is a hose diaphragm pump, in which pumping action is provided by a hydraulic fluid squeezing on the outside of an elastomeric hose. Pumping of sewage sludge concentrations of at least 15 wt% to an SCWO system is reported in the literature.[87] Other examples of slurries or readily slurried solids include industrial sludges, powdered activated carbon, ion exchange resins, heterogeneous catalysts, coal, sawdust, sediments, and soils. The upper pumpable limit for wood dust slurries also appears to be in the range of 10 to 15 wt%.[88] Above this concentration the slurry behaves like a damp solid. In certain cases, it may be advantageous to remove solids from a SCWO feed stream. For example, aluminum contained in a feedstock derived from surplus rocket motors has been removed via pH adjustment and filtration prior to treatment by SCWO in order to avoid reactor plugging.[35]

Purification of gases of toxic chemicals or pathogens is quite possible using SCWO. However, the high cost of compressing a gaseous feed stream makes the economics of such applications unlikely to be attractive.

Oxidant

Oxidant is generally supplied to SCWO systems at somewhat above the oxygen demand of the feed, 10 to 25% above the stoichiometric amount being typical. In general, the only economically practical oxidants for commercial SCWO plants are air, oxygen, or intermediate mixtures derived from air. As noted in the previous section, autogenic waste has a fairly high oxygen demand. If autogenic waste is processed in a SCWO system at 600°C with pure oxygen as the oxidant, oxygen will make up about 30 percent by weight of the process stream before reaction. In the case of air oxidant, about 70 wt% of the process stream before reaction will be air. For subautogenic feeds, the noncondensable gas content will be proportionately less.

Selection of air or oxygen as the oxidant typically involves an economic trade-off. Advantages to the use of oxygen include smaller equipment (as nitrogen does not have to be processed) and smaller plant capital cost since a large compressor need not be purchased. Advantages to the use of air include reduced operating cost (oxygen does not have to be purchased, although this is somewhat offset by much higher electricity usage), reduced plant safety concerns because a strong oxidant is not in use, and compatibility with titanium equipment. Titanium is a key material of construction for SCWO systems treating chlorine-containing materials.

Fuel

Virtually any type of high heating value liquid may be used as fuel for SCWO systems. In some applications, a site may have high heating value waste that can be processed, along with low-heating-value waste so that fuel costs can be minimized or eliminated. For applications where fuel must be purchased, kerosene is usually an economical choice. In applications where aqueous miscibility is important, the low-molecular-weight alcohols methanol, ethanol, and isopropanol have been used.

6.2 Reactor Type

Reactor types for SCWO fall into two broad categories—vessels, for which the process length-to-diameter ratio is less than about 20, and pipe reactors, for which the process length-to-diameter ratio is greater than about 100. Vessel reactors are typically mounted vertically, especially when solids are being handled, as gravity is used to help attain the desired solids disposition. Figure 10 shows a representative configuration for a commercial vessel reactor, referred to more specifically as a solid-wall, downflow reactor. In this design, feed material, oxidant, and any auxiliary streams such as fuel, water, or caustic, are injected through a single nozzle at the top of the reactor. Reaction occurs in the upper portion of the vessel, aided by a jet-induced recirculation flow that mixes incoming feed with the existing hot contents of the reactor. (The recirculation flow is associated with the feed nozzle design and is not inherent to the use of a vessel reactor.) Due to the backmixing effect, feed materials may be introduced to a vessel reactor without preheating if desired, a feature that allows the problems of preheater corrosion and scaling to be avoided. Salts, if any, precipitate in the backmix zone and are transported to the bottom of the reactor. As indicated by the parallel streamlines in Figure 10, a plug-flow zone is located below the backmixing zone. This section of the reactor helps to assure high destruction efficiency without the short

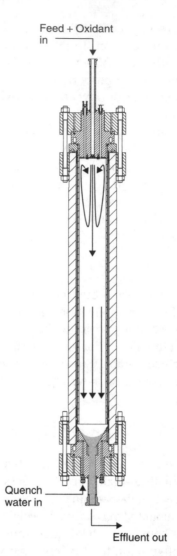

Feed + Oxidant
in

Quench
water in

Effluent out

Figure 10   Downflow vessel reactor configuration.

circuiting of incompletely reacted organics. A quench stream is introduced at the bottom of the reactor to bring the process stream to subcritical temperature. At this temperature, any precipitated salts are redissolved and can thus be removed from the reactor without causing plugging.

An important feature of the vessel reactor is the ability to incorporate a corrosion-resistant but replaceable liner, which can also be combined with a thermally insulating sleeve. In addition to allowing operation with highly corrosive feeds, the liner assembly allows the pressure shell to be fabricated of a lower-alloy material and to operate at temperatures much lower than that of the process. This feature reduces reactor cost and facilitates the fabrication of large vessels.

In contrast to vessel reactors, the relatively small diameter of pipe reactors makes them vulnerable to plugging when high-salt concentrations are encountered. For this reason, application of pipe reactors has to date been limited to waste salt contents of 1 wt% or less. Pipe reactors have the advantage of simple pressure envelope design, with many of the components available as standard piping designs (albeit using specialty materials), which helps to reduce reactor cost. Due to their great length, typically hundreds of meters, pipe reactors are generally coiled and mounted horizontally. Figure 11 shows a typical configuration for a SCWO pipe reactor, with the coiling omitted for clarity.[89] Due to their small diameter, pipe reactors cannot take advantage of direct backmixing of incoming feed with hot reactor contents. Pipe reactors with recycle loops have been proposed in numerous patents, but to our knowledge have not been successfully implemented.[90,91] Thus, pipe reactors are generally associated with preheating of the feed. Another feature of pipe reactors shown in Figure 11 is the use of multiple points of oxidant injection. Because the feed to a pipe reactor is typically preheated, oxidation of all but low-heat-content feed streams can lead to reactor overheating. For this reason, oxidant addition is frequently staged, with interstage heat loss eventually allowing the injection of additional oxidant. Alternatively, staged addition of quench water or feed can also be used to prevent reaction overtemperature.[85]

Feeds with salt concentrations in excess of 0.1 wt% are generally more suited to a vessel reactor than a pipe reactor. Feeds with high inert solids content are compatible with either a pipe or vessel reactor, but have to date been commercially processed only with pipe reactors. Table 5 in Section 4 summarized the types of feeds that are currently being processed commercially or are currently in commercial design, along with the reactor type employed.

To date, the largest capacity plants have utilized pipe reactors due to the relatively straightforward design of the pressure envelope. Large, high-pressure vessel reactors exist in industry for ammonia synthesis (2.4 to 3 m diameter × 18

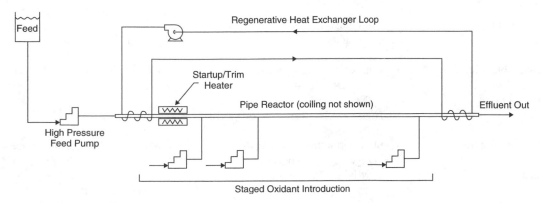

Figure 11 Pipe reactor configuration (coiling not shown).

to 27 m high, operating conditions up to 550°C and 300 bar)[92] and wet oxidation (2 m diameter × 18.5 m high, operating conditions up to 330°C and 250 bar),[93,94] but such large vessels have not yet been employed for SCWO. The SCWO-related technology of wet oxidation is primarily implemented using vessel-type reactors, which may indicate that SCWO will reach a similar state as the technology matures.

Table 9 provides a comparison of the suitability of the two reactor types with regard to major SCWO design factors. Potential clients should work with the SCWO vendor, as key features such as reactor temperature, pressure, and residence time will vary for the different reactor types as well as for the feed material.

6.3 Reactor Conditions

Residence time in SCWO reactors varies from seconds to minutes, depending on the reactor design and operating conditions. For a vessel reactor operating at

Table 9 Comparison of Pipe and Vessel Reactor Features

Design Factor	Pipe Reactor	Vessel Reactor
Pressure boundary design	Simpler; standard pipes and fittings may be used	More complex; pressure vessel code must be applied
Pressure boundary material	High temperature alloy required	High-temperature alloy not necessarily required
Feed preheating	Required; potential preheater corrosion problems	Optional; preheater corrosion problems may be avoided
High heating value feeds	Limited due to need for preheat	Preheat optional, not limiting
Low heating value feeds	Compatible	Compatible
Oxidant introduction	Complex; multiple injection points typically required	Simple; single point injection
Heat conservation	Difficult due to large surface area	Simple due to small surface area
Corrosion protection	Difficult due to large surface area	Simple due to small surface area (e.g., corrosion-resistant liner, transpiring wall)
Salt handling	Not compatible; sticky salts cause plugging	Compatible; various approaches available to move sticky salts through reactor
Non-salt solids handling	Compatible; high velocity facilitates solids transport through reactor	Compatible; solids fall through reactor
Erosion	High velocities; erosion a concern	Low velocities; erosion not a concern
Destruction efficiency	Short-circuiting of unoxidized organic not a concern	Short-circuiting of unoxidized organic possibly a concern
Capacity	May be preferable for high-capacity plants	Fabrication of large diameter high pressure vessels is a specialized capability

650°C, a typical residence time is 15 seconds. For a pipe reactor operating at 600°C, a typical residence time is 60 seconds. Given a target residence time, the approximate process volume required for a SCWO reactor may be calculated if the process fluid density is known. Density of the SCWO reactor fluid may be estimated by adding the densities of the gas and liquid at reactor temperature and at their respective partial pressures. When process simulator software is used, recommended equations of state are Redlich-Kwong-Soave (RKS, generally available) and SR-POLAR (for ASPEN).[95]

Consideration of Figures 1(a) and 1(b) indicates that for water at a typical SCWO operating temperature of 600°C, there are no sudden transitions as pressure is lowered from a normal operating pressure of 235 bar. This fact also holds true for oxidant-containing SCWO reaction fluids, and suggests that comparable high oxidation efficiencies may be obtained at subcritical pressures. As previously noted, due to the significant fraction of oxidant in SCWO systems, much of the fluid behavior is akin to subcritical conditions in any case. High organic oxidation efficiencies at subcritical pressures have indeed been verified.[96] While operating at lower pressures can facilitate design, fabrication, and procurement of equipment, cost reduction is not proportional as larger equipment becomes necessary at lower pressures (due to lower fluid densities) in order to maintain the necessary residence times. Studies on the effect of pressure on reaction kinetics have found that rates may increase or decrease, depending on the particular compound,[97] although a rate increase with pressure may be more common.[98]

6.4   Solubility and Phase Behavior

Solubility and phase behavior directly affect many of the key phenomena occurring within SCWO systems, including reaction kinetics, salt precipitation, and corrosion. A generic description of phase behavior and solubility in supercritical water was given previously in Section 2. This section provides a more specific discussion of phase behavior and solubility in SCWO systems.

It is important to note that while SCWO is formally defined in terms of the critical point of pure water, addition of any other constituents to the water will alter the critical point, and the system may or may not be supercritical with respect to this mixture critical point. Rather than a single critical point, for a binary system a critical curve exists that in the simplest cases joins the critical point of pure water to the critical point of the second substance across the composition space. For ternary mixtures the critical curve becomes a critical surface, and so on. In general, mixtures of water with higher volatility substances such as noncondensable gases or liquid organics will remain supercritical, while mixtures of water with lower-volatility substances such as salts will become subcritical and liquid or solid phases will precipitate from the vapor/ gas phase.

Figure 12(a) shows the phase behavior for the oxygen-water system at 250 bar, with operating lines at representative compositions.[99–101] A SCWO reactor processing autogenic feed requires about 30 wt% oxygen in the process stream when pure oxygen is used as the oxidant. As shown in Figure 12(a), the oxygen and water become a homogeneous vapor/gas phase above about 329°C, significantly below the mixture critical temperature of 373°C. Similarly, a SCWO reactor processing autogenic feed with air as the oxidant requires about 70 wt% air in the process stream. As an approximation to the phase behavior at this condition, Figure 12(a) shows an operating line at 70 wt% oxygen (air has solubility behavior comparable to oxygen). The air and water become a homogeneous vapor/gas phase above about 263°C. When processing subautogenic feeds, noncondensable gas content will be lower than the operating lines shown in Figure 12(a). In all cases, however, complete miscibility is attained below, and frequently well below, water's critical temperature of 374°C.

As oxidation proceeds through the SCWO reactor, oxygen is replaced by carbon dioxide. Figure 12(b) shows the very similar behavior of carbon dioxide to oxygen, indicating that single fluid phase conditions are maintained over the course of the reaction.[102–103]

Behavior of organic substances that are fluid phase and of sufficient thermally stability is generally similar to the behavior shown in Figures 12(a) and 12(b). In other words, organic substances will generally become miscible with water as system temperature is raised to water's critical temperature and, depending on the system composition, miscibility may occur well below water's critical temperature. Figure 12(c) illustrates this point for the system benzene-H_2O at 250 bar.[104–107] A system containing 10 percent benzene in water attains miscibility at a temperature of only 271°C.

The high noncondensable gas content of SCWO reactor fluids effectively reduces the partial pressure of water in the system to levels as low as about 100 bar, reducing any solvation effects due to water. This has a minor effect on organic solubility, as shown in Figure 12(d) for the system benzene-H_2O at 100 bar.[104,108] 10 percent benzene miscibility is now attained at 309°C. As a further illustration of the similar behavior between gases and organics in SCWO systems, Table 10 gives mixture critical temperatures and compositions for water with noncondensable gases and several representative organics at 250 bar. Complete miscibility is expected even for very heavy hydrocarbons with molecular weights of at least 400.[109]

Figure 12(e) shows the much more complex phase behavior observed when water is mixed with low volatility salts, in this case the sodium chloride–water system at 250 bar.[110,111] Due to the high critical temperature of sodium chloride, estimated at 3900K and 260 bar, this system is subcritical at SCWO reactor conditions.[112] Figure 12(e) shows an operating line for 10 wt% NaCl at 250 bar. As this brine is heated from ambient temperature, a split into vapor and liquid phases occurs in the vicinity of 400°C. As temperature increases to 450°C, NaCl

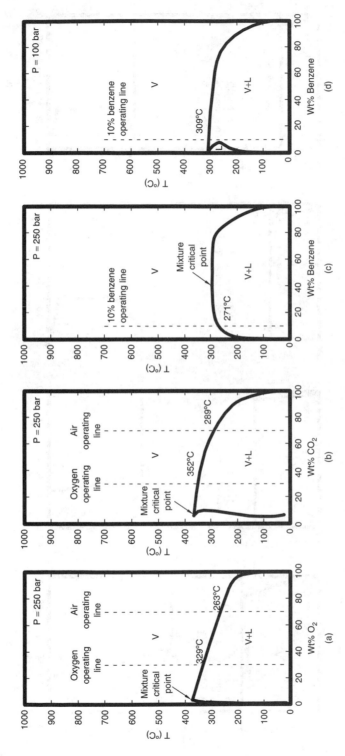

Figure 12 Solubility and phase behavior in supercritical water oxidation systems. (a) O_2-H_2O at 250 bar (based on data from Refs. 99 and 100 at 250°C and above, other data from Ref. 101); (b) CO_2-H_2O at 250 bar (based on data from Ref. 102 at 110°C and above, other data from Ref. 103; (c) Benzene-H_2O at 250 bar (based on data from Ref. 104 below 100°C, Refs. 105 and 106 from 287–295°C, other data from Ref. 107); (d) Benzene-H_2O at 100 bar (based on data from Ref. 104 up to 250°C, other data from Ref. 108). (e) NaCl-H_2O at 250 bar (vapor phase compositions from Ref. 110; other data from Ref. 111; (f) NaCl-H_2O at 100 bar (vapor phase compositions from Ref. 110, other data from Ref. 111); (g) Na_2SO_4-H_2O at 250 bar (based on data from Ref. 116 at 320°C and above; other data from Ref. 117); (h) Na_2SO_4-H_2O at 100 bar (based on data from Ref. 116 at 320°C and above, other data from Ref. 117 and Ref. 118.)

427

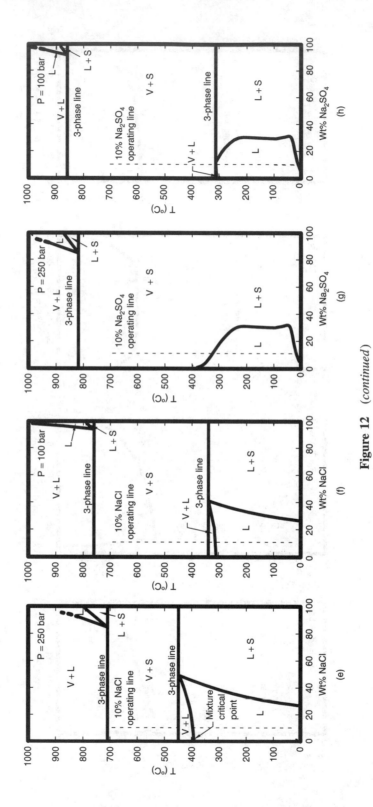

Figure 12 (*continued*)

428

Table 10 Gas and Organic Mixture Critical Points at 250 Bar

Compound	Formula	Mixture Critical T,°C	Wt% Compound at Mixture Critical Point	Reference
Oxygen	O_2	373	3.5	Japas and Franck, 1985 (Ref. 100)
Nitrogen	N_2	372	2.6	Japas and Franck, 1985 (Ref. 99)
Carbon dioxide	CO_2	367	6.1	Takenouchi and Kennedy, 1964 (Ref. 102)
Benzene	C_6H_6	296	40.3	Alwani and Schneider, 1967 (Ref. 107)
n-Heptane	C_7H_{16}	~356	~25	Connolly, 1966 (Ref. 105)
Naphthalene	$C_{10}H_8$	315	~50	Alwani and Schneider, 1969 (Ref. 151)
Biphenyl	$C_{12}H_{10}$	336	~50	Bröllos et al., 1970 (Ref. 152)

content of the vapor phase remains in the range of 200 to 300 ppm by weight, while NaCl content of the liquid brine phase increases up to about 40 wt%. At 450°C, the liquid phase disappears and a region of steam in equilibrium with solid NaCl is entered. As temperature is increased above 450°C, solubility of NaCl in the steam decreases to a minimum of about 150 ppm at 550 to 600°C and then begins to increase. At typical SCWO reactor temperatures of 600° to 650°C, NaCl solubility is in the range of 150 to 200 ppm. If temperature is increased beyond about 715°C, a liquid phase reappears. The liquid phase has a high concentration of sodium chloride and is more akin to a melt than a brine. Beyond the NaCl melting point of ~800°C (there is little effect of pressure on the melting temperature), solid NaCl is absent and vapor-liquid equilibrium dominates the composition space.

It is important to note that both the liquid phase below 450°C and that above 715°C contain NaCl in dissolved form and are potentially highly corrosive. A rule of thumb stated earlier is that SCWO corrosion is highest at subcritical conditions because of ionic corrosion reactions, while at SCWO conditions ionic dissociation is largely absent and corrosion reactions are correspondingly diminished. This generalization does not hold for salt-water systems that are in a region of brine or melt formation, even though temperatures may be well above the critical temperature of water.

As already noted, SCWO systems frequently have a high content of noncondensable gas, and the presence of this gas can significantly affect the phase behavior of the salt constituents. Little experimental data are available for supercritical water–gas–salt systems of relevance to SCWO, the only notable exception being the system water-carbon dioxide-sodium chloride.[113–115] Approximations may be made, however, by noting that the complete miscibility of the noncondensable gas and steam has the effect of reducing the partial pressure of water in the system. For an SCWO system using air oxidant and processing autogenic feed, the partial pressure of water is about 100 bar, and salt solubility will be more similar to the phase behavior at this pressure than at the total system pressure of ~250 bar. Figure 12(f) shows the phase behavior for the sodium chloride–water system at 100 bar.[110,111] Qualitatively, the behavior is similar to that at 250 bar, but the location of phase transitions and compositions has shifted significantly. For the 10 wt% NaCl operating curve, a vapor-liquid equilibrium region is entered at about 310°C, transitioning to a region of vapor-solid equilibrium at about 340°C. The vapor-solid region persists up to about 770°C, and is thus much broader than the vapor-solid region at 250 bar. This is consistent with the notion that the presence of the noncondensable gas has "dried out" the system, evaporating liquid phases in favor of the solid phase. At typical SCWO reactor conditions of 600° to 650°C, NaCl solubility is the range of 5 to 10 ppm. Above 770°C, meltlike liquids again make their appearance as the melting point of sodium chloride is approached. Since SCWO systems will operate with noncondensable gas content from a minimal value to that corresponding to autogenic operation, Figures 12(e) and 12(f) define approximately the range of NaCl phase behavior that will be encountered.

For high temperature and pressure phase behavior of salt-water systems, two classes of phase behavior exist. Type 1 systems, which include $NaCl$-H_2O, have a high salt solubility in the region of water's critical point. Type 2 systems have a low salt solubility, typically in the ppm range in the vicinity of water's critical point. An important example of a Type 2 system is Na_2SO_4-H_2O, for which the phase behavior at 250 bar is shown in Figure 12(g).[116,117] Following the operating curve for 10 percent Na_2SO_4, the salt remains in solution until a temperature of about 324°C, at which point a solid phase begins to precipitate out. Sodium sulfate solubility reaches ppm levels in the vicinity of water's critical point and the brine phase has transitioned to a low-density vapor phase. Parts-per-million salt solubilities persist through typical SCWO operating temperatures of 600° to 650°C up to about 830°C, at which point meltlike liquids make their appearance as the melting point of sodium sulfate (884°C) is approached. Figure 12(h) shows Na_2SO_4-H_2O phase behavior at 100 bar, with a similar drying-out effect as for $NaCl$-H_2O.[116–118]

Many SCWO applications involve mixtures of different salts. In such cases eutectic-type phenomena can occur, with broader regions of liquid phase behavior than with water-single salt systems. Further discussion of supercritical water systems containing mixed salts may be found in Hodes et al.[50]

6.5   Corrosion Handling

The elevated temperature and pressure of SCWO reactors, combined with the presence of water, oxygen, acids, bases, and salts, makes for a very aggressive environment in many cases. Fabrication of hot components from 316 stainless steel is suitable for relatively benign wastes containing only carbon, hydrogen, oxygen, and nitrogen, although even here high-nickel alloys are frequently preferred for their increased strength at higher temperatures. Also, materials other than stainless steel are preferred for applications where occasional contamination by heteroatoms such as chlorine may be encountered.

By way of example, Tables 11 and 12 illustrate several important features of corrosion in SCWO systems using sample corrosion data for the commonly

Table 11   Corrosion Results in High-Acid Chloride and Sulfate Environment

Zone	Temp. °C	HCl/H$_2$SO$_4$ mg/kg	Material	Exposure Hours	Corrosion rate mil/year
1 Reaction	620	9000/1600	Inconel 625	240	383
			Hastelloy C276	60	4563
			Hastelloy C22	240	466
			Inconel 686	60	355
2 Reaction	610	9000/1600	Inconel 625	240	495
			Hastelloy C276	120	1283
			Hastelloy C22	240	498
			Inconel 686	240	490
			Titanium Gr. 12	240	77
			Titanium Gr. 18	240	162
			Titanium Gr. 9	240	155
			Platinum	120	2
			Platinum-10 Iridium	240	3
			Platinum-30 Rhodium	240	1
3 Reaction	600	9000/1600	Inconel 625	240	235
			Hastelloy C276	120	946
			Hastelloy C22	240	314
			Inconel 686	240	342
4 Quench	400	5000/900	Inconel 625	240	366
			Hastelloy C276	240	747
			Hastelloy C22	240	380
			Inconel 686	240	449
			Platinum	60	45
			Platinum-10 Iridium	120	92
			Platinum-30 Rhodium	120	190
5 Neutralization	350–380	0/0 (converted to salts)	Inconel 625	60	550
			Hastelloy C276	60	480
			Hastelloy C22	240	270
			Titanium Gr. 18	240	0
			Titanium Gr. 9	240	0
			Inconel 686	240	258

Note: From Ref. 119.

432 Supercritical Water Oxidation

Table 12 Corrosion Results in Low-Acid Chloride and Sulfate Environment

Material	Temp. °C	Exposure Hours	Mass Change %	Corrosion rate mil/year	Comments
Inconel 625		434	−2.3	25	Pitting and crevice corrosion
Inconel 718		434	−1.0	12	Pitting and crevice corrosion
Inconel 718 U-bend		344	NA	NA	Pitting and crevice corrosion, no SCC
Inconel X750		434	−10.4	124	Pitting and crevice corrosion
Titanium Gr. 2		318	+1.1	Mass gain	Pitting only
Titanium Gr. 12	600 w/solid deposits	344	+1.0	Mass gain	Minor oxide spallation
Titanium beta-C		215	+0.4	Mass gain	Extensive oxide spallation
Platinum		434	+0.2	Mass gain	No evident attack
Platinum-10 Iridium		434	−0.5	0.4	No evident attack
Platinum-30 Rhodium		434	+1.3	Mass gain	No evident attack
ZrO_2 Tube		434	+0.08	Mass gain	No evident attack
TiO_2 Sheet		400	0	0	Pitting
TiO_2 Cast		344	−0.71	4.7	Pitting
Al_2O_3 Tube		434	−0.07	0.3	Pitting
Inconel 625		446	−0.4	5	Pitting and crevice corrosion
Inconel 625 weld coupon		446	−0.6	5	Pitting and crevice corrosion, no weld effects
Inconel 625 U-bend	550 no deposits	446	−0.3	1	Pitting and crevice corrosion, no SCC
Inconel 718		446	−0.2	2	Pitting and crevice corrosion
Inconel X750		446	−0.4	4	Pitting and crevice corrosion

Note: From Refs. 119, 120.

encountered SCWO environment containing both chlorine and sulfur. The data were obtained using an operating SCWO pilot plant (as opposed to a corrosion testing rig) and so closely represent actual operating conditions.[119,120]

Figure 13 shows the set-up used for a corrosion study at high acid levels, with specimens mounted in a chamber in a vertical downflow reactor configuration.[119] The chamber was nominally divided into five equal-length zones, with each zone being defined by the presence of a thermocouple. Oxidation of organic

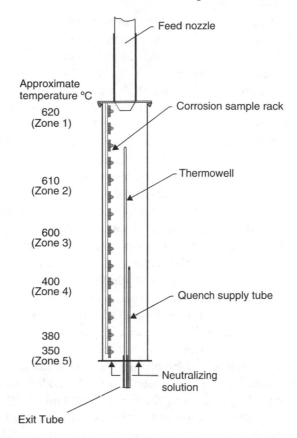

Figure 13   Configuration for high-acid corrosion testing.

chlorine and sulfur species in the top three zones results in an environment containing 9,000 mg/kg of HCl, 1,600 mg/kg of H_2SO_4, and 1,200 mg/kg of Na_2SO_4, equivalent to a pH of 0.6 at normal conditions. Quench water was introduced in Zone 4, reducing the acid and salt content by nearly half. In Zone 5 the stream was neutralized by mixing with a sodium hydroxide or sodium bicarbonate solution.

Table 11 shows the high-acid testing results. The high-nickel alloys have generally unacceptable corrosion rates at both reaction and quench temperatures. At reaction temperatures, titanium alloys have significantly lower corrosion rates than the high-nickel alloys. In the quenched and neutralized Zone 5, titanium alloys have excellent corrosion resistance. Platinum alloys give good performance at reaction temperatures, but become vulnerable and, given their high cost, unsuitable at quench temperatures.

Due to the high rates of general corrosion in this testing, specimen examination for localized corrosion was not conducted. Thermally sprayed ceramic coatings were simultaneously tested with the alloys shown in Table 11. The coating/substrate combination of titania on titanium showed excellent resistance in both the reaction and quench zones with 120 hours of exposure, but extended testing has not been carried out to date.[121,122]

Table 12 provides corrosion data for a lower-acid SCWO environment characterized by 350 mg/kg HCl, equivalent to a pH of 2 at normal conditions, and including a mixture of sodium chloride and sodium sulfate salts.[119,120] Such an environment could be attained by adding sodium hydroxide to the reactor feed of the preceding example (Table 11). Table 12 shows two different locations at nominal reaction conditions—one upstream of a hot filter, where salt deposits were present, and one downstream of the hot filter where salt deposits were absent. Under these relatively low-acid conditions, at reactor temperatures of 550° to 600°C, the high-nickel alloys 625 and 718 have acceptable general corrosion rates, although localized attack by pitting or crevice corrosion is a potential concern. Corrosion rates are higher at the higher temperature and in the presence of solid deposits. Two U-bend samples were used to test for stress corrosion cracking (SCC), which was not observed. In general, corrosion rates are far lower than for the high-acid environment.

The data in Tables 11 and 12 illustrate two common approaches to corrosion control in SCWO systems, i.e., selection of an alternate material of construction and pH neutralization. The problem of acid corrosion in SCWO systems is frequently addressed by adding alkaline material to the feed streams entering the reactor. The alkaline material may be added using a separate pump, or may be blended directly into the waste feed tank. With pH neutralization, salts rather than acids will form in the reactor, with much reduced although not insignificant corrosion rates. In fact, even neutral salts can be highly corrosive under SCWO conditions. In some applications with highly corrosive salts, no material with high resistance has been identified and replaceable sacrificial liners are used.[35]

In cases where a superior material of construction is not available, an alternative approach is to protect the reactor wall with a clean stream of water.[23,123,124] This *transpiring wall* approach has shown promise during pilot-scale testing of reducing corrosion and avoiding salt buildup in the reactor, but brings concerns of more complicated reactor design and possible reaction quenching by the cooler purge stream.[35] Transpiring wall reactors are of the vessel type.

As previously noted, the most aggressive corrosion environment in SCWO systems frequently occurs in preheaters or heat exchangers, where elevated but subcritical temperatures and ionic conditions occur. Approaches to dealing with preheater corrosion include avoidance of preheating (vessel reactor) and corrosion-resistant lining. Preheater/heat exchanger linings that have proven advantageous include titanium and tantalum.

6.6   Salts and Inert Solids Handling

Marrone et al. have described the numerous approaches that have been proposed for handling salts and inert solids in SCWO systems.[125] In dealing with salts in general, one can either (1) allow salts to precipitate and remove after surface accumulation (e.g., reactor flushing or scraping); (2) allow salts to precipitate but not accumulate (e.g., high velocity flow, feed additives, or transpiring wall reactor); or (3) avoid salt precipitation altogether by operating at conditions where salts remain dissolved (e.g., extreme high pressure). Table 13 summarizes the proposed approaches, which fall into the two basic categories of reactor designs and operating techniques. As noted earlier, salts are, in general, sticky—that is, they tend to adhere to process surfaces. Inert solids, by contrast, may settle or cause erosion, but have little or no tendency to stick to process surfaces. Inert solids are generally metal oxides or hydroxides. Some common examples are silica (sand), alumina, iron oxide, and clay. The final column of Table 13 notes the techniques that are currently in commercial use.

Table 13   Commercial Approaches to SCWO Salt and Solids Handling

Approach	Method	Comments	Tested at Pilot Scale?	Commercial Status
Reactor Designs	Reverse flow vessel reactor with brine pool		Yes	Practiced by Organo
	Transpiring wall reactor		Yes	
	Adsorption/reaction on fluidized solid phase		Yes	Practiced by Mitsubishi
	Reversible flow, tubular reactor	Not applicable for high salt	Yes	
	Centrifuge reactor	Not applicable for high salt	No	
Operating Techniques	High velocity flow	Not applicable for high salt	Yes	Practiced by Chematur
	Mechanical brushing	Not applicable for high salt	Yes	
	Rotating scraper		Yes	
	Reactor flushing/quenching		Yes	Practiced by Chematur and General Atomics
	Feed additives		Yes	Practiced by General Atomics
	Filtration	Not applicable for high salt	Yes	
	Inertial separation		Yes	
	Extreme pressure operation	Not economic for Type 2 salts	Yes	

6.7 Effluent Quality

Effluents from a SCWO system may be in the form of gases, liquids, and/or solids. This section discusses the characteristics of each of these streams.

SCWO Products

Table 14 shows typical reaction products resulting from the SCWO process. As noted previously, carbon, hydrogen, and oxygen form the conventional oxidation products carbon dioxide and water. Organic residuals, if any, will most commonly be as acetic acid or acetone.

Table 14 SCWO Products

Common Elements	SCWO Products	Reference
Carbon	CO_2, CO_3^{2-}	
Hydrogen	H_2O	
Oxygen	CO_2, H_2O	
Nitrogen	N_2, N_2O	Killilea et al., 1992 (Ref. 9)
Sulfur	SO_4^{2-}	
Chlorine	Cl^-	
Fluorine	F^-	
Phosphorus	PO_4^{3-}	
Iron	Fe_2O_3 or Fe_3O_4	
Silicon	SiO_2	
Aluminum	Al_2O_3	
Titanium	TiO_2	
Sodium	$NaHCO_3$ or salts	
Potassium	$KHCO_3$ or salts	
Calcium	$CaCO_3$ or salts	
Magnesium	$MgCO_3$ or salts	
Chromium	CrO_4^{2-} or $Cr_2O_7^{2-}$	
Lead	PbO	
Mercury	Cation or metallic	Swallow et al., 1990 (Ref. 153)
Nickel	Oxide or cation	
Zinc	ZnO	Swallow et al., 1990 (Ref. 153)
Zirconium	ZrO_2	
Cerium	CeO_2	
Tantalum	Ta_2O_5	
Tungsten	WO_2	
Hafnium	HfO_2	
Yttrium	Y_2O_3	
Molybdenum	MoO_4^{2-}	
Copper	Oxide or cation	
Silver	Cation or metallic	
Gold	Cation or metallic	
Palladium	PdO	Anonymous, 2003 (Ref. 154)
Platinum	Oxide or cation	Anonymous, 2003 (Ref. 154)
Rhodium	RhO_2	Anonymous, 2003 (Ref. 154)

Killilea et al. have reported on nitrogen chemistry in SCWO.[9] Both oxidized (e.g., nitro) and reduced (e.g., amino) forms of nitrogen are converted to N_2 and a minor fraction of N_2O in SCWO. Ammonia is a relatively refractory compound, requiring on the order of 700°C for complete destruction when not accompanied by organic material. In the presence of organic material, ammonia can be completely oxidized at temperatures below 650°C, an effect believed to be due to common reaction intermediate species. Significant residual ammonia in pH-neutral SCWO effluent will appear as a white precipitate of ammonium bicarbonate. Nitric acid or nitro moieties in the SCWO feed will act as oxidants in SCWO systems, provided a reduced material such as organic carbon is present.

For many of the metallic constituents, the product form in the SCWO liquid effluent is dependent to some degree on pH. As the liquid effluent is near ambient temperature, general chemistry rules may be applied. Acidic conditions can lead to higher dissolved levels of certain metals. A common example is provided by nickel, which forms nickel oxide in the SCWO reactor due to feed oxidation or corrosion. When excess hydrochloric or sulfuric acid is present in the SCWO effluent, some of the nickel oxide dissolves to yield dissolved nickel. Many of the entries in Table 14 have not been reported in the literature, but are based on unpublished observations by MODAR and General Atomics.

Gas Effluent

Table 15 shows a comparison of SCWO gaseous effluent with the U.S. hazardous waste incinerator air emission standards.[126] While incinerators require multiple clean-up steps to meet the regulations, SCWO inherently meets all but the particulate standard. The SCWO particulate carryover is due to aerosols from the gas–liquid separator, and (assuming a SCWO system was required to meet the incinerator standards) should be readily removed through the use of a coalescing filter. Not shown in Table 15 is the fact that compared to incineration, SCWO produces far lower levels of the acid rain precursor SO_2.

Liquid Effluent

The SCWO process converts waste feed to its mineral constituents. The cleanliness of the liquid effluent is such that, with the inclusion of a high-temperature solids separator, near-drinking-quality water can be produced.[20] For industrial applications, implementation of hot solids separation is not generally practical, and the liquid effluent will contain the soluble salt and suspended mineral constituents derived from the feed stream. SCWO liquid effluent is typically dischargeable to a publicly owned treatment works (POTW).[127] In one instance, SCWO liquid effluent was approved for use for lawn irrigation.[128]

As an indicator of SCWO liquid effluent cleanliness, total organic carbon levels <10 ppm, and frequently <1 ppm, are typically achieved. SCWO effluent

Table 15 Hazardous Waste Incinerator Standards Compared to Typical SCWO Performance

Type of Emissions in Effluent Gas Stream	Standards for New Incinerators	Typical Incinerator Emissions Controls Needed to Meet Standards	SCWO Inherent Performance* (No Gas Clean-up)
Dioxins/difurans, ng/DSCM (TEQ)	<0.2	Rapid quench, powdered activated carbon (PAC) with fabric filter baghouse	0.0028
Mercury, μg/DSCM	<8.1	Fabric filter baghouse or electrostatic precipitator	5.4
Particulate Matter, mg/DSCM	<3.4	Fabric filter baghouse or electrostatic precipitator	15
Toxic Metals, μg/DSCM	<10 for Cd + Pb <23 for Sb+As+Be+Cr	Wet electrostatic precipitator	1 for Cd+Pb, 4 for Sb+As+Be+Cr
DRE,** %	>99.99	Afterburner	>99.99
HCl + Cl_2, ppmv	<21	Packed tower wet scrubber	0.7
NO_x, ppm	Depends on air district—can be <100 ppm	Only local regulations apply. Ammonia or urea injection may be required.	<1
CO, ppm	<100	Afterburner	<10
Hydrocarbons, ppm	<10	Afterburner	<10

Notes: From Ref. 126.
*From General Atomics unpublished data.
**DRE = Destruction and Removal Efficiency.

is saturated with carbon dioxide, imparting a pH in the range of 3.5 when other mineral acids and salts are not present at significant levels. For some applications, pH adjustment of the effluent may be required. Also for some applications and SCWO system material selections, corrosion metals such as nickel and chromium may appear in dissolved form in the liquid effluent.

Solid Residue

Solid residue from the SCWO process is largely a function of the waste feed. Depending on the content and leachability of toxic metals, stabilization or disposal in a hazardous waste landfill may be required. For some applications and SCWO system material selections, corrosion metals such as nickel and chromium may add to the burden of toxic metals. One of the advantages of titanium as a material of construction is that the corrosion product is nontoxic titanium dioxide. In some applications, the solid residue is a valuable byproduct (see Section 8).

7 SCWO PERMITTING

SCWO is frequently evaluated as an option when permitting of an incinerator is difficult or impossible. To our knowledge, permitting of SCWO systems has in all cases to date been easily obtained. Table 16 summarizes the environmental permitting experience for a number of SCWO installations. Two cases (DSI and

Table 16   SCWO Permitting

Project	Location	Application	SCWO Vendor	Permit Description	Year	Reference
DSI (not built)	Texas	Various chemical wastes	Modar	RCRA Subpart B	1987	General Atomics unpublished
Huntsman	Texas	Laboratory wastewater	Ecowaste	liquid discharge to POTW or lawn irrigation	1994	McBrayer and Griffith, 1995 (Ref. 127); Wofford and Griffith, 2001 (Ref. 128)
EST	Texas	Chemical agent breakdown products	General Atomics	RCRA Subpart B	2000	General Atomics unpublished
NECDF (not built)	Indiana	Chemical agent breakdown products	GA	RCRA Subpart B, effluent acceptable for POTW	2000	General Atomics unpublished
Harlingen	Texas	Sewage sludge	HydroProcessing	No air discharge permit required, liquid discharge to POTW	2001	Wofford and Griffith, 2001 (Ref. 128)
Brimsdown	UK	Spent catalyst recovery	Chematur	Nonincineration process, no special discharge permits required	2002	Burgess, 2006 (Ref. 155)
Tokyo Bay	Japan	PCBs	SRI/Mitsubishi	Approved for PCB disposal, no additional permits required	2005	Jayaweera, 2006 (Ref. 162)

NECDF) have been included in which permits were obtained but the plant was not built.

8   SCWO ECONOMICS

The cost of SCWO processing varies greatly, depending on the waste stream being treated and to some degree on the treatment goals. Simpler waste streams that require little or no provision for corrosion or solids handling can, of course, be treated at lower cost. The cost benefit of simpler wastes is largely due to reduced capital expense, as the primary operating costs of oxidant, labor, and utilities do not vary strongly with waste type.

For incineration, air is free, but fuel must be purchased. By contrast, for SCWO fuel is free (at least the heating value provided by the waste), but air or other oxidant must be paid for. Air costs include a relatively expensive high-pressure air compressor, the operation of which entails the dominant fraction of the plant electricity usage, and frequently the dominant fraction of the plant operating costs. Except in very unusual cases, the only economic options to air are air derivatives including partially purified oxygen from an air separation plant and pure oxygen, typically supplied in cryogenic liquid form. The preferred oxidant is site and application specific.

Important factors for the most favorable economics include:

- Automated operation will reduce the operations staff requirements.
- Installation at a multiprocess site. Sharing labor with other local processes can help reduce operations staff requirements. In some cases, a common control room is advantageous.
- Recovery of SCWO products. Examples of valuable products include precious metal oxides,[27] phosphate,[129] paper fillers,[85] iron and aluminum coagulants,[28] and carbon dioxide.[128–130]
- For large-capacity plants with subautogenic feeds, regenerative heat recovery is very important to minimize the need for supplementary fuel and/or additional oxidant.
- Recovery of SCWO byproduct heat. This depends on the existence and needs of nearby processes and facilities. As an example, the HydroProcessing sewage sludge plant was to supply heat to a nearby clothing factory.[128]

Recovery of SCWO byproduct energy has not been included in this list, as it has not yet to our knowledge been demonstrated. Energy recovery from CHON waste streams should be relatively straightforward. A significant DOE-funded program was carried out by MODAR in the late 1980s to evaluate the possibility of energy recovery from salt-containing waste streams. Ceramic filters were developed to provide fluid clean-up prior to entry into a prime mover. Even given 100 percent removal of particles, residual solubility levels in the SCWO product fluid are orders of magnitude higher than operating levels in supercritical steam power plants. For example, steam purity recommendations for supercritical steam boilers are <3 ppb for sodium and <3 ppb for chloride.[16] As noted in Section 6.4, sodium chloride solubility at typical SCWO reactor conditions is on the order of 5 to 200 ppm. The presence of noncondensable gases in the SCWO effluent reduces power cycle efficiency, as compared to pure steam. For wet oxidation, energy recovery is not worthwhile until the plant size has a compressor horsepower of at least 400 (300 kW).[131]

A number of cost estimates for SCWO processes have appeared in the literature over the years. Table 17 provides a sampling of these estimates. To provide a more equal comparison, costs have been inflated by 3 percent per year since the date of publication. To help equalize the comparison, credits for feed tolls or material or energy byproducts have not been included in the table. However, no attempt has been made to equalize the other assumptions made, in particular the significant variables of capital recovery and staffing. The unit costs ($U.S./ton) shown in Table 17 include capital recovery and correlate fairly consistently with system size (wet tpd), with features such as reactor and oxidant type imparting some variation. Major contributors to processing costs are capital recovery, maintenance, operating labor, and oxidant (for oxygen) or electricity (for air oxidant). Depending on the application, caustic and residual solids disposal can also be major operating costs. The vessel reactor systems included in Table 17 are more

Table 17 SCWO Economics

Application	Wet* tpd	Feed Wt%	Ox	Reactor Type	Vendor	Installed Cost $MM	Installed Cost $K/tpd	O&M** Cost $MM/yr	Unit Cost $/ton	Year	2006 Installed Cost $K/tpd	2006 Unit Cost $/ton	Reference
Aqueous organic waste	10	14	Air	Vessel	MODAR				200	1984		431	Thomason and Modell, 1984 (Ref. 156)
Aqueous organic waste	22	10	O₂	Pipe	EWT	2.1	98	0.3	93	1994	144	137	Schmieder and Abeln, 1999 (Ref. 3)
Automobile painting waste	26	10	Air	Vessel	FZK				437	2005		450	Abeln et al., 2005 (Ref. 158)
Sewage sludge	33	15	O₂	Pipe	Modell	1.6	47	0.5	67	2001	54	77	Modell and Svanstrom, 2001 (Ref. 157)
Pulp mill sludge	50	10	O₂	Pipe	MODEC	3.1	62	0.6	62	1990	99	100	Modell, 1990 (Ref. 130)
Drinking water sludge	60	12	O₂	Pipe	Chematur/Feralco	6.5	108	0.7	55	2005	111	56	Stendahl and Stenmark, 2005 (Ref. 29)
Sewage sludge	67	15	O₂	Pipe	Modell	1.9	29	0.8	46	2001	33	53	Modell and Svanstrom, 2001 (Ref. 157)
Food processing waste	83	12	O₂	Pipe	MODAR	3.8	46	0.8	43	1989	75	71	Bettinger, 1989 (Ref. 16)

(continued overleaf)

441

Table 17 (continued)

Application	Wet* tpd	Feed Wt%	Ox	Reactor Type	Vendor	Installed Cost $MM	Installed Cost $K/tpd	O&M** Cost $MM/yr	Unit Cost $/ton	Year	2006 Installed Cost $K/tpd	2006 Unit Cost $/ton	Reference
Chlorinated pharmaceutical waste	83	17	O_2	Vessel	MODAR	5.0	60	1.6	74	1989	99	123	Bettinger, 1989 (Ref. 16)
Chlorinated hazardous waste	83	10	O_2	Vessel	MODAR	6.0	72	1.4	78	1989	119	128	Bettinger, 1989 (Ref. 16)
Pulp mill sludge	100	10	O_2	Pipe	MODEC	4.8	48	1.1	44	1990	78	70	Modell, 1990 (Ref. 130)
Aqueous organic waste	100	14	Air	Vessel	MODAR				40	1984		86	Thomason and Modell, 1984 (Ref. 156)
Sewage sludge	133	15	O_2	Pipe	Modell	2.5	19	1.2	33	2001	22	38	Modell and Svanstrom, 2001 (Ref. 157)
Sewage sludge	150	7	O_2	Pipe	Hydro Processing	3.5	23	0.6	18	2001	27	21	Wofford and Griffith, 2001 (Ref. 128)
Sewage sludge	220	15	O_2	Pipe	Chematur/ Feralco	12.8	58	1.6	43	2003	64	47	Stendahl and Jäfverström 2004 (Ref. 28)
Pulp mill sludge	1000	10	O_2	Pipe	MODEC	20.2	20	3.8	15	1990	32	24	Modell, 1990 (Ref. 130)

*tpd = tons per day (U.S.).
**O&M Cost = Operating and Maintenance Cost.

expensive than the pipe reactor systems but are generally handling wastes that are not treatable with pipe reactors. In addition, the vessel plants are generally handling higher heating value wastes and thus incur higher oxidant costs.

It must be emphasized that the costs shown in Table 17 are projected economics. To our knowledge, no actual processing cost data have been published to date. Until more operating history is gained and actual costs become available, the economics provided in Table 17 must be considered preliminary. While SCWO costs are believed to be competitive with incineration, SCWO is a less mature technology and therefore involves a higher element of risk.

9 SUMMARY

By taking advantage of the unique properties exhibited by water above its critical point, supercritical water oxidation has proven to be an effective technology for waste destruction. Compared to standard or more established technologies such as incineration, SCWO is applicable to a wider variety of chemical species, with minimal or no toxic byproducts. It has shown to be much easier to permit, meets or exceeds most regulatory constraints on effluent, has a more favorable public perception, and can be economically competitive with certain types of feeds. Limitations such as corrosion and salt/solids handling have slowed its development, but several decades of research, both in industry and academia, have now resulted in system designs where these issues are considered manageable. SCWO is less mature than the more-established competitive technologies, and this can result in a higher risk or cost for some applications. This picture, however, is gradually changing, as witnessed by the existence of a number of full-scale plants built by several different companies currently operating or under construction around the world for a variety of feeds of increasing complexity. Interest in the United States is still primarily limited to government agencies, but more commercial interest has developed internationally, as indicated by the distribution and location of the current full-scale plants.

Any of the current SCWO vendors listed in this chapter can provide more detailed information and assistance for addressing the key design and operating issues in applying SCWO to a specific feed or application. Considering expected future increases in waste quantities and its overall performance thus far, SCWO can play an important role as a viable technology for clean, acceptable, and effective waste destruction.

REFERENCES

1. P. Kritzer and E. Dinjus, "An Assessment of Supercritical Water Oxidation," *Chem. Eng. J.*, **83**, 207–214, 2001.
2. R. W. Shaw and N. Dahmen, "Destruction of Toxic Organic Materials Using Supercritical Water Oxidation: Current State of the Technology," *Supercritical*

Fluids—Fundamentals and Applications, E. Kiran, P. G. Debenedetti and C. J. Peters (eds.), Dordrecht: Kluwer—Academic, 2000, pp. 425–437.

3. H. Schmeider and J. Abeln, "Supercritical Water Oxidation: State of the Art," *Chem. Eng. Technol.*, **22**, 903–908 (1999).

4. E. F. Gloyna and L. Li, "Supercritical Water Oxidation for Wastewater and Sludge Remediation," in *Encyclopedia of Environmental Analysis and Remediation*, R. A. Meyers (ed.), John Wiley, New York, 1998.

5. E. F. Gloyna and L. Li, "Waste Treatment by SupercriticalWater Oxidation," in *Encyclopedia of Chemical Processing and Design*, pp. 272–304, Marcel Dekker, Inc., New York, 1998.

6. T. Tassaing, Y. Danten and M. Besnard, "Infrared Spectroscopic Study of Hydrogen-bonding in Water at High Temperature and Pressure," *J. Mol. Liq.*, **101**(1–3), 149–158 (2002).

7. W. Schilling and E. U. Franck, "Combustion and Diffusion Flames at High Pressures to 2000 Bar," *Ber. Bunsenges. Physik. Chemie*, **92**, 631–636 (1988).

8. R. R. Steeper, S. F. Rice, M. S. Brown and S. C. Johnston, "Methane and Methanol Diffusion Flames in Supercritical Water," *J. Supercrit. Fluids*, **5**, 262–268 (1992).

9. W. R. Killilea, K. C. Swallow and G. T. Hong, "The Fate of Nitrogen in Supercritical Water Oxidation," *J. Supercrit. Fluids*, **5**(1), 72–78 (1992).

10. V. S. Mishra, V. V. Mahajani and J. B. Joshi, "Wet Air Oxidation," *Ind. Eng. Chem. Res.*, **34**(1), 2–48 (1995).

11. D. S. Ross, I. Jayaweera, D. Bomberger, "On-site Disposal of Hazardous Waste via Assisted Hydrothermal Oxidation," *Rev. High Pressure Sci. Technol.*, **7**, 1386–1388 (1998).

12. M. Modell, R. C. Reid, S. I. Amin, "Gasification Process," U.S. Patent #4,113,446, September 12, 1978.

13. E. Dinjus and A. Kruse, "Hot Compressed Water—A Suitable and Sustainable Solvent and Reaction Medium," *J. Phys. Condens. Matter*, **16**, S1161–S1169, 2004.

14. P. A. Marrone, G. T. Hong and M. H. Spritzer, "Developments in Supercritical Water as a Medium for Oxidation, Reforming, and Synthesis," *J. Adv. Oxid. Technol.*, **10**(1), 157–168 (2007).

15. G. T. Hong, W. R. Killilea and T. B. Thomason, "Supercritical Water Oxidation: Space Applications," ASCE Proceedings of Space '88, Albuquerque, NM, August 29–31, 1988.

16. J. A. Bettinger, "Assessment and Development of an Industrial Wet Oxidation System for Burning Waste and Low-Grade Fuels," Final Report DOE/ID/12711-1, September 1989.

17. C. M. Barnes, "Evaluation of Tubular Reactor Designs for Supercritical Water Oxidation of U.S. Department of Energy Mixed Waste," Idaho National Engineering Laboratory Report INEL-94/0223, December, 1994.

18. G. T. Hong, "Hydrothermal Oxidation: Pilot Scale Operating Experiences," presented at the 56th Annual Water Conference, Engineers' Society of Western Pennsylvania, Pittsburgh, October, 1995.

19. M. Modell, J. Larson and S. F. Sobczynski, "Supercritical Water Oxidation of Pulp Mill Sludges," *Tappi J.*, **75**(6), 195–202 (1992).

20. G. T. Hong, P. K. Fowler, W. R. Killilea and K. C. Swallow, "Supercritical Water Oxidation: Treatment of Human Waste and System Configuration Tradeoff Study," SAE Technical Paper, No. 871444, 1987.

21. A. Shanableh and E. F. Gloyna, "Supercritical Water Oxidation—Wastewaters and Sludges," *Wat. Sci. Tech.*, **23**, 389–398 (1991).

22. National Research Council (Committee on Alternative Chemical Demilitarization Technologies; Board on Army Science and Technology), "Alternative Technologies for the Destruction of Chemical Agents and Munitions," National Academy Press, Washington DC, 1993.

23. P. J. Crooker, K. S. Ahluwalia, Z. Fan and J. Prince, "Operating Results from Supercritical Water Oxidation Plants," *Ind. Eng. Chem. Res.*, **39**(12), 4865–4870 (2000).

24. M. H. Spritzer, D. A. Hazlebeck and K. W. Downey, "Supercritical Water Oxidation of Chemical Agents and Solid Propellants," *J. Energetic Mater.*, **13**, 185–212 (1995).

25. P. Svensson, "Look, No Stack: Supercritical Water Destroys Organic Wastes," *Chemical Technology Europe*, 16–19 (January/February 1995).

26. R. N. McBrayer, J. W. Griffith and A. Gidner, "Operation of the First Commercial Supercritical Water Oxidation Industrial Waste Facility," Proceedings of International Conference on Oxidation Technologies for Water and Wastewater Treatment, 1996.

27. P. Grumett, "Precious Metal Recovery from Spent Catalysts," *Platinum Metals Rev.*, **47**(4), 163–166 (2003).

28. K. Stendahl and S. Jäfverström, "Recycling of Sludge with the Aqua Reci Process," *Water Sci. Tech.*, **49**(10), 233–240, 2004.

29. K. E. Stendahl and L. Stenmark, "The Use of Supercritical Water Oxidation for Recovery of Coagulants from Water Works Sludge," Presented at *11th International Conference on Advanced Oxidation Technologies for Treatment of Water, Air, and Soil*, Chicago, October 23–27, 2005.

30. J. Dahlin, "Oxidation of De-inking Sludge in Supercritical Water in Practice," Presented at Workshop on *Managing Pulp and Paper Process Residues*, Barcelona, Spain, May 30–31, 2002 (paper is also available for download at www.chematur.se).

31. I. Jayaweera, "Supercritical Water Oxidation Technology," in *Chemical Degradation Methods for Wastes and Pollutants* (Environmental Science Pollution Control Series 26), Chapter 3, 121–163, 2003.

32. N. Urai, Organo Corp., personal communication, October, 2005.

33. J. Han, Hanwha Chemical Corp., personal communication, June 2006.

34. N. Bonnaudin, "A Solution for the Treatment of Wastewater from Food and Pharmaceutical Industries: Hydrothermal Oxidation," Presented at *The Role of Supercritical Fluid Technology in Pharmaceutical/Nutraceutical/Food Processing*, Supermat Network Seminar, July 6–8, 2005 (www.univ-pau.fr/supermat/OP6.pdf).

35. P. A. Marrone, S. D. Cantwell and D. W. Dalton, "SCWO System Designs for Waste Treatment: Application to Chemical Weapons Destruction," *Ind. Eng. Chem. Res.*, **44**(24), 9030–9039 (2005).

36. J. W. Tester, H. R. Holgate, F. A. Armellini, P. A. Webley, W. R. Killilea, G. T. Hong and H. E. Barner, "Supercritical Water Oxidation Technology," in *Emerging Technologies for Hazardous Waste Management III*, W. D. Tedder and F. G. Pohland (eds.), American Chemical Society Symposium Series, Vol. 518, Washington DC, 35–76, 1993.

37. W. J. Lamb, G. A. Hoffman and J. Jonas, "Self-diffusion in Compressed Supercritical Water," *J. Chem. Phys.*, **74**(12), 6875–6880 (1981).

38. K. Yoshida, C. Wakai, N. Matubayasi and M. Nakahara, "A New High-Temperature Multinuclear-Magnetic-Resonance Probe and the Self-diffusion of Light and Heavy Water in Sub- and Supercritical Conditions," *J. Chem. Phys.*, **123**, (164506-1)–(164506-10) (2005).

39. K. H. Dudziak and E. U. Franck, "Messungen der Viskosität des Wassers bis 560°C und 3500 Bar," *Ber. Bunsenges. Physik. Chemie*, **70**(9–10), 1120–1128, 1966.

40. J. T. R. Watson, R. S. Basu and J. V. Sengers, "An Improved Representative Equation for the Dynamic Viscosity of Water Substance," *J. Phys. Chem. Ref. Data*, **9**(4), 1255–1290 (1980).

41. E. U. Franck, "Physicochemical Properties of Supercritical Solvents," *Ber. Bunsenges. Physik. Chemie*, **88**(9), 820–825, 1984.

42. A. I. Abdulagatov, G. V. Stepanov, I. M. Abdulagatov, A. E. Ramazanova and G. S. Alisultanova, "Extrema of Isochoric Heat Capacity of Water and Carbon Dioxide," *Chem. Eng. Commun.*, **190**(11), 1499–1520 (2003).

43. M. Uematsu and E. U. Franck, "Static Dielectric Constant of Water and Steam," *J. Phys. Chem. Ref. Data*, **9**(4), 1291–1306 (1980).

44. W. L. Marshall and E. U. Franck, "Ion Product of Water Substance, 0–1000°C, 1–10,000 Bars, New International Formulation and Its Background," *J. Phys. Chem. Ref. Data*, **10**(2), 295–304 (1981).

45. Z. Xu and S. I. Sandler, "Temperature-Dependent Parameters and the Peng-Robinson Equation of State," *Ind. Eng. Chem. Res.*, **26**(3), 601–605 (1987).

46. H. Touba and G. A. Mansoori, "Structure and Property Prediction of Sub- and Super-critical Water," *Fluid Phase Equilib.*, **150–151**, 459–468 (1998).

47. P. J. Smits, I. G. Economou, C. J. Peters and J. de Swaan Arons, "Equation of State Description of Thermodynamic Properties of Near-critical and Supercritical Water," *J. Phys. Chem.*, **98**, 12080–12085 (1994).

48. M. C. Kutney, V. S. Dodd, K. A. Smith, H. J. Herzog and J. W. Tester, "A Hard-Sphere Volume-Translated van der Waals Equation of State for Supercritical Process Modeling: 1. Pure Components," *Fluid Phase Equilib.*, **128**, 149–171 (1997).

49. W. Wagner and A. Pruβ, "The IAPWS Formulation 1995 for the Thermodynamic Properties of Ordinary Water Substance for General and Scientific Use," *J. Phys. Chem. Ref. Data*, **31**(2), 387–535 (2002).

50. M. Hodes, P. A. Marrone, G. T. Hong, K. A. Smith and J. W. Tester, "Salt Precipitation and Scale Control in Supercritical Water Oxidation—Part A: Fundamentals and Research," *J. Supercrit. Fluids*, **29**, 265–288 (2004).

51. P. A. Webley, J. W. Tester and H. R. Holgate, "Oxidation Kinetics of Ammonia and Ammonia-Methanol Mixtures in Supercritical Water in the Temperature Range 530–700°C at 246 Bar," *Ind. Eng. Chem. Res.*, **30**(8), 1745–1754 (1991).

52. P. A. Webley and J. W. Tester, "Fundamental Kinetics of Methane Oxidation in Super-critical Water," *Energy Fuels*, **5**, 411–419 (1991).

53. H. R. Holgate, P. A. Webley, J. W. Tester and R. K. Helling, "Carbon Monoxide Oxidation in Supercritical Water: The Effects of Heat Transfer and the Water-Gas Shift Reaction on Observed Kinetics," *Energy Fuels*, **6**, 586–597 (1992).

54. H. R. Holgate and J. W. Tester, "Fundamental Kinetics and Mechanisms of Hydrogen Oxidation in Supercritical Water," *Combust. Sci. and Tech.*, **88**, 369–397 (1993).

55. J. W. Tester, P. A. Webley and H. R. Holgate, "Revised Global Kinetic Measurements of Methanol Oxidation in Supercritical Water," *Ind. Eng. Chem. Res.*, **32**(1), 236–239 (1993).

56. J. C. Meyer, P. A. Marrone and J. W. Tester, "Acetic Acid Oxidation and Hydrolysis in Supercritical Water," *AIChE J.*, **41**(9), 2108–2121 (1995).

57. J. R. Katzer, "Solubilization and Wet Oxidation of Suspended and Dissolved Organics," U.S. Department of the Interior, Office of Water Research and Technology Report A-027-DEL, 1975.

58. P. A. Marrone, T. A. Arias, W. A. Peters and J. W. Tester, "Solvation Effects on Kinetics of Methylene Chloride Reactions in Sub- and Supercritical Water: Theory, Experiment, and *Ab-Initio* Calculations," *J. Phys. Chem. A*, **102**, 7013–7028 (1998).

59. F. Vogel, K. A. Smith, J. W. Tester and W. A. Peters, "Engineering Kinetics for Hydrothermal Oxidation of Hazardous Organic Substances," *AIChE J.*, **48**(8), 1827–1839 (2002).

60. J. M. Ploeger, P. A. Bielenberg, J. L. DiNaro-Blanchard, R. P. Lachance, J. D. Taylor, W. H. Green and J. W. Tester, "Modeling Oxidation and Hydrolysis Reactions in Supercritical Water—Free Radical Elementary Reaction Networks and Their Applications," *Combust. Sci. and Tech.*, **178**, 363–398 (2006).

61. Z. Y. Ding, M. A. Frisch, L. Li and E. F. Gloyna, "Catalytic Oxidation in Supercritical Water," *Ind. Eng. Chem. Res.*, **35**(10), 3257–3279 (1996).

62. P. E. Savage, J. B. Dunn and J. Yu, "Recent Advances in Catalytic Oxidation in Supercritical Water," *Combust. Sci. and Tech.*, **178**, 443–465 (2006).

63. F. J. Armellini and J. W. Tester, "Solubility of Sodium Chloride and Sulfate in Sub- and Supercritical Water Vapor from 450–550°C and 100–250 Bar," *Fluid Phase Equilib.*, **84**, 123–142 (1993).

64. M. M. DiPippo, "Phase Behavior of Inorganic Salts in Sub- and Supercritical Water," Doctoral Thesis, Massachusetts Institute of Technology, Cambridge, MA, 1997.

65. A. Anderko and K. S. Pitzer, "Equation of State Representation of Phase Equilibria and Volumetric Properties of the System NaCl H$_2$O Above 573K," *Geochim. Cosmochim. Acta*, **57**, 1657 1680 (1993).

66. M. T. Reagan, J. G. Harris and J. W. Tester, "Molecular Simulations of Dense Hydrothermal NaCl-H$_2$O Solutions from Subcritical to Supercritical Conditions," *J. Phys. Chem. B*, **103**, 7935–7941 (1999).

67. J. J. Kosinski and A. Anderko, "Equation of State for High-Temperature Aqueous Electrolyte and Nonelectrolyte Systems," *Fluid Phase Equilib.*, **183–184**, 75–86 (2001).

68. F. J. Armellini, J. W. Tester and G. T. Hong, "Precipitation of Sodium Chloride and Sodium Sulfate in Water from Sub- to Supercritical Conditions: 150 to 550°C, 100–300 Bar," *J. Supercrit. Fluids*, **7**, 147–158 (1994).

69. S. N. Rogak and P. Teshima, "Deposition of Sodium Sulfate in a Heated Flow of Supercritical Water," *AIChE J.*, **45**(2), 240–247 (1999).

70. M. Hodes, P. Griffith, K. A. Smith, W. S. Hurst, W. J. Bowers and K. Sako, "Salt Solubility and Deposition in High Temperature and Pressure Aqueous Solutions," *AIChE J.*, **50**(9), 2038–2049 (2004).

71. N. Boukis, N. Claussen, K. Ebert, R. Janssen and M. Schacht, "Corrosion Screening Tests of High Performance Ceramics in Supercritical Water Containing Oxygen and Hydrochloric Acid," *J. European Ceramic Soc.*, **17**, 71–76 (1997).

72. N. Boukis, C. Friedrich and E. Dinjus, "Titanium as Reactor Material for SCWO Applications. First Experimental Results," Corrosion '98, National Association of Corrosion Engineers, March 1998.

73. D. B. Mitton, J. H. Yoon, J. A. Cline, H. S. Kim, N. Eliaz and R. M. Latanision, "Corrosion Behavior of Nickel-Based Alloys in Supercritical Water Oxidation Systems," *Ind. Eng. Chem. Res.*, **39**(12), 4689–4696 (2000).

74. P. Kritzer, N. Boukis and E. Dinjus, "Corrosion of Alloy 625 in Aqueous Solutions Containing Chloride and Oxygen," *Corrosion*, **54**(10), 824–834 (1998).

75. J. H. Yoon, K. S. Son, B. Mitton, R. Latanision, Y. R. Yoo and Y. S. Kim, "Corrosion Behavior of 316L Stainless Steel in Supercritical Water Environment," *Materials Science Forum*, **475–479**, 4207–4210 (2005).

76. L. B. Kriksunov and D. D. Macdonald, "Potential-pH Diagrams for Iron in Supercritical Water," *Corrosion*, **53**(8), 605–611 (1997).

77. D. D. Macdonald and L. B. Kriksunov, "Probing the Chemical and Electrochemical Properties of SCWO Systems," *Electrochim. Acta*, **47**, 775–790 (2001).

78. C. Liu, S. R. Snyder and A. J. Bard, "Electrochemistry in Near-critical and Supercritical Fluids. 9. Improved Apparatus for Water Systems (23–385°C). The Oxidation of Hydroquinone and Iodide," *J. Phys. Chem. B*, **101**, 1180–1185 (1997).

79. P. Kritzer, "Corrosion in High-Temperature and Supercritical Water and Aqueous Solution: A Review," *J. Supercrit. Fluids*, **29**, 1–29 (2004).

80. L. B. Kriksunov and D. D. Macdonald, "Corrosion in Supercritical Water Oxidation Systems: A Phenomenological Analysis," *J. Electrochem. Soc.*, **142**(12), 4069–4073 (1995).

81. R. M. Latanision, "Corrosion Science, Corrosion Engineering, and Advanced Technologies," *Corrosion*, **51**(4), 270–283 (1995).

82. M. Modell and M. Svanström, "Comparison of Supercritical Water Oxidation and Incineration for Treatment of Sewage Sludges," 6th Conference on Supercritical Fluids and Their Applications, September 9–12, Maiori, Italy, 2001.

83. L. Stenmark, "Aqua Critox®, the Chematur AB Concept for SCWO," in *Supercritical Water Oxidation—Achievements and Challenges in Commercial Applications*, A. Nazeri, R. W. Shaw and P. A. Marrone (eds.), Proceedings of Workshop held August 14–15, 2001, in Arlington, Virginia, Strategic Analysis, Inc., 2001.

84. A. Suzuki, "Commercialization of the First Supercritical Water Oxidation Facility for Semi-conductor Manufacturing Wastes," in *Supercritical Water Oxidation—Achievements and Challenges in Commercial Applications*, A. Nazeri, R. W. Shaw and P. A. Marrone (eds.), Proceedings of Workshop held August 14–15, 2001, in Arlington, Virginia, Strategic Analysis, Inc., 2001.

85. A. V. Gidner, L. B. Stenmark and K. M. Carlsson, "Treatment of Different Wastes by Supercritical Water Oxidation," Proceedings of the International Conference on Incineration and Thermal Treatment Technologies, Philadelphia, PA, May, 2001 (paper is also available for download at www.chematur.se).

86. R. N. McBrayer Jr., J. G. Swan and J. S. Barber, "Method and Apparatus for Reacting Oxidizable Matter with Particles," U.S. Patent #5,620,606, 1997.

87. A. Gidner and L. Stenmark, "Supercritical Water Oxidation of Sewage Sludge—State of the Art," IBC Conference on Sewage Sludge and Disposal Options, Birmingham, England, March 2001 (paper is also available for download at www.chematur.se).

88. G. T. Hong and M. H. Spritzer, "Supercritical Water Partial Oxidation," Proceedings of the 2002 U.S. DOE Hydrogen Program Review, NREL/CP-610-32405, 2002.

89. F. Cansell, "Method for Treating Waste by Hydrothermal Oxidation," U.S. Patent #6,929,752, August 16, 2005.

90. M. Modell, "Processing Methods for the Oxidation of Organics in Supercritical Water," U.S. Patent #4,338,199, July 6, 1982.

91. J. M. Eller, R. N. McBrayer, Jr., R. D. Peacock, J. S. Barber, W. H. Stanton, F. Applegath and G. H. Lovett, "Heating and Reaction System and Method Using Recycle Reactor," U.S. Patent #6,001,243, December 14, 1999.

92. C. M. Barnes and C. H. Oh, "Chemical Reactor Design for Supercritical Water Oxidation of U.S. DOE Waste," First International Conference on Advanced Oxidation Technologies for Water and Air Remediation, London, Ontario, Canada, June 25–30, 1994.

93. F. J. Zimmermann, "New Waste Disposal Process," *Chem. Eng.*, **65**, 117–120 (1958).

94. H. Perkow, R. Steiner and H. Vollmüller, "Wet Air Oxidation—a Review," *German Chemical Engineering*, **4**, 193–201 (1981).

95. M. Modell, A. Z. Panagiotopoulos and M. C. Kutney, "Modeling and Simulation of the Thermodynamic Properties of Water and Supercritical Water Mixtures," Joint 6th International Symposium on Hydrothermal Reaction and 4th International Conference on Solvo-thermal Reactions, Kochi, Japan, 2000.

96. G. T. Hong, "Process for Oxidation of Materials at Supercritical Temperatures and Subcritical Pressures," U.S. Patent #5,106,513, 1992.

97. R. R. Steeper and S. F. Rice, "Optical Monitoring of the Oxidation of Methane in Supercritical Water," in *Physical Chemistry of Aqueous Systems*, Begell House, New York, 1995, pp. 652–659.

98. H. R. Holgate and J. W. Tester, "Oxidation of Hydrogen and Carbon Monoxide in Sub- and Supercritical Water: Reaction Kinetics, Pathways, and Water Density Effects. 1. Experimental Results," *J. Phys. Chem.*, **98**, 800–809 (1994).

99. M. L. Japas and E. U. Franck, "High Pressure Phase Equilibria and PVT-Data of the Water-Nitrogen System to 673 K and 250 MPa," *Ber. Bunsenges. Phys. Chem.*, **89**, 793–800 (1985).

100. M. L. Japas and E. U. Franck, "High Pressure Phase Equilibria and PVT-Data of the Water-Oxygen System Including Water-Air to 673 K and 250 MPa," *Ber. Bunsenges. Phys. Chem.*, 89:1268–1275 (1985).

101. H. A. Pray, C. E. Schweickert and B. H. Minnich, "Solubility of Hydrogen, Oxygen, Nitrogen and Helium in Water At Elevated Temperatures," *Ind. and Eng. Chem.*, **44**(5), 1146–1151 (1952).

102. S. Takenouchi and G. C. Kennedy, "The Binary System H_2O-CO_2 at High Temperatures and Pressures," *Am. J. of Sci.*, **262**, 1055–1074 (1964).

103. R. Wiebe, "The Binary System Carbon Dioxide-Water Under Pressure," *Chem. Rev.*, **29**, 475–481 (1941).

104. A. Seidell, "Solubilities of Inorganic and Organic Compounds," Vol. II, 3rd ed., p. 368, Van Nostrand, New York, 1940.

105. J. F. Connolly, "Solubility of Hydrocarbons in Water Near the Critical Solution Temperatures," *J. Chem. and Eng. Data*, **11**, 13–16 (1966).

106. T. M. O'Grady, "Liquid-Liquid Equilibria for the Benzene-n-Heptane-Water System in the Critical Solution Region," *J. Chem. and Eng. Data*, **12**, 9–12 (1967).

107. Z. Alwani and G. M. Schneider, "Druckeinfluß auf die Entmischung flüssiger Systeme," *Ber. Bunsengesellschaft*, **71**, 633–638 (1967).

108. C. J. Rebert and W. B. Kay, "The Phase Behavior and Solubility Relations of the Benzene-Water System," *AIChE J.*, **5**, 285–289 (1959).

109. N. D. Sanders, "Visual Observation of the Solubility of Heavy Hydrocarbons in Near-Critical Water," *Ind. Eng. Chem. Fundam.*, **25**, 169–171 (1986).

110. K. S. Pitzer and R. T. Pabalan, "Thermodynamics of NaCl in Steam," *Geochim. Cosmochim. Acta*, **50**, 1445–1454 (1986).

111. S. Sourirajan and G. C. Kennedy, "The System H_2O- NaCl at Elevated Temperatures and Pressures," *Am. J. of Sci.*, **260**, 115–141 (1962).

112. K. S. Pitzer, "Ionic Fluids," *J. Phys. Chem.*, **88**, 2689–2697 (1984).

113. S. Takenouchi and G. C. Kennedy, "The Solubility of Carbon Dioxide in NaCl Solutions at High Temperatures and Pressures," *Am. J. of Sci.*, **263**, 445–454 (1965).

114. M. Gehrig, H. Lentz and E. U. Franck, "The System Water-Carbon Dioxide-Sodium Chloride to 773 K and 300 MPa," *Ber. Bunsengesellschaft*, **90**, 525–533 (1986).

115. T. S. Bowers and H. C. Helgeson, "Calculation of the Thermodynamic and Geochemical Consequences of Nonideal Mixing in the System H_2O-CO_2-NaCl on Phase Relations in Geologic Systems: Equation of State for H_2O-CO_2-NaCl Fluids at High Pressures and Temperatures," *Geochim. Cosmochim. Acta*, **47**, 1247–1275 (1983).

116. M. I. Ravich and V. Y. Borovaya, "Phase Equilibria in the Sodium-Sulphate Water System at High Pressures and Temperatures," *Russ. J. of Inorg. Chem.*, **9**(4), 520–532 (1964).

117. W. C. Schroeder, A. Gabriel and E. P. Partridge, "Solubility equilibria of sodium sulfate at temperatures of 1508C to 3508C. I. Effect of sodium hydroxide and sodium chloride," *J. Amer. Chem. Soc.*, **57**, 1539 (1935).

118. N. B. Keevil, "Vapor pressures of aqueous solutions," *J. Am. Chem. Soc.*, **64**, 841–850 (1942).

119. G. T. Hong, D. W. Ordway and V. A. Zilberstein, "Materials Testing in Supercritical Water Oxidation Systems," First International Workshop on Supercritical Water Oxidation, Amelia Island, FL, February 1995.

120. V. A. Zilberstein, J. A. Bettinger, D. W. Ordway and G. T. Hong, "Evaluation of Materials Performance in a Supercritical Wet Oxidation System," Corrosion '95, National Association of Corrosion Engineers, Orlando, March 1995.

121. K. M. Garcia and R. Mizia, "Corrosion Investigation of Multilayered Ceramics and Experimental Nickel Alloys in SCWO Process Environments," First International Workshop on Supercritical Water Oxidation, Amelia Island, Florida, February, 1995.

122. G. T. Hong, "Ceramic Coating System for Water Oxidation Environments," U.S. Patent #5,545,337, August 13, 1996.

123. P. J. Crooker, K. S. Ahluwalia and Z. Fan, "Supercritical Water Oxidation of Chemical Weapon Agents by Transpiring Wall Reactor," Proceedings of the International Conference on Incineration and Thermal Treatment Technologies, Philadelphia, PA, May, 2001.

124. T. G. McGuinness, "Supercritical Oxidation Reactor," U.S. Patent #5,384,051, January 24, 1995.

125. P. A. Marrone, M. Hodes, K. A. Smith and J. W. Tester, "Salt Precipitation and Scale Control in Supercritical Water Oxidation—Part B: Commercial/Full-scale Applications," *J. Supercrit. Fluids*, **29**, 289–312 (2004).

126. U.S. Environmental Protection Agency, "National Emission Standards for Hazardous Air Pollutants: Final Standards for Hazardous Air Pollutants for Hazardous Waste Combustors," *Federal Register*, Vol. 70, No. 196, October 12, 2005.

127. R. N. McBrayer Jr. and J. W. Griffith, "Operation of the First Supercritical Water Oxidation Industrial Waste Facility," 56th Annual Water Conference, Engineers' Society of Western Pennsylvania, Pittsburgh, October, 1995.

128. W. T. Wofford and J. W. Griffith, "Commercial Application of SCWO to the Treatment of Municipal Sludge," in *Supercritical Water Oxidation—Achievements and Challenges in Commercial Applications*, A. Nazeri, R. W. Shaw, and P. A. Marrone (eds.), Proceedings of Workshop held August 14–15, 2001, in Arlington, Virginia, Strategic Analysis, Inc., 2001.

129. M. Svanström, K. Johansson and K. Stendahl, "Supercritical Water Oxidation of Sewage Sludge in Combination with Phosphate Recovery," presentation at the Leading-Edge Conference on Drinking Water and Wastewater Treatment Technologies, Prague, 2004.

130. M. Modell, "Treatment of Pulp Mill Sludges by Supercritical Water Oxidation," Final Report DOE/CE/40914-T1, U.S. Department of Energy, July, 1990.

131. G. H. Teletzke, "Wet Air Oxidation," *Chem. Eng. Prog.*, **60**, 33–38 (1964).

132. J. W. Griffith and D. H. Raymond, "The First Commercial Supercritical Water Oxidation Sludge Processing Plant," Proceedings of the International Conference on Incineration and Thermal Treatment Technologies, Philadelphia, PA, May 2001.

133. J. Abeln, M. Kluth, G. Petrich and H. Schmieder, "Supercritical Water Oxidation (SCWO): A Process for the Treatment of Industrial Waste Effluents," *High Pressure Research*, **20**, 537–547 (2001).

134. B. Wellig, K. Lieball and P. R. von Rohr, "Operating Characteristics of a Transpiring-wall SCWO Reactor with a Hydrothermal Flame as Internal Heat Source," *J. Supercrit. Fluids*, 34: 35–50, 2005.

135. P. Teshima, S. Rogak, D. Fraser, E. Hauptmann, S. Gairns and J. Lota, "Fouling of Supercritical Water Oxidation Reactors," Proceedings of the Air and Waste Management Association's Annual Meeting & Exhibition, June 8–13, 1997, Toronto, Ontario, Canada, 97–RA133.03.

136. M. J. Cocero, E. Alonso, R. Torio, D. Vallelado, T. Sanz and F. Fdz-Polanco, "Supercritical Water Oxidation (SCWO) for Poly(ethylene terephthalate) (PET) Industry Effluents," *Ind. Eng. Chem. Res.*, **39**(12), 4652–4657 (2000).

137. F. Vogel, J. L. DiNaro-Blanchard, P. A. Marrone, S. F. Rice, P. A. Webley, W. A. Peters, K. A. Smith and J. W. Tester, "Critical Review of Kinetic Data for the Oxidation of Methanol in Supercritical Water," *J. Supercrit. Fluids*, **34**, 249–286 (2005).

138. J. Schanzenbächer, J. D. Taylor and J. W. Tester, "Ethanol Oxidation and Hydrolysis Rates in Supercritical Water," *J. Supercrit. Fluids*, **22**, 139–147, (2002).

139. D. Lee and E. F. Gloyna, "Hydrolysis and Oxidation of Acetamide in Supercritical Water," *Environ. Sci. Technol.*, **26**(8), 1587–1593 (1992).

140. L. Li, J. R. Portela, D. Vallejo and E. F. Gloyna, "Oxidation and Hydrolysis of Lactic Acid in Near-critical Water," *Ind. Eng. Chem. Res.*, **38**(7), 2599–2606 (1999).

141. J. L. DiNaro, J. W. Tester, J. B. Howard and K. C. Swallow, "Experimental Measurements of Benzene Oxidation in Supercritical Water," *AIChE J.*, **46**(11), 2274–2284 (2000).

142. R. Li, P. E. Savage and D. Szmukler, "2-Chlorophenol Oxidation in Supercritical Water: Global Kinetics and Reaction Products," *AIChE J.*, **39**(1), 178–187, 1993.

143. K. S. Lin, H. P. Wang and M. C. Li, "Oxidation of 2,4-Dichlorophenol in Supercritical Water," *Chemosphere*, **36**(9), 2075–2083 (1998).

144. C. J. Martino, P. E. Savage and J. Kasiborski, "Kinetics and Products from o-Cresol Oxidation in Supercritical Water," *Ind. Eng. Chem. Res.*, **34**(6), 1941–1951 (1995).

145. N. Crain, S. Tebbal, L. Li and E. F. Gloyna, "Kinetics and Reaction Pathways of Pyridine Oxidation in Supercritical Water," *Ind. Eng. Chem. Res.*, **32**(10), 2259–2268 (1993).

146. N. Crain, S. Tebbal, L. Li and E. F. Gloyna, "Comments on 'Kinetics and Reaction Pathways of Pyridine Oxidation in Supercritical Water,'" *Ind. Eng. Chem. Res.*, **34**, 1499 (1995).

147. G. Anitescu and L. L. Tavlarides, "Oxidation of Biphenyl in Supercritical Water: Reaction Kinetics, Key Pathways, and Main Products," *Ind. Eng. Chem. Res.*, **44**(5), 1226–1232 (2005).

148. G. Anitescu and L. L. Tavlarides, "Oxidation of Arochlor 1248 in Supercritical Water: A Global Kinetic Study," *Ind. Eng. Chem. Res.*, **39**(3), 583–591 (2000).

149. P. A. Sullivan and J. W. Tester, "Methylphosphonic Acid Oxidation Kinetics in Supercritical Water," *AIChE J.*, **50**(3), 673–683, 2004.

150. S. Gopalan and P. E. Savage, "A Reaction Network Model for Phenol Oxidation in Supercritical Water," *AIChE J.*, **41**(8), 1864–1873 (1995).

151. Z. Alwani and G. M. Schneider, "Phasengleichgewichte, kritische Erscheinungen und PVT—Daten in binären Mischungen von Wasser mit aromatischen Kohlenwasserstoffen bis 420°C und 2200 bar," *Ber. Bunsengesellschaft*, **73**, 294–301 (1969).

152. K. Bröllos, K. Peter and G. M. Schneider, "Fluide Mischsysteme unter hohem Druck. Phasengleichgewichte und kritische Erscheinungen in den binären Systemen Cyclohexan-H_2O, n-Heptan-H_2O, Biphenyl-H_2O und Benzol-D_2O bis 420°C und 3000 bar," *Ber. Bunsengesellschaft*, **74**, 682–686 (1970).

153. K. C. Swallow, W. R. Killilea, G. T. Hong and H. W. Lee, "Behavior of Metal Compounds in the Supercritical Water Oxidation Process," SAE Technical Paper No. 901314, 1990.

154. Anonymous, "Recovery of Precious Metal Catalysts with Supercritical Water Oxidation," *Filtration and Separation* (June 2003).

155. A. Burgess, Chematur, personal communication, 2006.

156. T. B. Thomason and M. Modell, "Supercritical Water Destruction of Aqueous Wastes," *Hazardous Waste*, **1**, 453–467 (1984).

157. M. Modell and M. Svanström, "Economics and Full-Scale Application," in *Supercritical Water Oxidation—Achievements and Challenges in Commercial Applications*, A. Nazeri, R. W. Shaw, and P. A. Marrone, (eds.), Proceedings of Workshop, August 14–15, 2001 Arlington, Virginia, Strategic Analysis, Inc., 2001.

158. J. Abeln, M. Kluth and M. Pagel, "Results and Rough Cost Estimation for SCWO of Painting Effluents Using a Transpiring Wall and a Pipe Reactor," *J. Adv. Oxid. Technol.*, **10**(1), 169–176 (2007).

159. A. Botti, F. Bruni, M. A. Ricci and A. K. Soper, "Neutron Diffraction Study of High Density Supercritical Water," *J. Chem. Phys.*, **109**(8), 3180–3184 (1998).

160. A. G. Kalinichev and J. D. Bass, "Hydrogen Bonding in Supercritical Water. 2. Computer Simulations," *J. Phys. Chem. A*, **101**, 9720–9727 (1997).

161. T. Oe, A. Suzuki, H. Suzugaki and S. Kawasaki, "Commercialization of the First Supercritical Water Oxidation Facility for Semiconductor Manufacturing Wastes," Proceedings of 1998 Semiconductor Pure Water and Chemicals Conference, 399–407, 1998.

162. I. Jayaweera, SRI International, Personal Communication, 2006.

Index

A

Acidithiobacillus ferrooxidans, 42
Acid mine drainage (AMD), 42
Acid neutralization, 284–285
Acid-rock drainage (ARD), 41–43
Acquisition prices, 165
Activated sludge process, 229
Adsorption, 226–227, 289–290
  theoretical considerations, 289–290
Advanced oxidation process (AOP), 222
Agglomeration. *See* Product
AHO. *See* Assisted hydrothermal
  oxidation
Air concentrators, 327–329
Air cyclones, 330–331
Air pollution, 252–253
  equivalence index, 198
Air stripping, 213–214
Allocation methodology, 189
Alluvial mining, 16
Alternative building materials,
    environmental life-cycle analysis
  introduction, 180
  references, 205–206
Aluminum
  material, 244
  packages, recycling, 264
Ambient conditions, temperature/pressure
    (increase), 389–390
American Plastics Council (APC), study,
    365
Annex V of the International Convention
    for the Prevention of Pollution from
    Ships (MARPOL), 252
Apron feeders, 342–343
Aquatic mining, 16–20
Aqueous processing
  potential/limitations, 281
  unit operations, 281–296
Aqueous systems, unit activity, 51
Arbitrary scaling, 198
Arrhenius-form rate constant, 410
Aseptic packaging, 265
ASR. *See* Automotive shredder residue
Assisted hydrothermal oxidation (AHO),
    398

Association of Plastics Manufacturers in
    Europe (APME), 262
Athena EIE software, 202
Athena Sustainable Materials Institute,
    182
Attrition force, 312
Audit exercise, planning, 135–139
Audit programs
  department, informing, 139
  development, 137–138
Audit team, formation, 134–135
Audit workplan/checklists, preparation,
    138
Automotive shredder residue (ASR),
    374–375

B

Backfilling/grading, 25
Balers, 351–352
Baling, 351–352
Ballistic separation, 341
Baseline information, collection, 139–141
Basic oxygen furnaces (BOF), 45–46
Bed depth service time (BDST) method,
    227
Belt conveyors, 345–346
Bench-scale laboratory, 99
Best available technology (BAT), 209
Best practicable control technology
    (BPT), 209
Biodegradable plastics, 247
Biodegradable polymers, disposal,
    266–267
Biological aerated filters (BAFs),
    231–232
Biological methods, 291
Biological oxygen demand (BOD), 252
Biological waste treatment, 228–232
  treatment processes, 228–229
Blast furnaces, feedstock recycling use,
    260
Blister copper, 47
Bottle bill
  passage, 368
  states, 369

Briquetting, 349
Bucket elevators, 347–348
  capacity, 347

C

Cake filters, 216
Carbon, adsorption, 290
Carbon dioxide, 48–50
Casting sands, 46
Catalysis, 414
Categorical pretreatment standards, 209
Cationic species, removal, 295
Cellophane composites, 246
Cementation, phenomenon, 286
Centralized RPSs, 165–167
Chematur Engineering AB, 404
Chemical metallurgy wastes, 43–57
Chemical oxidation/reduction, 222–224
Chemical oxygen demand (COD), 145, 252
Chemical plant operations, waste reduction, 89
  introduction, 90–93
  references, 121–124
Chemical plants, life cycles
  economic features, 61–63
  environmental features, 63–67
Chemical processing plants, life-cycle evaluation, 59
  references, 87–88
Chemical processing plants, life cycles, 59–61
  phases, 60
Chemical recycling, 377
Chemical treatment technologies, 219–227
Clarification, 214
Clarifying filters, 216
Classification, 318–323
Clay minerals, adsorption, 290
Clean Air Act of 1977, 47
  Amendments of 1990, 367
Cleaner production
  concept, 64–65
  economics, 78–81
  global applicability, 65
  initiatives, integration, 64
  waste elimination/reduction, 64
Clean Water Act (1987), 207
Closure, activities, 26–29
Collection/cleaning/separation. *See* Plastics; Recycling
Collection/processing networks, interdependence/viability/growth, 157–159

Commingled plastics, 375
Commingled waste, liberation, 323
Compaction, 351–352
Compactors, 351
Composites, material, 245–247
Compressions, comminution force, 312
Consortium for Research on Renewable Industrial Materials (CORRIM), 185
Copper electrorefining plant, electrolyte usage, 55
Corrosion, 415–417
  studies, 100
Cross-flow filtration, 216
Cubing, 349–350

D

DARPA. *See* U.S. Defense Advanced Research Projects Agency
Decentralized decision-making framework, 165
Decentralized RPSs, 163–165
Degradable plastics, 247–248
  recovery/recycling, 378
Density separation, 327–331
Development, activities, 8–12
Diffusivity, 386
Dipole-dipole interactions, 389
Disc screens, 322–323
Dissolved air flotation (DAF), 216
Distillation, 219
Downstream boundary tier entities, 162
Downstream tier sites, 165
Dredging, 16
  operational cycle, 17–19
Drift/adit/shaft/slope, usage, 14
Duales System Deutschland (DSD), 258
DuPont, Ten-Step Method of Engineering Evaluations of Pollution Prevention, 105, 107–108
Dust, particulates, 46–47

E

Eco-efficiency, 80–81
Economic incentive instruments, 80
Economic indicators, 82
EC Packaging Directive, 263
Eddy-current separation, 338–340
  applications, 338
  operation, principles, 338–340
Eddy-current separators, performance, 339–340
Effluent, quality, 436–438
Electrical/electronic products, 376–377
Electrochemical process, 225–226

Electrokinetic remediation, 288–289
Electrolysis, 288
Electroosmosis, 288–289
Electrophoresis, 288–289
Electrorefining (ER)/electrowinning
    (EW), spent electrolytes, 54–56
Electrostatic separation, 331, 340–341
Elementary reaction network modeling,
    409–414
End-of-pipe treatment, 128
Energy
    conversation, advantages, 128
    recycling, industrial ecology
        (extension), 85
    savings/recovery, 366–368
    use, 250–251
Environmental costs
    assessment/quantification, frameworks,
        105
    evaluation, AIChE-CWRT TCA
        method usage, 105–107
Environmental data, 140
Environmental engineering, 280
Environmental externalities, 79–80
Environmental impact reduction, cost
    linkage, 76–78
Environmentally conscious practices, 3–4
Environmental planning/permitting,
    10–12
Environmental protection, 280–281
    issues/challenges, 280–281
Environmental protection, aqueous
        processing
    introduction, 279–280
    references, 296–305
Environmental stewardship, 35
Equilibrium constant, 53
Equipment selection/specification, 9
Evaporation, 218–219
Exploration, activities, 4–8
Externalities, 79
Extraction, activities, 12–20
Extractive inputs, contrast, 190–191

F

Feeders, 342–343
Feeding systems, 342–345
Feed materials, 417–421
    chemical composition, 418–420
Feeds, salt concentration, 423
Feedstock recycling, 260–262
Fe(III) compounds, precipitation, 285
Films, recycling, 377
Filtration, 216
Financial data, 140

Flail mill, usage, 316
Flotation, 216–217
Flowsheet analysis, hierarchical design
    procedures, 115
Fluid clean-up, providing, 440
Foam separation processes, 294–296
    theoretical considerations, 295–296
Franklin Associates, Ltd., research, 380
Franklin Database, usage, 193
Froth flotation, 342

G

Gas effluents, 437
Gaseous emissions, measurement, 147
Gaseous wastes, 47–50
Gasification, feedstock recycling use, 261
General Atomics, 405, 408
Geophysical surveys, 6–8
Gibbs adsorption equation, 295–296
Glass
    material, 243
    recycling, 263–264
Global kinetics, 409–414
Global material flows, uncertainties,
    159–161
Granular activated carbon (GAC), 227
Gravity separation, 327–331
Green chemistry, 96
Ground granular blast furnace slag
    (GGBFS), 46

H

Hammer mills, usage, 313–315
Handling system, 342, 345–349
Hand sorting
    calculation, data (usage), 324
    electronic sorting, contrast, 323–326
    line, design criteria, 323–325
Hanwha Chemical Corp., 404
Hazard and Operability (HAZOP)
    analysis/studies, 114
HDPE. See High density polyethylene
Heat exchangers, fouling rates, 100
Heating value/concentration, 417–418
Heat integration, 117
Heavy media separators, 331
High-acid testing results, 433
High density polyethylene (HDPE), 358
    bottle recycling, 365
Higher heating value (HHV), 418
High heating value liquid, usage, 421

High-value recycled plastics, production, 382
High-water-content feeds, minimization, 394
House designs, comparison, 201–204
Household packaging, contamination, 361
Housekeeping processes, audit, 136
Hydraulicking, 3, 16
  precautions, 19–20
Hydraulic retention times (HRTs), 229
Hydrogenation, feedstock recycling use, 261–262
Hydrometallurgical processing wastes, 50–57
Hydrometallurgical wastes, 39–41

I

IAPWS. *See* International Association for the Properties of Water and Steam
Impact, size-reduction force, 312
Impact assessment, 182
Impact categories, development, 196–197
Impactors, usage, 313–315
Improvement analysis, 201–204
Incremental sampling, 311
Industrial ecology, 83–85, 109
  Australian example, 84–85
  example, 85f
Industrial pollution, perception (change), 127
Industrial sectors, materials integration, 109
Industrial waste auditing, 125
  overview, 125–126
  references, 153
Industry data, 140
Inert solids, handling, 434–435
Infrastructure/surface facilities design/construction, 9–10
Injection-molded products, 375
Input-oriented categories, output-oriented categories (contrast), 197–198
International Association for the Properties of Water and Steam (IAPWS), 409
International commitments, change, 127
International Organization for Standards (IOS), 182
Intrinsic viscosity (IV), decrease, 370
Inventory analysis, 75
Inventory processes, audit, 136
Ion exchange, 226–227, 291–293
  theoretical considerations, 292–293
Ion flotation, 294
Island Copper Mine, case study, 29–31

J

Jet-induced recirculation flow, 421

K

Kinetic studies, trends, 413–414
Klobbie-based intrusion processes, 375–376
Knife shredders, 317–318
Kraton, 376

L

Lamination, 265
Land acquisition, 8–9
Land-use analysis, 22
Lenz's rule, 338
Life cycle, concept, 60
Life-cycle analyses (LCA), 108
  software, 193
  strengths/weaknesses, 204–205
  studies, findings, 184–187
Life-cycle assessment (LCA), 109–110, 196–201
  application, 73
  basics, 180–187
  methodology, 72–78
    development/application, 78
  practical applications, 182–184
  steps, 74–76
Life-cycle impact assessment (LCIA), 182
Life-cycle inputs/outputs, 190–196
Life-cycle inventory (LCI), 180–182, 272
  cross-sectional information, 204–205
  data collection, 187–188
  dynamic inventory, 204–205
  framework design, 187–190
  industry description, 188
  measures, development process, 200
  methods, 191
  strengths/weaknesses, 204–205
  unit processes, 188–189
Light fraction, composition, 327
Liquid effluents, 437–438
Liquid packaging board (LPB), 267
Litter, 251–252
Load sampling, truck (usage), 310–311
Logistic consumption data, 140
Loss prevention, 269

M

Magnetic drum, 334–335
Magnetic field, achievement, 332
Magnetic head pulley, 333–334
Magnetic rotor, power, 339

Magnetic separation, 331–338
Magnets
classification, 332–333
configurations/arrangements, 333–337
in-line configuration, 336
Maintenance, waste minimization, 71–72
Management data, 140
Marine mining, 16–20
Mass exchange network (MEN)
synthesis, 118–120
Material balance, 143–148
evaluation/refinement, 147
information sources, 144–145
preparation, steps, 146–147
Material data, 140
Material flows, 189
analysis, 111, 114
Material form, 420
Material recovery facilities (MRFs),
369–370
Materials, industrial ecology (extension),
85
Mechanical recycling, feedstock
recycling (comparison), 262–263
Mediated electrochemical oxidation
(MEO), 226
Melt-blown process, 371
Membrane bioreactors, 230–231
Membrane processes, 293–294
theoretical considerations, 294
Metal-contaminated acid-rock drainage,
treatment, 284
Metal hydroxides, precipitation, 284–285
Metal ions, concentration, 52–53
Metal ion spills, treatment, 284
Metals manufacturing, waste reduction,
33
references, 57–58
Metal sulfides, precipitation, 285
Mined resources, classifications, 2t
Mine facility removal, 27
Mine planning, 12
Mineral extraction, engineering, 1
references, 31–32
Mine sites
maintenance, 28
wastes, 35–43
Mine waste disposal sites, reclamation,
26
Mining activities, range, 1–2
Mining practices, 2–3
Mitsubishi Heavy Industries Ltd., 404
Mixed board/flexible packaging (MBFP)
material, 267
Mixed plastic (MP), 375–377
waste, 267

MODAR, 440
initiation, 398–399
MODEC, joining, 399
MRFs. *See* Material recovery facilities
Multimaterial packages, recycling,
264–265
Municipal solid waste (MSW)
combusting, 366
disposal, 357
waste packaging, 255–256

N

National Pollution Discharge Elimination
System (NPDES) permit, 207–208
National Pretreatment Program, discharge
control, 208
National Renewable Energy Laboratory,
182, 183
National Research Council (NRC),
incineration review, 399
Network/collection recruitment, growth,
160–161
Network design models. *See* Reverse
logistics
Neutralization, 219
Non-bottle bill states, 369
Nongovernmental organizations (NGOs),
157
Nonpolar organic compounds, 391–392

O

Oil/grease removal, 217–218
Operations, waste minimization, 71–72
Optimization techniques, 113
Opto-electronic sorting, 325–326
Ore, metal value, 40
Organic substances, behavior, 426
Organizational data, 140
Output categorization, 192–195
Output mass balance, 191–192
Oversize/overflow, 318
Oxidant, supply, 420–421
Oxides, adsorption, 290
Ozone depleting substances (ODS), 126

P

Packages
recycled materials, safety, 265–266
recycling, 258–259
Packaging
advertising function, 241
distribution function, 240
environmental impacts

Packaging (Continued)
    introduction, 238–239
        references, 274–278
    functions, 239–242
    household function, 240
    image-component function, 241
    intermediate function, 240
    materials, 242–248
        consumption, 248–250
        environmental assessment, 253–255
    processes, audit, 136
    systems, 268–272
        LCA, 272–273
    value-forming function, 241
    waste-reduction function, 241–242
Packing tower, height, 213
Paper/board, material, 242–243
Paper-plastic composites, 246
PE. See Polyethylene
PEF. See Processed-engineered fuels
Pellet fuels, production, 367
Pelletizing, 349–350
Percolation leaching, 40
Persistent organic pollutants (POPs), 126
PET. See Polyethylene terephtalate
Photodegradable plastics, 248
Physical property determination, 408–409
Physical treatment technologies, 213–219
Pickling liquors, 56–57
Pilot-plant-scale development/testing,
    99–100
Pinch analysis/technology, 113
Pinning, 333
Placer mining, 16
Plant walkthrough survey, 141–142
Plastic film, 377
Plastic packages, recycling, 259–260
Plastic-plastic composites, 246–247
Plastics
    biodegradation, 248
    collection/cleaning/separation, 368–369
    material, 245
    postconsumer recycling, 368–378
    recycling, 264
    strapping, 371–372
PMMA. See Polymethylmethacrylate
Pneumatic jig, operation, 329
Pneumatic systems, 348–349
Pollution, packaging (role), 251–253
Pollution prevention (P2)
    definition, 92–93
    economics, 105–108
    flowsheet analysis, 111–120
    initiatives, 89, 94
    macroscale, 108–110
    mesoscale, 110–120

    microscale, 120–121
    opportunities
        flowsheets/processes, examination
            (frameworks), 114–115
        identification, 93–102
    programs
        development, 93–105
        integrated methodology, 102–105
    qualitative methods, 114–117
    quantitative methods, 117–120
    tools/technologies/best practices,
        survey, 108–121
Polyethylene (PE), 253
Polyethylene terephtalate (PET // PETE),
    253, 266, 326, 358
    bottles, acceptance, 369
    number 1, resin recycling, 365
    recycling, 370–372
Polymethylmethacrylate (PMMA),
    pyrolysis, 378
Polyolefins, 372–373
Polypropylene (PP), 253
Polystyrene (PS), 253
    usage, 373–375
Polyurethane (PUR), 253
    obtaining, 377
Polyvinyl chloride (PVC)
    recycling, 374
    use, 253
Popcorn slag, 46
Postconsumer packaging
    economical aspects, 382
    LCA, 380–381
    prospects, future, 382
    recovery, 255–268
Postconsumer recycled plastics,
    processing
    background, 358–362
    introduction, 357–358
    references, 382–383
Postconsumer recycled plastics, products,
    378–379
Postmining liability, 28–29
Powdered activated carbon (PAC),
    227
PP. See Polypropylene
Precipitation, 219–222, 282–285
    overview, 50–54
    theoretical considerations, 282–284
Pregnant leach solution, 40
Preliminary material balance, preparation,
    147
Premine planning/permitting, 21–22
Price-flow contract, 163–164
Price-flow dependence, 170–171
Primary packaging, 242

Priority unit operation, selection, 145
Process data, 140
Processed-engineered fuels (PEF),
    production, 367
Process energy integration/heat
    integration, 117
Process flow diagrams, construction, 143
Processing scope/objectives/boundaries,
    189–190
Process inputs/outputs, 143–148
Process mass integration, 117–118
Process operations stage, P2 usage,
    100–101
Product
    agglomeration, 349–350
    data, 140
    functions, 72
    stability, 100
    storage, 352–353
Production processes, audit, 135
Product life cycle, 65–67
    cleaner production, applications, 64
Product/process engineering stage, P2
    usage, 100
Prohibited discharge standards, 209, 212t
Protection function, 239
Prussian blue precipitate, 37
PS. *See* Polystyrene
Publicly operated treatment work
    (POTW), 209, 437
PUR. *See* Polyurethane
Purchased inputs, contrast, 190–191
Purchased materials, recycled materials
    (contrast), 191
Pyrolysis, 377–378
    feedstock recycling use, 261
Pyrometallurgical processing wastes,
    44–50

Q

Quality protection, 269
Queensland fertilizer project, 85

R

Rapid small-scale column tests
    (RSSCTs), 227
Raw material data, 140
Reactive method, 128
Reactors
    effluent, 396
    mixing/feed distribution, impact,
        99–100
    performance, 67
    type, 421–424

waste, feedstock impurities, 68
waste minimization, 67–69
Readily implementable options,
    identification, 142
Reclamation
    activities, 20–26
    operations, 22–26
Reconnaissance, 4
Recycled, term, 364
Recycled metals, separation, 34
Recycled PET market, 371–372
Recycled resins, 372–373
Recycling
    approach, 364–366
    challenges, 367–368
    collection/cleaning/separation, 368–369
    importance, 308
    term, 364
Redlich-Kwong-Soave, 425
Reductive precipitation, 286–288
    theoretical considerations, 286–288
Refuse derived fuel (RDF), 267
Regulatory drivers, 86
Renewable inputs, contrast, 190–191
Reporting procedures, development, 138
Research and development (R&D) stage,
    P2 usage, 95–100
Resource conservation, necessity,
    308–309
Reusable plastic shipping containers
    (RPSCs), 363–364
Reuse
    approach, 363–364
    importance, 308–309
Revegetation, 25–26
Reverez process, 376
Reverse logistics, network design models,
    159–160
Reverse production systems (RPSs), 155
    background, 157–161
    decentralized framework, 160
    decision making, 160
    experimental comparisons, 167–171
    introduction, 155–157
    material flows, 157f
    references, 173–177
    strategic design models, 161–167
Rigid Plastics Packaging Container Law
    (California), 379
Rolling mills, 56
Rotary screens, 321–322
Rotary shear shredders, 316–317
RPSCs. *See* Reusable plastic shipping
    containers

S

Salts
  handling, 434–435
  nucleation/growth, 415
  phase equilibrium, 414–415
Sampling, 5–6
Sampling, waste characterization,
    310–311
Screw feeders, 344–345
SCWO. *See* Supercritical water oxidation
Secondary packaging, 242
Secondary reactions, minimization, 68
Sedimentation, 214
  rates, 100
Self-diffusion coefficient, 408
Separated plastics, processing, 370–375
Separation equipment, waste
    minimization, 69–70
Sequencing batch biofilm bioreactor
    (SBBR), 230
Sequencing batch reactors (SBRs),
    229–230
Shear force, 312
Shear shredders, 316
Shredders, usage, 315–318
Size reduction, 311–318
  equipment, selection factors, 311–313
Slags, 45–46
Sludge-retention times (SRTs), 229
Social indicators, 82
Society for Environmental Toxicology
    and Chemistry (SETAC), 182
Soil reconstruction, 25
Solid/liquid phase separation, 397
Solid residues, 438
Solids-containing feeds, 420
Solids removal,
    clarification/sedimentation, 214–215
Solid waste disposal/recycling
  introduction, 308–310
  references, 354–355
Solid waste-processing plants,
    classification (usage), 319
Solid-waste recycling, scientific basis,
    309–310
Solid wastes, 45–47, 253
  size reduction , equipment type,
    313–318
Solid waste separation, unit
    operation/equipment overview
  introduction, 308–310
  references, 354–355
Solvent extraction, 291–293
  theoretical consideration, 292–293
Source reduction, 363
Source-sink mapping, 118

SR-POLAR, 425
Starter dike, 38
Steel
  material, 243–244
  production, change, 44–45
  recycling, 264
Steel plate conveyors, 345
Steel Wall representation, 203
Sticky salts, 419
Stoners, 329–330
Straight run gas oil (SRGO), 75
Sulfur dioxide, 47–48
Supercritical fluid
  thermodynamic critical point, 386
  water, comparison, 388–389
Supercritical water
  properties, 386–392
  solvent, usage, 391–392
  temperatures/conditions, characteristics,
    393
Supercritical water oxidation (SCWO),
    224–225
  application, 395–396
  commercialization, 399–401, 404–405
  corrosion
    handling, 430–434
    research, conclusions, 416–417
    study, 432–433
  designs, 394–395
  economics, 439–443
  history/status, 398–408
  introduction, 385–386
  permitting, 438–439
  pipe reactor, configuration, 423
  pressure reduction, 397
  process, 392–398
    operating conditions, 393–394
  processing, aspects, 417–438
  products, 436–437
  reactor
    conditions, 390, 424–425
    types, 421–424
  references, 443–453
  research efforts, 399, 408–417
  system
    corrosion, 419
    erosion, 419
    noncondensable gas, high content,
      429–430
    solubility/phase behavior, 425–430
  technology, maturation, 401
  treatment, optimum, 418
Supply chain coordination, 160–161
Surface mining, 13–14
Surface Mining Control and Reclamation
    Act (SMCRA) of 1977, 21

Suspended magnet, 335–337
Sustainability, 81–83
    design, 381–382
    indicators, 82
    temporal characteristics, 83
Sustainable development, 308
System impacts, capture, 204

T

Tailings, definition, 37
Tailings impoundments
    considerations, 37–38
    planning/operating, 36–39
Target processes, preassessment, 139–142
Tertiary packaging, 242
Thermolysis, feedstock recycling use, 261
Tie compounds, selection, 145
Time value, 205
Total Cost Assessment (TCA) method,
    105
Total emissions, site-generated emissions
    (contrast), 195–196
Total suspended solids (TSS), 252
Toxin example, offense, 198–200
Trommels, 321–322
Tub grinders, 318

U

Uncapacitated case, material flow
    allocation, 170
Underground mines, sealing, 27–28
Underground mining, 14–16
    environmental considerations, 16
    operational cycle, 15
Undersize/underflow, 318
UN Group of experts on the Scientific
    Aspects of Marine Pollution
    (GESAMP), 252
Unit operations, listing, 143
Unit operations, P2 usage, 110–111
Upstream boundary tier entities, 161
U.S. Defense Advanced Research
    Projects Agency (DARPA), SCWO
    interest, 399
U.S. Environmental Protection Agency
    (EPA), three Rs, 363
Used packages, energy source, 267
Utility systems, waste minimization,
    70–71

V

Vapor-liquid equilibrium, limit, 386
Vertical airflow, 328

Vertical supply chain, double
    marginalization, 169–170
Vessel reactor, feature (importance),
    422–423
Vessel-type reactors, 396
Vibrating conveyors, 346
Vibrating feeders, 343–344
Vibrating screens, 320–321
Viscosity, 386
Volatile organic compounds (VOCs),
    213
Volatile suspended solids (VSS), 252

W

Waste
    auditing, 130–152
        assessment, 142–148
    construction materials, derivation, 268
    end-of-pipe treatment, 126
    feed, suitability, 417–418
    generation, audits, 109
    material recovery, 109
    reduction, definition, 92–93
    sources, 96t
    treatment processes, audit, 136
Waste audits
    management/staff involvement,
        131–134
    options, screening/selecting, 149–150
    phases, 131–152
    preliminary technical/economical
        evaluation, 150–151
    preparatory work, 131–138
    synthesis/preliminary analysis,
        148–152
Waste management
    audits, 109
    costs, 78–79
    hierarchy, 90
        preference order, 90
    options, selection, 257–258
Waste material production, 155
    background, 157–161
    experimental comparisons, 167–171
    introduction, 155–157
    references, 173–177
    strategic design models, 161–167
Waste minimization
    advantages, 128
    audits, 130
    cycle, 129–130
    implementation barriers, 128
    phases, 129–130
    programs, 127–128
    strategies, 67

Waste reduction
    action plan, development, 152
    approaches, 362–366
    options, preliminary prioritizing,
      151–152
    packaging, 256–257
Waste-to-energy (WTE) facilities,
    366
Wastewater, accounting, 147
Wastewater engineering
    introduction, 207–208
    references, 233–235
Wastewater treatment
    strategies/requirements,
      208–213
Water pollution, 252

Water usage, 394
    recording, 146
Weighted sum method, 149–150
Wet air oxidation, 224–225
Wood, material, 248
Wooden packaging, recovery, 268
Wood grinders, 318
World Business Council for Sustainable
    Development (WBCSD), 80
Worst-offending safety standard, 199–200
Worst-offending substance, 198
WTE. *See* Waste-to-energy

Z
Zero-valent iron (ZVI) barriers, 288